Communications in Computer and Information Science 2859

Series Editors

Rationale
The CCIS series is devoted to the publication of proceedings of computer science conferences. Its aim is to efficiently disseminate original research results in informatics in printed and electronic form. While the focus is on publication of peer-reviewed full papers presenting mature work, inclusion of reviewed short papers reporting on work in progress is welcome, too. Besides globally relevant meetings with internationally representative program committees guaranteeing a strict peer-reviewing and paper selection process, conferences run by societies or of high regional or national relevance are also considered for publication.

Topics
The topical scope of CCIS spans the entire spectrum of informatics ranging from foundational topics in the theory of computing to information and communications science and technology and a broad variety of interdisciplinary application fields.

Information for Volume Editors and Authors
Publication in CCIS is free of charge. No royalties are paid, however, we offer registered conference participants temporary free access to the online version of the conference proceedings on SpringerLink (http://link.springer.com) by means of an http referrer from the conference website and/or a number of complimentary printed copies, as specified in the official acceptance email of the event.

CCIS proceedings can be published in time for distribution at conferences or as post-proceedings, and delivered in the form of printed books and/or electronically as USBs and/or e-content licenses for accessing proceedings at SpringerLink. Furthermore, CCIS proceedings are included in the CCIS electronic book series hosted in the SpringerLink digital library at http://link.springer.com/bookseries/7899. Conferences publishing in CCIS are allowed to use our online conference service (Meteor) for managing the whole proceedings lifecycle (from submission and reviewing to preparing for publication) free of charge.

Publication process
The language of publication is exclusively English. Authors publishing in CCIS have to sign the Springer CCIS copyright transfer form, however, they are free to use their material published in CCIS for substantially changed, more elaborate subsequent publications elsewhere. For the preparation of the camera-ready papers/files, authors have to strictly adhere to the Springer CCIS Authors' Instructions and are strongly encouraged to use the CCIS LaTeX style files or templates.

Abstracting/Indexing
CCIS is abstracted/indexed in DBLP, Google Scholar, EI-Compendex, Mathematical Reviews, SCImago, Scopus. CCIS volumes are also submitted for the inclusion in ISI Proceedings.

How to start

To start the evaluation of your proposal for inclusion in the CCIS series, please send an e-mail to ccis@springer.com

Tolga Ensari · Akram Bennour ·
Bassem Bouaziz · Imad Rida · Walid Mahdi
Editors

Intelligent Systems and Pattern Recognition

5th International Conference, ISPR 2025
Hammamet, Tunisia, September 25–27, 2025
Revised Selected Papers, Part II

Editors
Tolga Ensari
Arkansas Tech University
Russellville, AR, USA

Bassem Bouaziz
University of Sfax
Sfax, Tunisia

Walid Mahdi
University of Sfax
Sfax, Tunisia

Akram Bennour
Larbi Tebessi University
Tebessa, Algeria

Imad Rida
University of Technology of Compiègne
Compiègne, France

ISSN 1865-0929 ISSN 1865-0937 (electronic)
Communications in Computer and Information Science
ISBN 978-3-032-21584-0 ISBN 978-3-032-21585-7 (eBook)
https://doi.org/10.1007/978-3-032-21585-7

This Springer imprint is published by the registered company Springer Nature Switzerland AG
The registered company address is: Gewerbestrasse 11, 6330 Cham, Switzerland

Preface

It is with great honor and pleasure that we present the proceedings of ISPR 2025, the Fifth International Conference on Intelligent Systems and Pattern Recognition. Building upon the strong scientific legacy of the ISPR conference series, ISPR 2025 provided a distinguished international forum for researchers, academics, and practitioners to share innovative ideas, cutting-edge methodologies, and recent advances in artificial intelligence and pattern recognition.

Organized under the auspices of the MIRACL Laboratory, Sfax University (Tunisia), and held in collaboration with prominent international academic partners, ISPR 2025 further strengthened the conference's position as a leading scientific venue endorsed by the International Association for Pattern Recognition (IAPR). This fifth edition reflected the growing international recognition and increasing scientific influence of the ISPR series.

ISPR 2025 received 180 submissions that successfully passed the initial screening and were forwarded to the peer-review process. Each manuscript was rigorously evaluated by at least three expert reviewers through a double-blind review process. Following this stringent evaluation, 65 papers were accepted for presentation, among which 59 were successfully registered and presented at the conference.

All presented papers were revised according to reviewers' comments. The authors further improved and extended their manuscripts based on feedback and discussions held during the event. The accepted papers cover a wide spectrum of contemporary research topics, including machine learning, deep learning, computer vision, data mining, intelligent systems, explainable artificial intelligence, pattern recognition methodologies, and multimedia analysis.

We sincerely hope that the ISPR 2025 proceedings will serve as a lasting and valuable reference for researchers and practitioners and will foster further collaboration and innovation within the global scientific community.

October 2025

Tolga Ensari
Akram Bennour
Bassem Bouaziz
Imad Rida
Walid Mahdi

Organization

General Chairs

Tolga Ensari	Arkansas Tech University, USA
Bassem Bouaziz	University of Sfax, Tunisia
Abdel-Badeeh Salem	Ain Shams University, Egypt
Akram Bennour	Larbi Tebessi University, Algeria

Organizing Chairs

Imad Rida	University of Technology of Compiègne, France
Ayush Dogra	Chitkara University, India
Carmen Bisogni	University of Salerno, Italy
Mohamed Elhoseny	University of Sharjah, UAE
Lucia Cascone	University of Salerno, Italy
Gökhan Bakal	Abdullah Gül University, Turkey
Walid Mahdi	University of Sfax, Tunisia
Moises Diaz	Universidad de Las Palmas de Gran Canaria, Spain
Najib Ben Aoun	Al Baha University, Saudi Arabia

Local Arrangement Committee

Bassem Bouaziz (Chair)	University of Sfax, Tunisia
Basma Jalloul	University of Sfax, Tunisia
Siwar Chaabene	University of Sfax, Tunisia
Amal Bouadya	University of Sfax, Tunisia
Tarek Zlitni	University of Sfax, Tunisia
Adham Bennour	Constantine 2 University, Algeria
Marzoug Soltan	Larbi Tebessi University, Algeria

Steering Committee

Faiez Gargouri	University of Sfax, Tunisia
KC Santosh	University of South Dakota, USA

Ahmed Bouridane	University of Sharjah, United Arab Emirates
Muhammad Attique Khan	Lebanese American University, Lebanon
Mohammed Al-Sarem	Aylol College University, Yemen
Mohammad Shabaz	Model Institute of Engineering and Technology, India
Nazife Çevik	İstanbul Arel University, Turkey
Slim Kallel	University of Sfax, Tunisia

International Advisory Board

José Ruiz-Shulcloper	University of Informatics Sciences, Cuba
Osamah Ibrahim Khalaf	Al-Nahrain University, Iraq
Johan Debayle	École Nationale Supérieure des Mines de Saint-Étienne, France
Deepak Gupta	Maharaja Agrasen Institute of Technology, India
Faiez Gargouri	University of Sfax, Tunisia
Takashi Matsuhisa	Karelia Research Centre, Russian Academy of Sciences, Russia
Sabah Mohammed	Lakehead University, Canada
Mostafa M. Aref	Ain Shams University, Cairo, Egypt
Gurinder Bawa Stellantis	Fiat Chrysler Automobiles, USA

Publicity Committee

Chintan M. Bhatt	Pandit Deendayal Energy University, India
Mohammed al-Chaabi	Tayba University, Saudi Arabia
Sean Eom	Southeast Missouri State University, USA
Ersin Sener	Kirklareli University, Turkey
Ayush Dogra	Chitkara University, India
Ali Ismail Awad	United Arab Emirates University, Abu Dhabi, UAE
Majid Banaeyan	TU Wien, Austria
Mustafa Dagtekin	Istanbul University - Cerrahpaşa, Turkey
Mohamed Hammad	Menoufia University, Egypt
Mustafa Ali Abuzaraida	Utara University, Malaysia
Jinan Fiaidhi	Lakehead University, Canada

Program Committee Chairs

Tolga Ensari	Arkansas Tech University, USA
Akram Bennour	Larbi Tebessi University, Algeria
Bassem Bouaziz	University of Sfax, Tunisia
Imad Rida	University of Technology of Compiègne, France
Walid Mahdi	University of Sfax, Tunisia

Scientific Committee

Varuna De Silva	Loughborough University, UK
Mohamed Elhoseny	University of Sharjah, UAE
Mahmood Seyyedzadeh	University of Tabriz, Iran
Karri Chiranjeevi	University of Porto, Portugal
Kamel Boukhalfa	National School of Artificial Intelligence, Algeria
Faiez Gargouri	University of Sfax, Tunisia
Linas Petkevičius	Vilnius University, Lithuania
Ersin Sener	Kirklareli University, Türkiye
Khalid El Khadiri	Chouaib Doukkali University, Morocco
Emrah Önder	Istanbul University, Turkey
Ayush Dogra	Chitkara University, India
Slim Kallel	University of Sfax, Tunisia
Mohammed Benmohammed	Constantine 2 University, Algeria
José Ruiz-Shulcloper	University of Informatics Sciences, Cuba
Muhammad Attique Khan	Lebanese American University, Lebanon
Kouloud Boukadi	University of Sfax, Tunisia
Mohammad Shabaz	Model Institute of Engineering and Technology, India
Deepak Gupta	Maharaja Agrasen Institute of Technology, India
Selçuk Topal	Gebze Technical University, Turkey
Hala Bezine	University of Sfax, Tunisia
Mohammed al-Sarem	Aylol College University, Yemen
Gökhan Bakal	Abdullah Gül University, Turkey
Afef Kacem	Université de Tunis - ENSIT, Tunisia
Hemam Sofiane Mounine	Université de Khenchela, Algeria
Chintan M. Bhatt	Pandit Deendayal Energy University, India
KC Santosh	University of South Dakota, USA
Mohamed Hammad	Menoufia University, Egypt
Sengul Bayrak Hayta	Istanbul Sabahattin Zaim University, Turkey
Osamah Ibrahim Khalaf	Al-Nahrain University, Iraq
Houari Aoued	Hassiba Benbouali University of Chlef, Algeria

Sam Zaza	Middle Tennessee State University, USA
Bechir Alaya	University of Gabès, Tunisia
Rachid Oumlil	École Nationale de Commerce et de Gestion Agadir, Morocco
Tolga Ensari	Arkansas Tech University, USA
Jerry Wood	Arkansas Tech University, USA
Toufik Sari	Badji Mokhtar University, Algeria
Sean Eom	Southeast Missouri State University, USA
Walid Mahdi	University of Sfax, Tunisia
Anna Maria Di Sciullo	*Université du Québec à Montréal, Canada*
Bhaskar Ghosh	Arkansas Tech University, USA
Abdelkader Nasreddine Belkacem	Osaka University, Japan
Faezeh Soleimani	Ball State University, USA
Robin Ghosh	Arkansas Tech University, USA
Indira Dutta	Arkansas Tech University, USA
Suzan Anwar	Philander Smith University, USA
Nassima Bouchareb	Constantine 2 University, Algeria
Shridhar Devamane	Global Academy of Technology, India
Chokri Ben Amar	Taif University, Saudi Arabia
Hafidi Mohamed	Cadi Ayyad University, Morocco
Sidahmed Benabderrahmane	Inria, France
Shahin Gelareh	University of Artois, France
Abdel-Badeeh Salem	Ain Shams University, Egypt
Xinli Xiao	Arkansas Tech University, USA
Weiru Chen	Arkansas Tech University, USA
Yudith Cardinale	Simón Bolívar University, Venezuela
Ankur Singh Bist	Towards Blockchain, India
El-Sayed M. El-Horbaty	Ain Shams University, Egypt
Abdelghani Ghomari	University of Oran 1, Algeria
Brahim Hnich	University of Monastir, Tunisia
Dragana Krstić	University of Niš, Serbia
Kechar Bouabdellah	University of Oran 1, Algeria
Krassimir Markov	Institute of Information Theories and Applications, Bulgaria
Dana Simian	Lucian Blaga University of Sibiu, Romania
Marina Nehrey	National University of Life and Environmental Sciences of Ukraine, Ukraine
Rossitsa Yalamova	University of Lethbridge, Canada
Vitalina Babenko	Kharkiv National University of Radio Electronics, Ukraine
Vera Meister	Brandenburg University of Applied Sciences, Germany

Paata Kervalishvili	Georgian Technical University, Georgia
Roumen Kountchev	Technical University of Sofia, Bulgaria
Nagwa Badr	Ain Shams University, Egypt
Natalya Shakhovska	Lviv Polytechnic National University, Ukraine
Anastasia Y. Nikitaeva	Southern Federal University, Russia
Francesco Sicurello	University of Milano-Bicocca, Italy
Maria Brojboiu	University of Craiova, Romania
Liliana Moga	Dunarea de Jos University of Galati, Romania
Nouhad Rizk	University of Houston, USA
Volodymyr Romanov	Glushkov Institute of Cybernetics of National Academy of Sciences of Ukraine, Ukraine
Romina Kountchev	Technical University of Sofia, Bulgaria
Mohamed Hammami	University of Sfax, Tunisia
Cornelia Aida Bulucea	University of Craiova, Romania
Elena Nechita	Vasile Alecsandri University of Bacău, Romania
Vicente Rodríguez Montequín	Universidad de Oviedo, Spain
Livia Bellina	MobileDiagnosis, Italy
Yukako Yagi	Harvard Medical School, USA
Abdelaziz Triki	Berlin School of Business and Innovation, Germany
Qinghan Xiao	Defence R&D Canada, Canada
Felix T. S. Chan	Macau University of Science and Technology, China
Mohamed Tmar	University of Sfax, Tunisia
Carmen Bisogni	University of Salerno, Italy
Lucia Cascone	University of Salerno, Italy
Redouane Tlemsani	University of Science and Technology of Oran, Algeria
Sami Saleh	Sains University, Malaysia
Djamel Samai	University of Ouargla, Algeria
Abdelaziz Kallel	Digital Research Center of Sfax, Tunisia
Paulo Batista	University of Évora, Portugal
Baskar Arumugam	Amrita Vishwa Vidyapeetham University, India
Loukas Ilias	National Technical University of Athens, Greece
Nabiha Azizi	Badji Mokhtar University, Algeria
Imran Mudassir	Air University, Pakistan
Abdellatif Rahmoun	Higher National School of Computer Science, Algeria
Rudresh Dwivedi	Netaji Subhas University of Technology, India
Mariagrazia Fugini	Politecnico di Milano, Italy
Akram Boukhamla	Badji Mokhtar University, Algeria
Arcangelo Castiglione	University of Salerno, Italy

Contents

Spine 3-D Representation from 2-D X-Ray Images and Measurement for Scoliosis Patient Follow-up

Zineb Hadjadj(✉), TinHinane Bencherif, Hayette Hadjar, and Abdelkrim Meziane

Research Center on Scientific and Technical Information (CERIST), Algiers, Algeria
hadjadj_zineb@yahoo.fr, {zhadjadj,bencherif,hhadjar,ameziane}@cerist.dz

Abstract. Scoliosis is a congenital disorder that causes the spine to be deformed from its normal shape, it greatly influences the patient's health and quality of life, therefore, optimal solutions for scoliosis patient's follow-up are needed.
The most common method for diagnosing scoliosis in humans is two-dimensional X-ray imaging. However, better exploration and analysis of the scoliosis spine may be possible with 3-D visualization. Moreover, it is much better to be able to manipulate or interact with the scoliosis spine when it is represented in three dimensions. Applications of Augmented Reality (AR) technology have recently showed promising results in supporting medical decisions, diagnosis, and therapy. In this paper, we propose a conceptual augmented reality framework for spinal disorders 3-D representation based on 2-D X-ray images.
Accurate scoliosis measurement in spinal radiographs is critical for clinical assessment, diagnosis, and treatment planning. Conventional manual methods are time-consuming and subject to inter-observer variability. Here, we propose an automatic scoliosis measurement method, tested on X-ray images from Bounaama Djilali Hospital (CHU Douera), Algeria. Results show that 3-D reconstruction from 2-D landmarks is feasible and that Cobb angle estimation validates the effectiveness of the approach.

Keywords: Scoliosis · Spine · 3-D representation · Scoliosis measurement · Augmented Reality (AR)

1 Introduction

1.1 Spine

The spine, commonly known as the backbone, is a flexible column composed of 33 superimposed vertebrae. In a healthy spine, the alignment of the vertebrae is straight in the frontal plane, while the sagittal plane shows the spine's natural flexibility and elasticity [1]. The spine presents two physiological curvatures with opposite convexities (Fig. 1):

T. Ensari et al. (Eds.): ISPR 2025, CCIS 2859, pp. 1–19, 2026.
https://doi.org/10.1007/978-3-032-21585-7_1

- Kyphosis: is a posterior deviation with a forward-directed concavity.
- Lordosis: is a prior deviation with a backward-directed concavity.

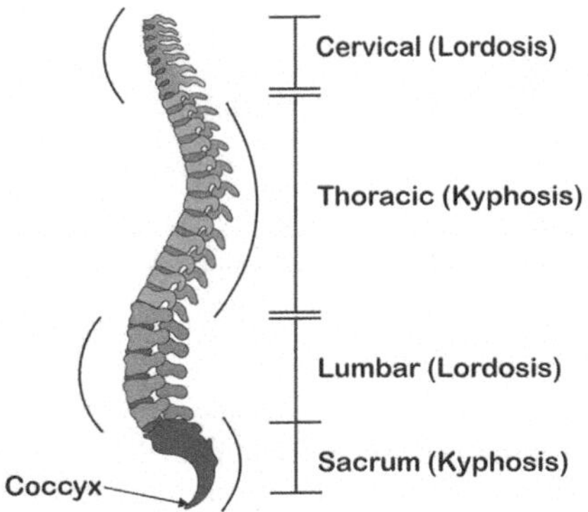

Fig. 1. Natural curvatures of the spine.

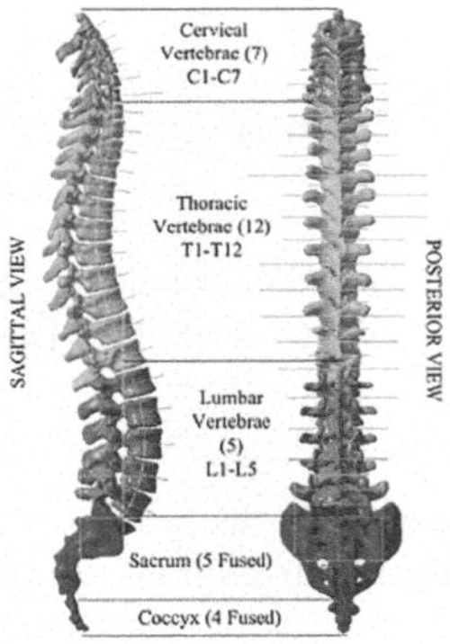

Fig. 2. The vertebral column.

Five spinal regions are formed by these curvatures [2], as Fig. 2 illustrates:

- Cervical region, which supports the head and consists of seven cervical vertebrae (C1-C7), exhibiting a lordotic curvature.
- Thoracic region, which corresponds to the chest and includes twelve thoracic vertebrae (T1-T12), characterized by a kyphotic curvature.
- Lumbar region, which forms the lower back and comprises five lumbar vertebrae (L1-L5), presenting a lordotic curvature.
- Sacral region, which constitutes the pelvic region and includes five welded vertebrae (S1-S5), with a kyphotic curvature.
- Coccygeal region, which consists of four welded vertebrae forming the coccyx.

1.2 Scoliosis

Scoliosis (from the Greek skolios, meaning crooked) is a spinal deformity characterized by a rotational displacement of the vertebrae, leading to three-dimensional deformation of the spine in the frontal, axial, and sagittal planes [3]. In individuals with scoliosis, the spine typically presents a C- or S-shaped curvature, as illustrated in Fig. 3. According to the Algerian press, scoliosis affects approximately 2–5% of the population in Algeria [4].

1.3 Diagnosis of Scoliosis

Scoliosis is typically confirmed through physical examination and medical imaging techniques such as spinal radiography (X-ray), computed tomography (CT),

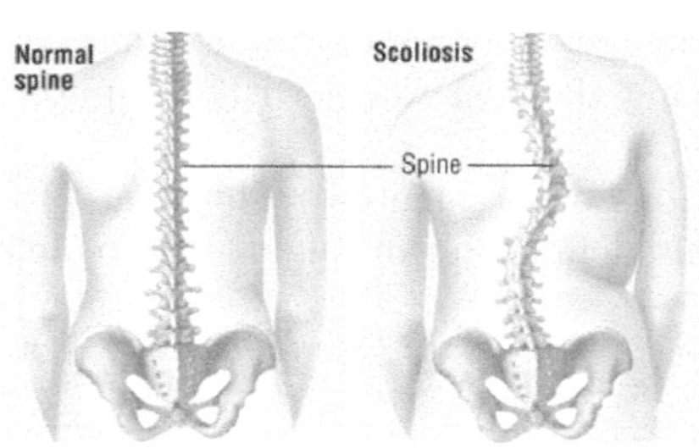

Fig. 3. Normal spine and scoliosis [5].

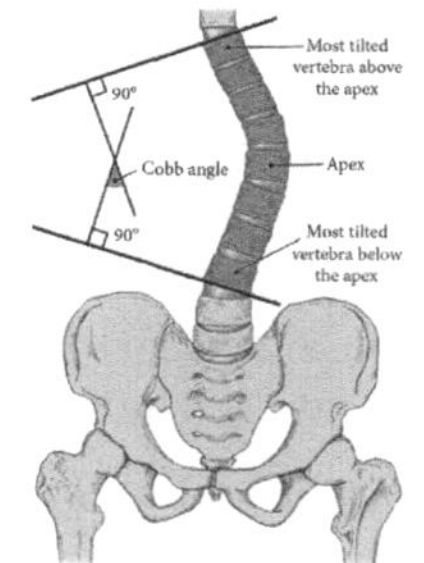

Fig. 4. Measurement using the Cobb method [7].

or magnetic resonance imaging (MRI). Among these modalities, radiography is the most widely used for clinical assessment and measurement of scoliosis due to its high availability, relatively low cost, and rapid data acquisition.

The Cobb angle, first introduced by the American orthopedic surgeon John Robert Cobb, is the most widely used metric for assessing scoliosis severity [6]. It is calculated as the angle between two lines drawn perpendicular to the spinal curve at the inflection points. Fig. 4 illustrates the procedure for Cobb angle measurement and the identification of the relevant vertebrae.

The classification of scoliosis severity is detailed in Table 1. A Cobb angle below 10°C is generally considered indicative of spinal curvature rather than scoliosis [8]. Angles between 10 and 20°C correspond to mild scoliosis, values between 20 and 40°C indicate moderate scoliosis, and angles exceeding 40°C are classified as severe scoliosis [9].

Table 1. Classification of scoliosis severity [9].

Cobb angle	Definition
$0° - 10°$	Normal Spinal curvature
$10° - 20°$	Mild scoliosis
$20° - 40°$	Moderate scoliosis
$> 40°$	Severe scoliosis

Recent studies have demonstrated that emerging technologies such as virtual reality and augmented reality can provide promising outcomes in supporting medical decision-making, diagnosis, and therapeutic interventions.

1.4 Augmented Reality

Three-dimensional visualization enables a deeper understanding of the presented data. Moreover, interacting with or manipulating information displayed in three dimensions provides a more intuitive and effective user experience. Within the medical field, augmented reality (AR) is increasingly employed for interactive three-dimensional visualization, enhancing the real world with computer-generated information.

One of the most widely accepted definitions of augmented reality was proposed by Ron Azuma [10]. According to this definition, an augmented reality system must satisfy the following criteria:

1. Integration of real and virtual environments;
2. Real-time interactivity of the augmentations;
3. Accurate three-dimensional registration of virtual content within the real world.

According to [11], augmented reality technology represents a novel approach for supporting educational activities and assisting diagnosis, contributing to the enhancement of traditional learning and diagnostic practices.

The goal of our study is to apply new technologies to meet the needs of health professionals and facilitate patient follow-up. In this paper, we use augmented reality for the diagnosis and visualization of the scoliosis spine.

The remainder of this paper is structured as follows. Section 2 provides a brief review of three-dimensional spine reconstruction techniques and scoliosis measurement methods. Section 3 describes the proposed system, followed by the experimental results presented in Sect. 4. Finally, Sect. 5 concludes the paper and discusses avenues for future work.

2 Related Work

This section is organized into three main parts, with the first addressing 3D spine reconstruction from X-ray images.

2.1 3-D Spine Reconstruction Review

Due to its advantages, including low cost and minimal radiation exposure, X-ray imaging is the most frequently used modality in clinical practice. Clinicians typically acquire bi-planar X-ray images in anterior-posterior (AP) and lateral (LAT) views for the diagnosis of spinal disorders. However, a major limitation of X-ray imaging is that it produces only two-dimensional projections of three-dimensional structures, providing limited information that is generally insufficient for a complete 3-D representation of the spine. This limitation restricts clinical assessment and treatment planning to partial visual information and prior experience, potentially leading to unstable and error-prone decisions. In this context, reconstructing the three-dimensional structure of the spine from 2-D X-ray images is expected to significantly enhance clinical decision-making.

The literature reports many approaches for 3-D reconstruction:

- In [12], the scoliotic spine was modeled as a 3-D Cosserat rod, allowing reconstruction of its three-dimensional structure by minimizing the potential energy for optimal registration with X-ray images.
- Novosad et al. [13] introduced a 3-D spine reconstruction method from a single X-ray image using a prior vertebral model. Its primary drawback was the need for models constructed without bone growth, making landmark identification challenging.
- In [21], the authors calculate 3-D spine models in real-time using augmented reality to support the diagnosis and treatment of scoliosis.
- In [11] a framework for spinal disorders diagnosis is presented as well as a full 3-D spinal model.

The second part of Sect. 2 focuses on the diagnosis of scoliosis.

2.2 Scoliosis Measurement Review

Manual measurement of Cobb angles remains the standard for scoliosis diagnosis and treatment planning. However, anatomical variability, low X-ray contrast, and line intersections extending beyond image boundaries often lead to inter- and intraobserver variability. This motivates the development of automated computer-based methods for reliable and robust scoliosis assessment.

Several studies have proposed methods for measuring scoliosis:

- In [7], an end-to-end model was proposed for fully automatic and reliable vertebrae segmentation in scoliosis measurement.
- Horng et al. [9] introduced an automatic measurement approach using object detection with a deep learning model, showing strong agreement (r = 0.89) with expert manual measurements.
- Kusuma et al. [14] introduced a K-means and curve-fitting method for Cobb angle estimation, requiring multiple pre-processing steps.
- Moura et al. [15] developed methods to isolate the spine, detect vertebral positions using a progressive threshold, and identify lateral vertebral boundaries. A tree structure was used to remove redundant information and merge small regions. These boundaries were subsequently used to measure the Cobb angle.
- Okashi et al. [16] introduced a fully automatic method for spine segmentation and curvature quantification from X-ray images, consisting of region of interest preparation, segmentation, and curvature measurement. The method, however, depends on complex image processing steps.
- Regression-based approaches that directly identify vertebral corners have also been proposed [17–20]. Despite their potential, these methods are limited in clinical use due to reduced accuracy and poor explainability.

The lack of novel research and reliable solutions in this field is what motivated us to use new technologies, such as augmented reality, for spinal diagnosis and visualization.

2.3 Medical Augmented Reality Applications Review

According to Gartner's hype cycle, augmented reality (AR) has passed from the emerging technology class to a more mature state. It may be the key technology to enable the visualization of 3-D patient-specific models in the physical environment, such as a 3-D spinal model superimposed over the patients' dorsal surface [21].

The interest in medical augmented reality applications has grown in recent years, mainly due to the recent hardware and software development. PubMed database is one of the most accurate sources where the popularity of AR applied for medical purposes can be observed (Fig. 5). Among 7336 papers with "Augmented Reality" in the title, 323 of them discuss the use of AR for spinal affections and treatment procedures, mainly focused on surgical procedures.

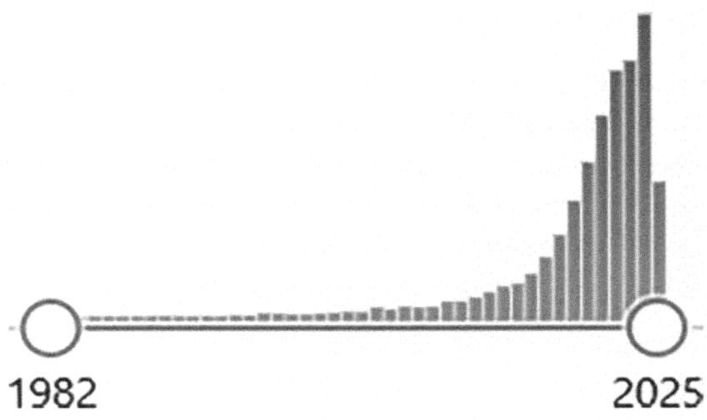

Fig. 5. PubMed database AR results by year [22].

In the field of spinal disorders, AR applications are mainly developed to support surgical procedures. For the spinal deformity evaluation and diagnosis, a few AR-based studies were found:

- In [21], the authors employed augmented reality to generate three-dimensional spine models in real time, supporting both the diagnosis and treatment of scoliosis.
- In [11] an AR framework for spinal disorders diagnosis is presented as well as a full 3-D spinal model.

In the proposed method, we use augmented reality for the diagnosis and visualization of the scoliotic spine.

3 Proposed Method

This section presents the proposed system and is divided into three subsections.

3.1 Scoliotic Spine 3-D Representation

The proposed method for 3D spine reconstruction is based on augmented reality (AR), an innovative technology that enhances data visualization and augments human perception. A flowchart of the proposed system is presented in Fig. 6. The system comprises four main steps: X-ray labeling, thresholding-based landmarks detection, binarization, and connected component labeling-based vertebrae coordinates detection, and moving from 2-D to 3-D. Each step is described below.

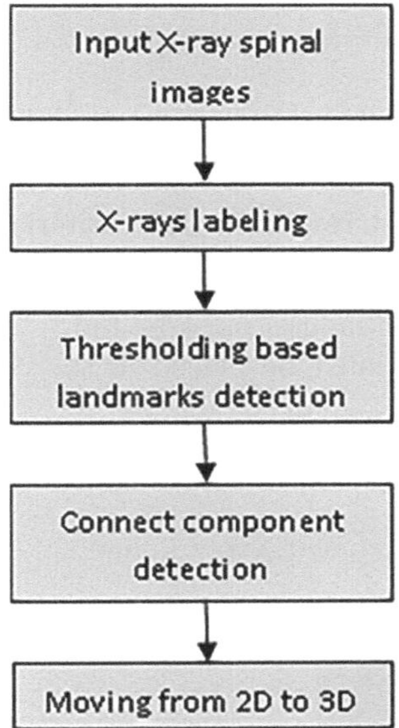

Fig. 6. Flowchart of the proposed scoliosis spine 3-D representation method.

X-Ray Image Scanning. The X-ray images used in this study were provided by the Physical Medicine and Rehabilitation Service of Bounaama Djilali Hospital (CHU Douera) in Algeria, the images are in the form of film photography (hard copies). Therefore, a negatoscope and a camera are used to take pictures of the films in order to numerically register them. The films are first cropped, and any zones that include the patient's name, address, or place of birth are removed.

X-Ray Image Labeling. Labeling of the resulting X-ray image consists of adding manually landmarks on the vertebrae of the posteroanterior and lateral views that will help represent the spine in three dimensions in the next steps. We use a filled red rectangle as a landmark on each vertebra (Fig. 7).

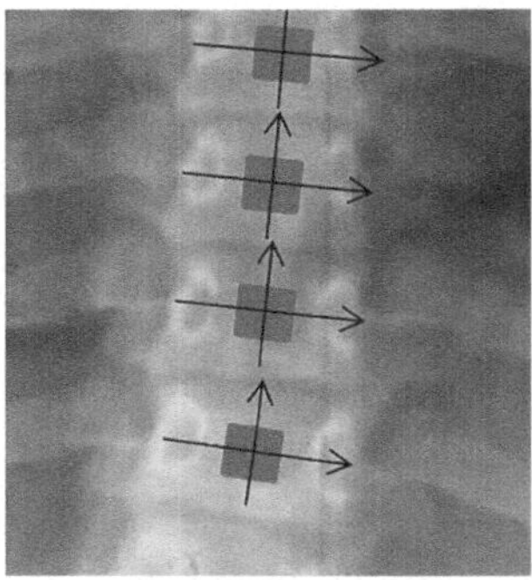

Fig. 7. Placing a red landmark on each vertebra. (Color figure online)

The filled rectangles must respect certain criteria which are summarized as follows:

- Color: the color of the landmarks must be red.
- Position: each landmark must be placed in the center of the corresponding vertebra.
- Rotation: each landmark must be rotated according to the rotation of the corresponding vertebra.

Figure 8 shows the final labeled X-ray image.

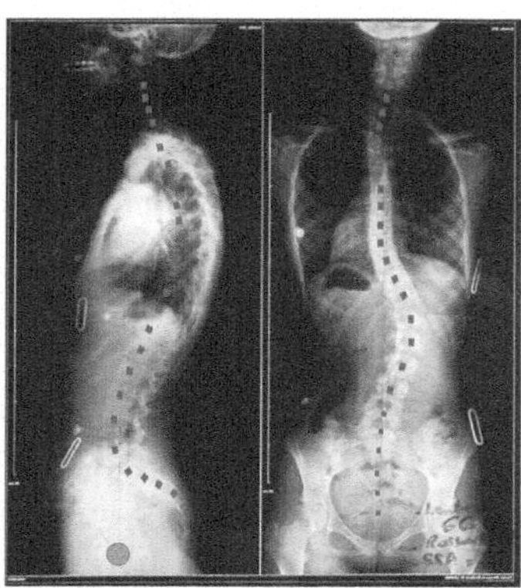

Fig. 8. Final labeled X-ray image.

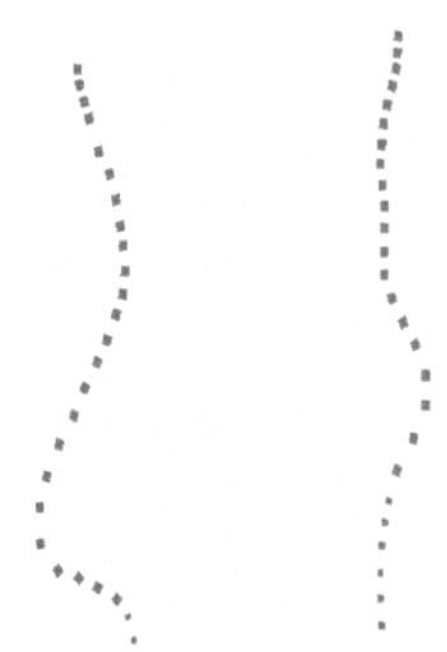

Fig. 9. Extracted landmarks from the image in Fig. 8 using thresholding.

The primary advantage of these images is the uniform color of the landmarks. This consistency facilitates segmentation, as landmarks can be extracted using thresholding techniques. The landmark detection process is described in the following paragraph.

Thresholding-Based Landmarks Detection. Since the landmarks of the X-ray images were added manually using a unified color, a simple thresholding of the image in the HSV space allows us to obtain these landmarks. The global thresholds were obtained as follows:

- Initially, the image is transformed into the HSV color space.
- Subsequently, a sliding window is employed to compute thresholds for all three channels.
- To decide on the thresholds to choose, thresholding is applied using the selected thresholds, and the result is visually analyzed.
- After fixing the thresholds, the image is segmented using global thresholding.

The results are shown in Fig. 9. This step is used to compute the global positions of the vertebrae in the image.

Binarization and Connected Component Labeling-Based Vertebrae Coordinates Detection. Connected Component Labeling (CCL) is a computer vision concept, it is applied to binary images and consists of finding groups of connected pixels forming objects in the image, called connected components. The principle of CCL is to group all the connected pixels under the same label. Depending on the needs, it is then possible to provide as a product, a labeled image or a set of characteristics (connected component analysis) such as the bounding box, and the mass center of each connected component in the input image. The main goal of this step is to extract the information necessary to construct the corresponding 3-D spine model: the first information is the position of the vertebrae centers which will help to construct the vertical format of the 3-D spine model, the second information is the degree of rotation of each vertebra, which will be used to create the horizontal format. The results are displayed in Fig. 10. Rotation angles are represented by green and yellow color (Fig. 10.b, Fig. 10.a respectively) and position coordinates (x, y) in blue (Fig. 10.b), position z in orange (Fig. 10.a).

After collecting the necessary information in this section, the reconstruction of the 3-D model will be the objective of the next step.

Moving from 2-D to 3-D. The basic model: we have an adult male spine model (Fig. 11) with seven cervical vertebrae, twelve thoracic vertebrae, five lumbar vertebrae and an object bringing together the sacral and coccygeal part. (Model link: [23]).

This model will be adjusted based on the patient's information (Age, weight, and sex) in order to estimate its length. Using the data extracted after segmentation, we assemble the spatial coordinates X, Y, and Z by stacking the vertebrae. Each vertebra will therefore have a rotation, in order to reproduce with maximum similarity the shape of the spine of the scoliosis subject.

The 3-D spine reconstruction was implemented using Unity 3D and Vuforia (see Sect. 4 for implementation details).

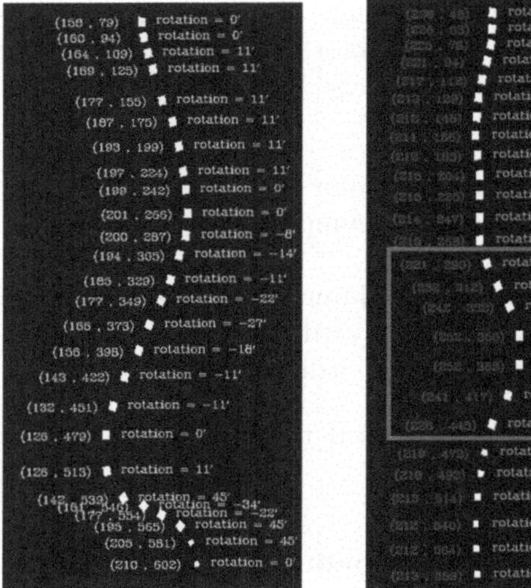

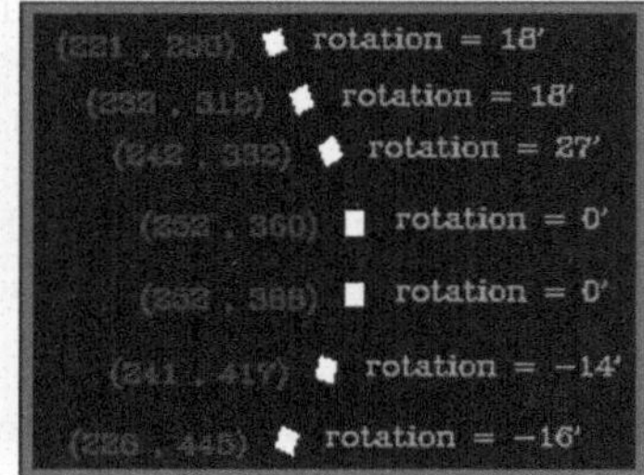

Fig. 10. Results of calculation of rotation angles and positions of each vertebra.

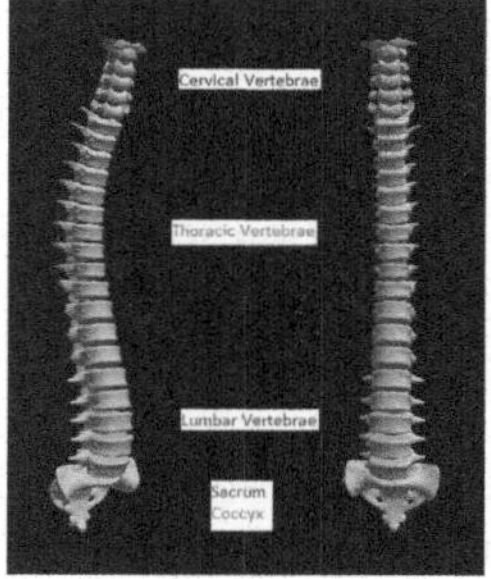

Fig. 11. The used model of the human spinal column [23].

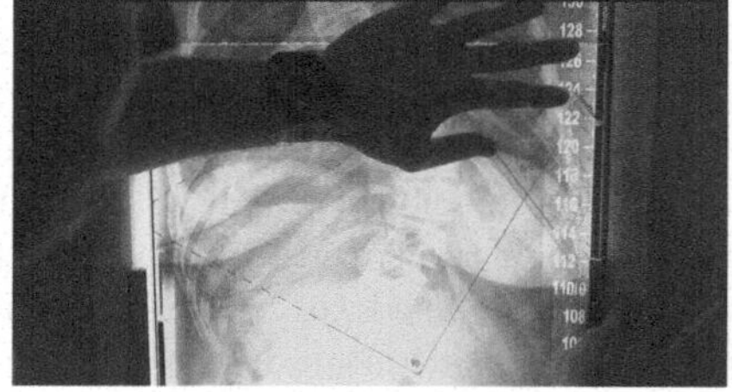

Fig. 12. Manual method for measuring the Cobb-angle using a protractor.

3.2 Scoliosis Measurement

Traditional spine image analysis relies on manual labor and hand-crafted feature extraction to measure scoliosis (Fig. 12). This process takes a lot of time and effort, and it is accompanied by issues like intraobserver and interobserver variations (Fig. 12).

In this section, we suggest using X-ray images to automatically calculate the curvature of the spine. The suggested method's flowchart is displayed in Fig. 13. The proposed method includes three stages: Apex vertebra identification, determination of the most tilted vertebrae above and below the apex, and spine curvature quantification. The steps are described in the following.

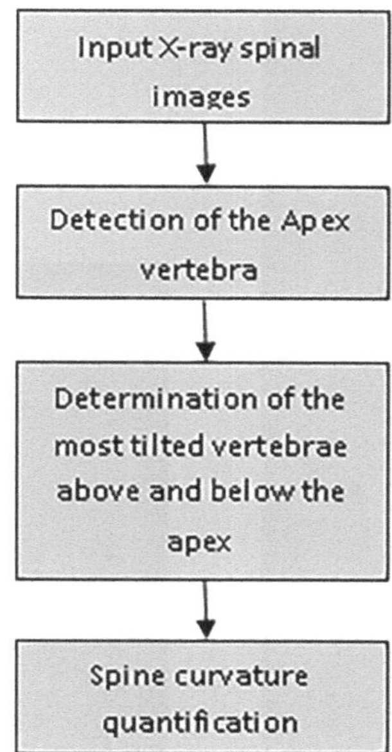

Fig. 13. Flowchart of the proposed Cobb angle measurement method.

Apex Vertebra Identification. For calculating the Cobb angle in a spine X-ray, we have to identify the relevant vertebrae. This first step is very important to extract the vertebra which represents the apex vertebra, a single apex in the case of thoracic, lumbar, or thoracolumbar type scoliosis and two apex in the case of combined type (two curves in different directions). Initially, locate the central line, which is the line that descends from the first vertebra (the orange line in Fig. 14 represents this line). Then, we will search both the left and right sides of this line for the vertebra that is furthest along the X-axis. In Fig. 14, we can observe that there are two vertebrae with the greatest distance on the left and right; the vertebra with the green line is considered as the apex. Remember that the cervical vertebrae (from C1 to C7) do not have any form of scoliosis; hence, the vertebra with the red line will not be included in the computation.

Determination of the Most Tilted Vertebrae Above and Below the Apex. After identifying the apex vertebra in the first part, we will now determine the upper and lower limit vertebrae which are the most important elements for calculating the Cobb angle in our method.

The most tilted vertebra above the apex: Starting at the point of the apex, we evaluate the rotational angle of each landmark while moving higher. We stop when we reach the angle with the largest value; this vertebra is the upper limit vertebra (Fig. 15).

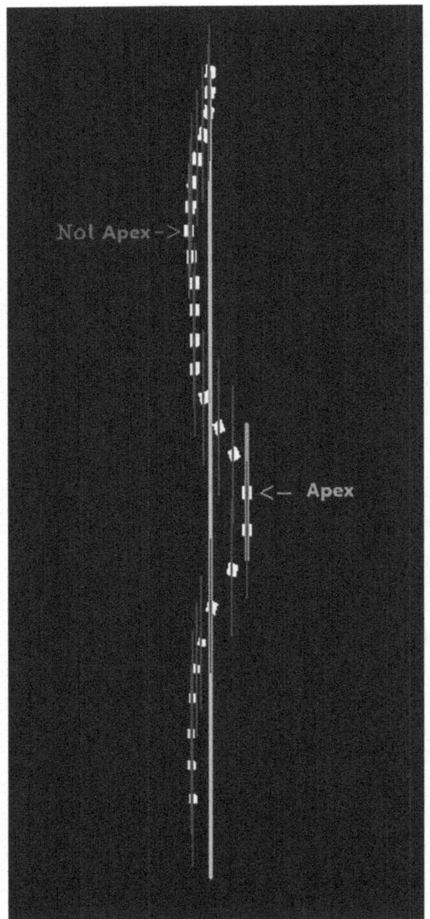

Fig. 14. Apex vertebrae determination.

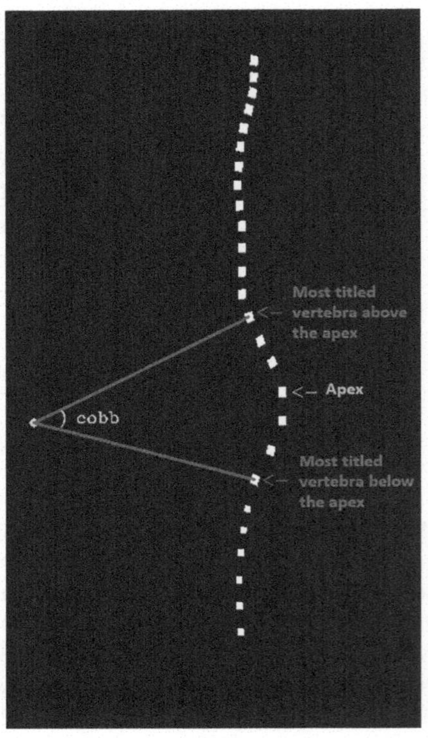

Fig. 15. Determination of the most tilted vertebrae above and below the apex.

The most tilted vertebra below the apex: Similarly, to find this vertebra, we will descend and always search for the angle that has the highest value (Fig. 15).

Once the most tilted vertebrae above and below the apex are identified, tangents are drawn from the upper edge of the superior vertebra and the lower edge of the inferior vertebra. These tangents are then used to calculate the Cobb angles and classify the scoliosis.

Cobb Angle Measurement. The last step calculates the spinal curvature using the Cobb angle measurement criterion (Fig. 15). The Al-Kashi's theorem (law of cosines) mathematical formulas will be used to get the angle. The Al-

Kashi theorem (Fig. 16) is a generalization of the Pythagorean theorem 2 to any triangle, it connects the length of the sides using the cosine of one of the triangle's angles, it is expressed in Eq. (1):

$$c^2 = a^2 + b^2 - 2.a.b.cos(\gamma) \tag{1}$$

Knowing that the length of the three sides of the triangle is known, we can use Eq. (2) to compute the angle γ:

$$\gamma = arccos\left(\frac{a^2 + b^2 - c^2}{2.a.b}\right) \tag{2}$$

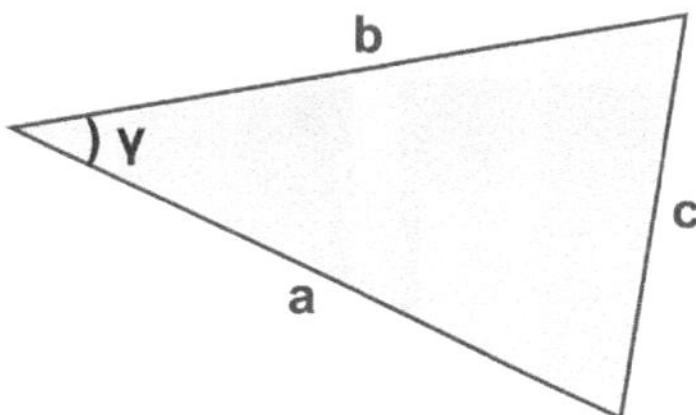

Fig. 16. Al-Kashi's triangle.

The implementation of the scoliosis measurement method described in this subsection is presented in Sect. 4.

4 Experimental Evaluation and Discussion

Experiments were carried out to assess the proposed system and identify the key inputs necessary for precise 3-D reconstruction and measurement of scoliotic spines. Qualitative evaluations were performed on paired PA and lateral X-ray images.

Following the description of our system in the previous section, we arrive at the last part of this article, it's the time to transform the completed study into a functional mobile application that facilitates user interaction.

4.1 Registration and Login

The doctor utilizing our "Spine 3D" application needs to register first and then authenticate using his email address and password (Fig. 17). To validate the registration and for security reasons, we send a validation code to the email address entered by the doctor, who must then enter it to confirm their registration.

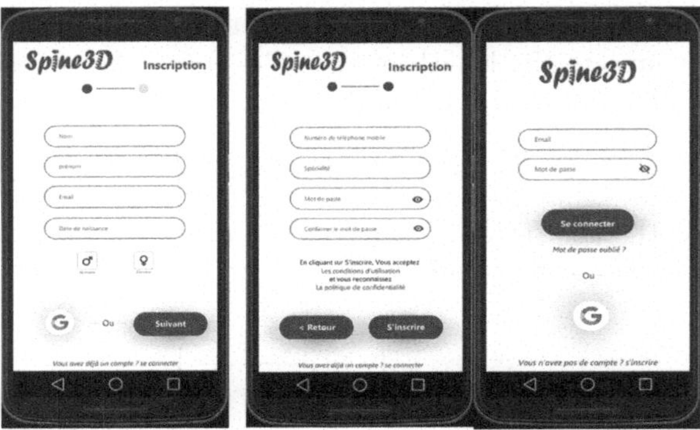

Fig. 17. Registration and login in "Spine 3D" application.

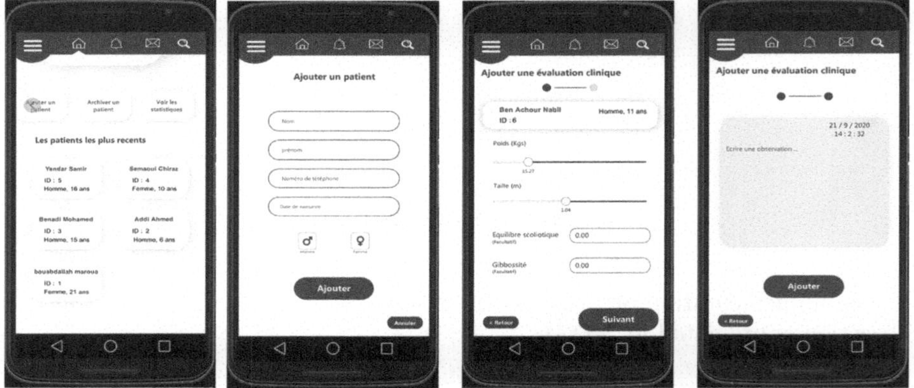

Fig. 18. Patient records management.

4.2 Patient Records Management

The doctor generates a patient profile including weight, height, scoliotic balance, and gibbosity, with optional observations (Fig. 18). The data is subsequently sent to a server and stored in the database.

4.3 X-Ray Images Labeling

After loading the image (Fig. 19), landmarks will be manually placed on each vertebra in the PA and lateral views (Fig. 20). The user will need to click the "Ajouter marqueur" button in order to accomplish this task. This will display a filled red rectangle that the user will place in the center of each vertebra, he will be able to change the rotation of the landmark according to that of the vertebra, increase its dimensions, move it or delete it, in addition, he will be able to zoom the image to manipulate it better. Moreover, the user can position a green marker in the center of the femoral heads which will be useful for calculating a number of clinical indices, including sagittal tilt (Fig. 20).

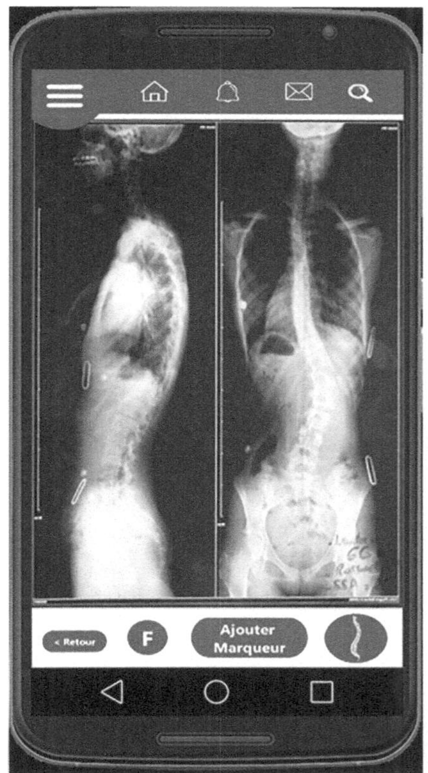

Fig. 19. 2D X-ray image loading.

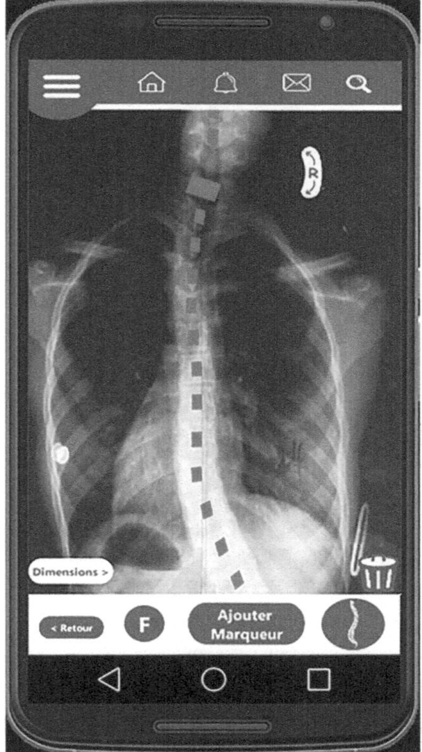

Fig. 20. Placing red landmark on each vertebra. (Color figure online)

The detected landmarks are then used to determine the 3D vertebrae positions and orientations.

4.4 Spine 3-D Representation and Cobb Angle Calculation

The proposed method was evaluated using X-ray images exhibiting various shapes and positions of spinal curvature. Most of the clinical images used in this study were provided by the Physical Medicine and Rehabilitation Service of Bounaama Djilali Hospital (CHU Douera), Algeria.

Once all the landmarks are placed on the frontal and the sagittal images, the image will then be augmented, and the 3-D spine will appear on the same scene and can be manipulated to see it in the three axes (Fig. 21). By clicking “Voir calculs" button, scoliosis measurements, including Cobb, Cyphose, and Lordose angles will be displayed (Fig. 22).

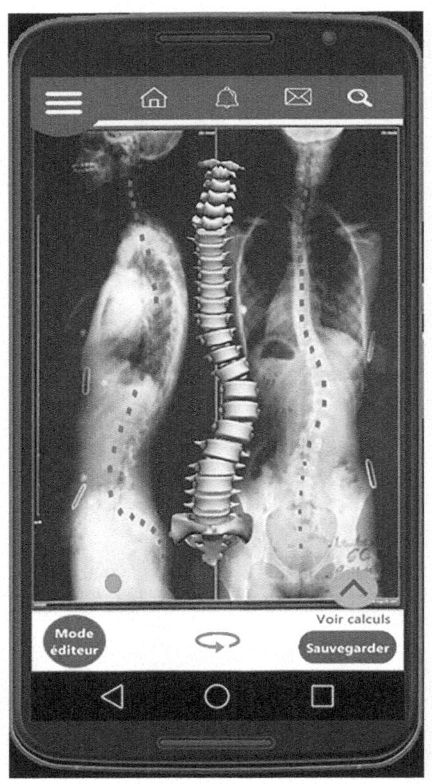

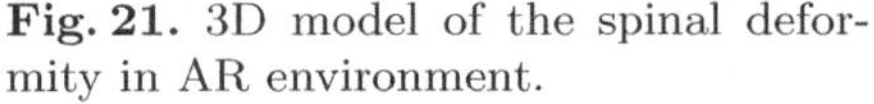

Fig. 21. 3D model of the spinal deformity in AR environment.

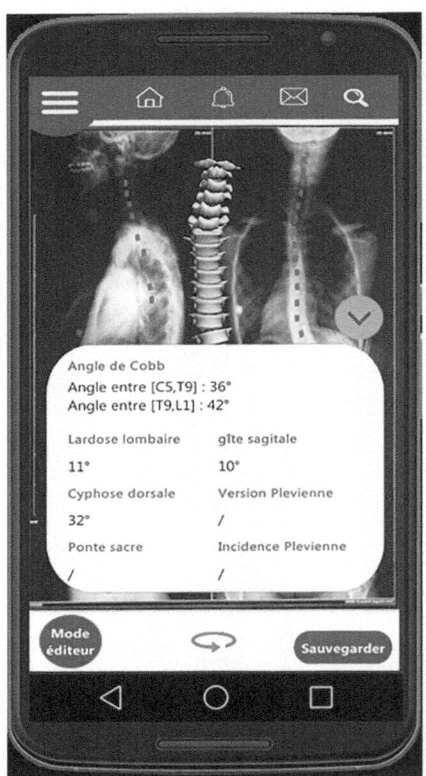

Fig. 22. Cobb angle calculation by the proposed method.

4.5 Quantitative Evaluation

In this experiment, our method was compared with expert manual measurements (Table 2) to evaluate its reliability and effectiveness.

As shown in Table 2, the two results are very close to each other. This confirms that the proposed method is reliable and can be effectively applied to scoliosis measurement from spinal X-ray images.

Table 2. Performance of the proposed method for Cobb angle calculation on selected X-ray images compared with expert manual measurements.

	Our method			Manual measurements of an expert		
X-ray image number	Upper Vert	Lower Vert	Cobb angle	Upper Vert	Lower Vert	Cobb angle
1	T6	T12	7,8	T6	L1	6,4
2	T3	L2	10,1	T4	L2	11,6
3	T12	L4	16,6	T11	L4	18,4
4	T9	L5	13,5	T10	L4	14,4
5	T11	L4	15,9	T10	L4	16,9
6	T9	L1	7,6	T6	L4	7,1

5 Conclusion and Future Perspectives

This paper presents an integrated system combining image processing techniques and augmented reality to generate an interactive 3-D model of the spine using only 2-D X-ray images, offering a practical and low-cost solution. The accompanying application provides a user-friendly interface for doctors to register patient data and perform semi-automated measurements.

Our first contribution is a three-dimensional spine representation system designed to enhance the perception of students and healthcare professionals, providing information that cannot be directly perceived through human senses or standard X-ray imaging.

The second contribution is a method for accurate and user-friendly Cobb angle measurement in clinical settings. The method was tested on real X-ray images obtained from a medical institution. The results showed very close alignment with manual measurements made by an expert, indicating the effectiveness of the proposed system in clinical applications and scoliosis follow-up.

However, the system relies on manual landmark placement, which introduces a high potential for user-related error and can affect measurement accuracy. Additionally, the number of images used in the experimental evaluation is limited reducing the generalizability of the results. These issues could be addressed in future work or an extended version of the study.

A statistical analysis to measure the deviation or confidence intervals of the results and comparisons with other automated methods in the field are also the objectives of future work or an extended version of the study.

Acknowledgement. The authors express their gratitude to Noureddine Aouaa for his contribution to this study.

References

1. Atlas of Clinical Gross Anatomy. Elsevier Health Sciences (2012)
2. Gesbert, J.-C.: Modélisation 3D du rachis scoliotique: fusion de donnes et personnalisation exprimentale. PhD thesis, IRMAR - Institut de Recherche Mathematique de Rennes, Université Rennes 1 (2013)
3. Janicki, J.A., Alman, B.: Scoliosis: review of diagnosis and treatment. Paediatr. Child Health **12**(9), 771–776 (2007). November
4. Benmansour, F.: La scoliose touche 2 5[percent] de la population en algrie. LIBERTE, 6:– (2014)
5. HHPublishing. Scoliosis (2016). https://www.health.harvard.edu/a_to_z/scoliosis-a-to-z
6. Cobb, J.: Outline for the study of scoliosis. Instr. Course Lect. **5**, 261–275 (1948)
7. Al-Zubaer Imran, A., et al.: Analysis of Scoliosis From Spinal X-Ray Images. CoRR abs/2004.06887 (2020)
8. Kim, H., et al.: Scoliosis imaging: what radiologists should know. Radiographics **30**(7), 1823–1842 (2010)
9. Horng, M.-H., et al.: Cobb angle measurement of spine from X-ray images using convolutional neural network. Comput. Math. Methods Med. **2019**, 6357171 (2019)
10. Azuma, R.T.: A survey of augmented reality. In: Presence: Teleoperators and Virtual Environments, Vol. 6, No. 4, pp. 355–385 (1997)
11. Sasa, C., et al.: Conceptual augmented reality framework for spinal disorders representation and diagnosis. In: 2nd Regional Conference Mechatronics in Practice and Education - MECHEDU2013, Subotica (2013)
12. Li, H., Leow, W.K., Huang, C.H., Howe, T.S.: Modeling and measurement of 3D deformation of scoliotic spine using 2D X-ray images. In: Computer Analysis of Images and Patterns: 13th International Conference, CAIP 2009, Münster, Germany, September 2–4: Proceedings 13, pp. 647–654. Springer, Berlin Heidelberg (2009)
13. Novosad, J., Cheriet, F., Petit, Y., Labelle, H.: Three dimensional (3-D) reconstruction of the spine from a single X-Ray image and prior vertebra models. IEEE Trans. Biomed. Eng. **51**(9), 1628–1639 (2004). September
14. Kusuma, B.A.: Determination of spinal curvature from scoliosis X-ray images using K-means and curve fitting for early detection of scoliosis disease. ICITISEE (2017)
15. Moura, D.C., Correia, M.V., Barbosa, J.G., Reis, A.M., Laranjeira, M., Gomes, E.: Automatic vertebra detection in x-ray images. In: Proceedings of the International Symposium CompIMAGE 2006, pp. 307–312, Coimbra, Portugal (2006)
16. Okashi, O.A., Du, H., AI-Assam, H.: Automatic spine curvature estimation from X-ray images of a mouse model. Comput. Methods Programs Biomed. **140**, 175–184 (2017)
17. Wu, H., Bailey, C., Rasoulinejad, P., Li, S.: Automatic landmark estimation for adolescent idiopathic scoliosis assessment using BoostNet. In: International Conference on Medical Image Computing and Computer-Assisted Intervention, pp. 127–135. Springer (2017)
18. Wu, H., Bailey, C., Rasoulinejad, P., Li, S.: Automated comprehensive adolescent idiopathic scoliosis assessment using MVC-Net. Med. Image Anal. **48**, 1–11 (2018)
19. Sun, H., Zhen, X., Bailey, C., Rasoulinejad, P., Yin, Y., Li, S.: Direct estimation of spinal cobb angles by structured multi-output regression. In: International Conference on Information Processing in Medical Imaging, pp. 529–540. Springer (2017)

20. Imran, A.-A.-Z., et al.: Bipartite distance for shape-aware landmark detection in spinal X-ray images. In: Medical Imaging Meets NeurIPS Workshop. Vancouver, Canada (2019). December
21. Cukovic, S., Petruse, R. E., Meixner, G., Buchweitz, L.: Supporting diagnosis and treatment of scoliosis: using augmented reality to calculate 3D spine models in real-time-ARScoliosis. In: 2020 IEEE International Conference on Bioinformatics and Biomedicine (BIBM) (pp. 1926–1931). IEEE (2020)
22. National Center for Biotechnology Information "PubMed" (2024). https://pubmed.ncbi.nlm.nih.gov. Accessed May 2025
23. The human spinal column. https://sketchfab.com/3d-models/the-human-spinal-column-bcd9eee09ce044ef98a69c315aa792e2

Stacked Deep Learning Models Leveraging Denoising Autoencoder-Based Features for Student Performance Prediction

Amani Khalifa[1](✉), Fatma BenSaid[2], and Yessine Hadj Kacem[1]

[1] ReDCAD Laboratory, Enis, University of Sfax, B.P. 1173, Sfax 3038, Tunisia
amani.khalifa@fsegs.usf.tn, yessine.hadjkacem@enis.tn

[2] REGIM Laboratory, Enis, University of Sfax, Soukra road km 4, Sfax 3038, Tunisia
fatma.bensaid@enis.usf.tn

Abstract. Accurately predicting student academic performance remains a persistent challenge in educational data mining, particularly when dealing with limited, imbalanced, and noisy datasets. The present study proposes hybrid models in which denoising autoencoders (DAEs) are employed for feature extraction, followed by a range of deep learning classifiers—namely recurrent neural networks (RNN), long short-term memory networks (LSTM), artificial neural networks (ANN), deep neural networks (DNN), gated recurrent units (GRU), and convolutional neural networks (CNN)—to improve classification accuracy across multiple performance levels. In addition, a Stacking ensemble strategy is employed to integrate the strengths of individual models and enhance overall robustness. The experimental evaluation was conducted on a real-world educational dataset, transformed into a multi-class classification task reflecting four academic performance categories. The results demonstrate that the Stacking model achieved the highest accuracy (98.45%), followed by the DAE_RNN (Denoising Autoencoder with Recurrent Neural Network) and DAE_GRU (Denoising Autoencoder with Gated Recurrent Unit) models, which also exhibited strong class-wise F1-scores. Recurrent models consistently outperformed feedforward and convolutional counterparts, highlighting the importance of temporal pattern modeling in academic data. The proposed approach demonstrates the effectiveness of combining denoising-based representation learning with deep neural architectures to address classification challenges in the educational domain, offering a scalable and robust solution for learning analytics applications.

Keywords: Student performance prediction · Denoising autoencoder · Stacking · Deep neural architectures

1 Introduction

Predicting student performance has become a key task in educational data mining, aiming to support early intervention strategies and enhance decision-making

T. Ensari et al. (Eds.): ISPR 2025, CCIS 2859, pp. 20–34, 2026.
https://doi.org/10.1007/978-3-032-21585-7_2

in academic environments. The increasing availability of educational data provides an opportunity to develop predictive models that can identify at-risk students and improve educational outcomes [14]. Traditional machine learning approaches often rely on handcrafted features and may struggle to capture the complex, non-linear patterns inherent in educational datasets. This motivates the use of deep learning architectures that can automatically learn meaningful representations from raw data. To address this limitation, deep learning techniques, particularly autoencoders, have gained attention due to their ability to learn compact and informative feature representations in an unsupervised manner [10]. In particular, the Denoising Autoencoder (DAE) has shown effectiveness in extracting robust features by learning to reconstruct clean inputs from their corrupted versions, thereby enhancing generalization capabilities. The integration of DAE with deep classifiers represents a novel approach for handling noisy and complex educational datasets. In the present study, we propose a hybrid predictive framework that integrates a DAE-based feature extraction module with a deep neural network classifier. The autoencoder component is trained to encode the input features into a compressed latent representation, which is then fed into a deep learning model for classification. This architecture is particularly suited for complex tabular educational data, where the relationships among features may be noisy or redundant. A critical challenge in educational datasets is class imbalance, where the number of students in each performance category is uneven. Such imbalance often biases the learning process toward the majority class, leading to suboptimal performance. To mitigate this issue, we employ the Synthetic Minority Over-sampling Technique (SMOTE) [6], a widely used oversampling method that generates synthetic samples of the minority class by interpolating between existing examples. By applying SMOTE prior to training, the proposed approach ensures a more balanced learning environment, improving the model's ability to generalize across all classes.

The effectiveness of the proposed pipeline is evaluated on a real-world student performance dataset collected from higher education, referred to as GCSDV1. Experimental results demonstrate that the combination of robust feature extraction, balanced data distribution, and deep learning-based classification leads to high predictive accuracy and strong generalization across all student performance categories. These aspects collectively highlight the novelty, sound methodology, and practical contributions of the proposed approach. The key contributions of the present study are:

- The proposal of an innovative predictive approach that combines denoising autoencoder-based feature extraction with a diverse set of deep learning architectures, including RNN, LSTM, ANN, DNN, GRU, and CNN, in order to effectively capture complex patterns within student performance data.
- The integration of a stacking ensemble strategy built upon the base models, aiming to improve predictive reliability and performance by exploiting the complementarity of individual learners and the strength of meta-level aggregation.

The remainder of this paper is organized as follows. Section 2 reviews related work. Section 3 presents the proposed methodology. Section 4 presents the experimental results and discussion, emphasizing the comparative performance of the proposed approach. Finally, Sect. 5 concludes the study and outlines potential directions for future work.

2 Related Work

Predicting student performance using deep learning has received growing attention due to its ability to handle sequential, high-dimensional, and noisy educational data.

Several recent contributions have highlighted the effectiveness of hybrid deep learning architectures for student performance prediction, particularly when combining spatial and temporal modeling capabilities. For instance, the study by [1] validated the advantage of hybrid CNN–LSTM architectures on academic performance datasets collected from two different educational institutions. Their approach achieved up to 98.9% accuracy, confirming the superiority of models that jointly exploit spatial and temporal information in educational data.

Building on this idea, [18] explored convolutional neural networks (CNN) combined with LSTMs in an ensemble framework to capture both spatial and temporal features from multivariate student activity logs. Using the KDD Cup 2015 dataset, their model reached 94% accuracy and an F1-score of 98%, demonstrating the effectiveness of convolutional layers in identifying local patterns alongside temporal dependencies for student performance prediction.

In a different learning environment, [2] developed a hybrid Artificial Neural Network and Long Short Term Memory model aimed at early dropout prediction in Massive Open Online Courses (MOOCs). Their approach leverages the temporal dependencies captured by LSTMs along with the ANN's capability for non-linear transformations. The model achieved approximately 70% accuracy at early checkpoints, outperforming standalone LSTM and Gated Recurrent Unit (GRU) architectures, thus highlighting the benefit of hybrid recurrent models in early intervention scenarios.

However, one common limitation across these works lies in the treatment of input data, which often contain noise or redundancy due to the nature of educational records. Addressing this challenge, [19] introduced denoising autoencoders (DAEs) as a method to reconstruct clean inputs from corrupted versions, thereby learning stable and robust feature representations. Although their application in educational data mining (EDM) remains limited, DAEs have shown promising results in various domains involving structured and tabular data. This suggests that DAEs can effectively reduce noise and redundancy in student records, potentially enhancing the predictive accuracy of models relying on complex educational datasets.

Although the individual contributions of denoising autoencoders, deep learning classifiers, and SMOTE [7] have been explored, few studies combine these components in an integrated methodological paradigm for student performance prediction. This paper aims to fill this gap by proposing a novel modeling approach that leverages robust feature extraction, data augmentation, and deep learning classification to address imbalanced educational data effectively.

Despite numerous advances in predictive modeling for student performance, existing approaches often fail to address multiple interrelated challenges in a unified manner. In educational datasets, noisy and incomplete data are common due to errors in data collection, inconsistencies in grading, and missing values, which can significantly hinder predictive accuracy. DAEs have demonstrated strong potential in mitigating such noisy inputs by learning robust feature representations that capture essential data patterns while filtering out corruption. However, despite their effectiveness, DAEs remain underutilized in the domain of student performance prediction. At the same time, models that achieve high predictive accuracy frequently lack robustness when facing imbalanced data quality. This fragmented treatment exposes a clear gap in current research. To bridge this gap, we introduce a comprehensive deep learning approach that combines DAE-based noise reduction, multiple deep neural models (RNN, LSTM, ANN, DNN, GRU, CNN), and SMOTE-based data augmentation.

3 Research Methodology

The methodology adopted in the present study is structured around a robust and innovative deep learning approach designed to address the complex challenge of student performance classification, particularly in the presence of significant class imbalance. The proposed approach is articulated in three main stages: data preprocessing, feature extraction via a Denoising Autoencoder (DAE), and the construction of a stacking ensemble classifier based on diverse deep learning models (see Fig. 1).

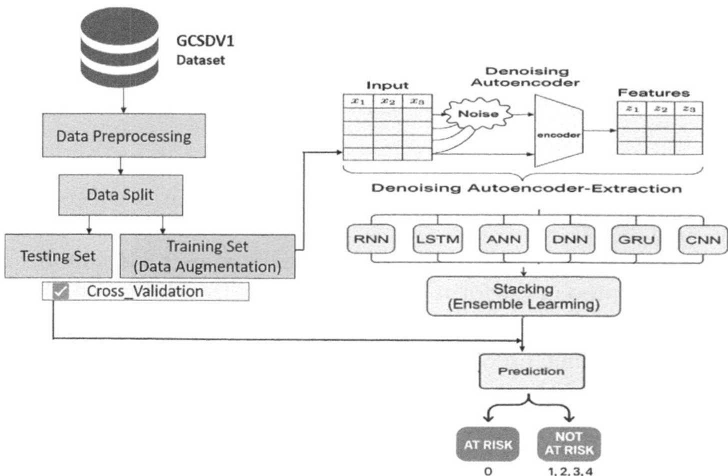

Fig. 1. Proposed methodology

3.1 Theoretical Grounding of the Proposed Methodology

Let the dataset be denoted by $\mathcal{D} = \{(x_i, y_i)\}_{i=1}^{N}$, where each $x_i \in \mathbb{R}^d$ denotes a d-dimensional observation associated with the i-th student, and $y_i \in \mathcal{Y}$ corresponds to the ground truth label. In order to ensure statistical coherence across features, each variable is rescaled using min-max normalization, as shown in Eq. (1):

$$x_i^{(j)} \leftarrow \frac{x_i^{(j)} - \min(x^{(j)})}{\max(x^{(j)}) - \min(x^{(j)})}, \quad \forall j \in \{1, \ldots, d\}. \tag{1}$$

To alleviate class imbalance, which is recurrent in real-world academic datasets, a synthetic oversampling method denoted $\mathcal{R}$ (SMOTE) is applied, yielding an enriched training set $\mathcal{D}' = \mathcal{R}(\mathcal{D})$ that preserves class distributional symmetry and reduces bias in model estimation.

Following training data augmentation, the representation space is transformed through the construction of a denoising autoencoder (DAE). This neural architecture is designed to capture robust latent structures by reconstructing original observations from their noise-corrupted versions. Formally, the encoder $\Phi : \mathbb{R}^d \rightarrow \mathbb{R}^k$ maps each perturbed input $\tilde{x}_i$ to a latent representation z_i, as defined in Eq. (2):

$$z_i = \Phi(\tilde{x}_i) = f(W_e \tilde{x}_i + b_e), \tag{2}$$

where $W_e \in \mathbb{R}^{k \times d}$ and $b_e \in \mathbb{R}^k$ are the encoder parameters, and f denotes a non-linear activation function such as ReLU. The decoder reconstructs the input from z_i according to Eq. (3):

$$\hat{x}_i = \Psi(z_i) = f(W_d z_i + b_d), \tag{3}$$

and the network is optimized by minimizing the mean squared reconstruction loss as shown in Eq. (4):

$$\mathcal{L}_{\text{rec}} = \frac{1}{N} \sum_{i=1}^{N} \|x_i - \hat{x}_i\|^2. \tag{4}$$

Upon convergence, the encoder Φ is retained as a feature extractor, generating a new representation space $\mathcal{F}_{\text{DAE}} = \{\Phi(x_i)\}_{i=1}^{N}$ that is subsequently used for classification.

To integrate multiple perspectives on the classification task, a model aggregation strategy is adopted in the form of stacked generalization (ensemble learning). Let $\{M_1, M_2, \ldots, M_K\}$ denote a collection of K distinct deep models trained on $\mathcal{F}_{\text{DAE}}$. Each base learner produces an individual prediction $\hat{y}_i^{(k)} = M_k(\Phi(x_i))$, and a meta-learner H is trained to combine these outputs. The final prediction is obtained as described in Eq. (5):

$$\hat{y}_i = H\left(\hat{y}_i^{(1)}, \hat{y}_i^{(2)}, \ldots, \hat{y}_i^{(K)}\right). \tag{5}$$

The hierarchical formulation enables the learning process to benefit from the complementary inductive biases of the base models while enhancing generalization performance through joint inference. The overall methodology thus combines robust representation learning with strategic model fusion to address the dual challenges of high-dimensionality and imbalanced class distribution in educational datasets.

3.2 GCSDV1 Description

[11] introduced the initial version of the dataset, referred to as GCSD, which contained 13 attributes and 358 academic and demographic records collected between 2014 and 2022. In the present study, we introduce an improved version, named GCSDV1, which expands the temporal coverage to span from 2010 to 2022 and includes a substantially larger number of records. The observation window was extended by four additional years, capturing data from 2010 to 2014. The computer science department was selected for its higher data availability compared to other departments. All records were sourced from the same governmental institution to ensure consistency and reliability. This study utilizes the *GCSDV1* dataset, a multi-class educational dataset specifically curated for academic performance prediction. It comprises 1,618 student records, each labeled according to one of four distinct performance categories (0, 2, 3, and 4), ranked from lowest to highest performance. Table 1 presents the attributes (features) used from the GCSDV1 dataset.

Table 1. GCSDV1 Dataset Features

Attribute	Description	Type
stdID	ID for each student	String
Batch	Year of first registration	Integer
Gender	Female/male	Integer
Math1	Mathematics score of the first semester	Real
Algo and Prog1	Algorithms and Programming score of the first semester	Real
OS and arch	Operating systems and architecture score of the first semester	Real
Logic and Multi	Logic and multimedia score of the first semester	Real
Communic Skill1	Transversal unit score of the first semester	Real
Math2	Mathematics score of the second semester	Real
Algo and Prog2	Algorithms and Programming score of the second semester	Real
OS and network	Operating systems and networks score of the second semester	Real
DB	Database Systems score of the second semester	Real
Communic Skill2	Transversal unit score of the second semester	Real

3.3 GCSDV1 Labelling

In the *GCSDV1* dataset, student performance was categorized based on the average score (AS) from 0 to 20. Students with AS below 10 were considered at-risk (class 0), while students with higher scores were grouped into four additional classes reflecting increasing performance levels, from "pass" to "very good". Specifically, class 4 corresponds to an AS between 10 and 12 ("pass"), class 3 to 12–14 ("fairly good"), class 2 to 14–16 ("good"), and class 1 to 16–20 ("very good"). It should be noted that the dataset contains no instances corresponding to class 1.

3.4 GCSDV1 Preprocessing and Class Rebalancing

The initial step involves rigorous data preparation. All input features are normalized using the MinMaxScaler [12]to ensure that their values lie within a common numerical range. Following normalization, the dataset is split into training and test subsets using a stratified holdout strategy [16] that preserves the original distribution of classes in both partitions. However, a major challenge lies in the strongly imbalanced nature of the GCSDV1 dataset. As illustrated in Fig. 2, class 0 (at-risk) overwhelmingly dominates with over 1,000 instances, whereas class 2 is severely underrepresented with only 33 samples. Classes 3 and 4 contain 217 and 347 instances, respectively. This imbalance introduces a strong bias in favor of the majority class, leading to skewed learning and degraded performance on minority classes.

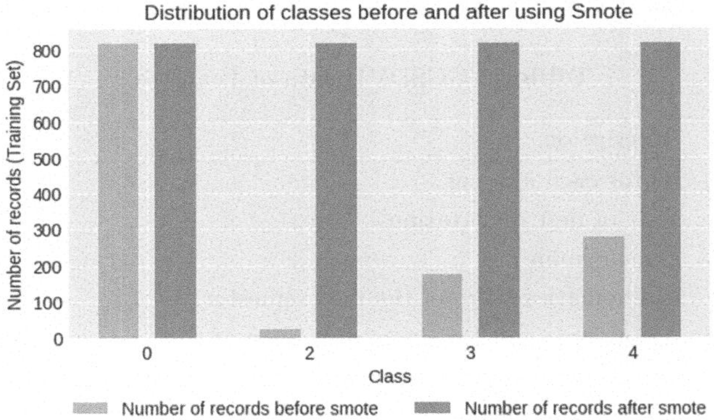

Fig. 2. GCSDV1: Distribution of classes before and after applying SMOTE

To mitigate this effect, the Synthetic Minority Over-sampling Technique (SMOTE) is employed. SMOTE generates synthetic data points by interpolating between existing instances from the minority classes, thereby enriching their representation without duplicating original records. After applying SMOTE to the training set, each class is balanced to contain 680 instances, yielding a uniform class distribution that allows the learning algorithms to equally attend to all categories. Importantly, SMOTE is applied exclusively to the training data to prevent information leakage into the test set. This strategy ensures that the models learn under balanced conditions, while evaluation remains faithful to the original class distributions, thereby preserving the validity of performance estimates.

3.5 Feature Extraction with Denoising Autoencoder

At the core of the proposed approach lies a Denoising Autoencoder (DAE), structured to learn compact and noise-invariant representations from input features. As illustrated in Fig. 3, the architecture follows a symmetrical encoder-decoder design. The encoder compresses the input data through successive dense layers of sizes 128, 64, and 32, producing a compact 32-dimensional latent representation. This bottleneck vector serves as the input to the decoder, which mirrors the encoder by expanding the data through layers of sizes 32, 64, and 128 to reconstruct the original input. Each hidden layer is regularized using L1 (1e-5) and L2 (1e-4) kernel penalties to encourage sparsity and improve generalization. In addition, Dropout layers with a rate of 0.2 are applied to mitigate overfitting. During training, Gaussian noise with a noise factor of 0.1 is injected into the input data to enhance robustness. The model is optimized using the Adam optimizer [21], and the training objective is to minimize the mean squared error (MSE) between the original and reconstructed inputs. Training proceeds for 100 epochs with a batch size of 64, and an early stopping strategy (patience = 15) is employed to monitor the validation loss and restore the best-performing weights.

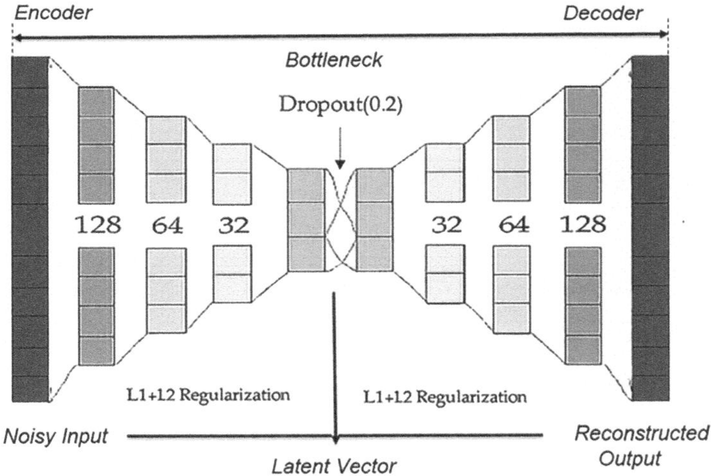

Fig. 3. Proposed denoising autoencoder (DAE) architecture

3.6 Stacked Hybrid Deep Learning Architecture

The encoded features are fed into a set of heterogeneous deep learning classifiers, each designed to capture distinct temporal, spatial, or hierarchical patterns within the data. As shown in Fig. 4, this ensemble includes ANN [4] for capturing basic non-linear feature interactions, DNN [5]for modeling complex abstractions, RNN [15] and GRU [8] for sequential dependencies, LSTM networks [9] for long-range pattern recognition, and CNN [13] for detecting local feature patterns.

Each model is architected with task-specific layers, activation functions such as ReLU, dropout mechanisms for regularization, and tuned hyperparameters. The final output layer employs a softmax activation function and the compilation is carried out using the Adam optimizer with a categorical cross-entropy loss function and accuracy as the evaluation metric. Furthermore, model evaluation is conducted using a 5-fold stratified cross-validation scheme with 50 training epochs, a batch size of 32, and training verbosity set to zero to ensure consistency and robustness across splits.

To leverage the complementarity of the individual models, a stacking ensemble approach [20] is employed. In contrast to hard and soft voting strategies [3], which aggregate predictions at the decision level, stacking facilitates the integration of meta-knowledge by employing a meta-learner trained to effectively combine the outputs of base models, thus enhancing the overall predictive accuracy. In this scheme, a *Random Forest* classifier serves as the meta-model to aggregate the predictions of the base learners. This choice is motivated by the Random Forest's robustness against overfitting, achieved through the ensemble of multiple decision trees. Consequently, the meta-model enhances the overall predictive accuracy of the ensemble.

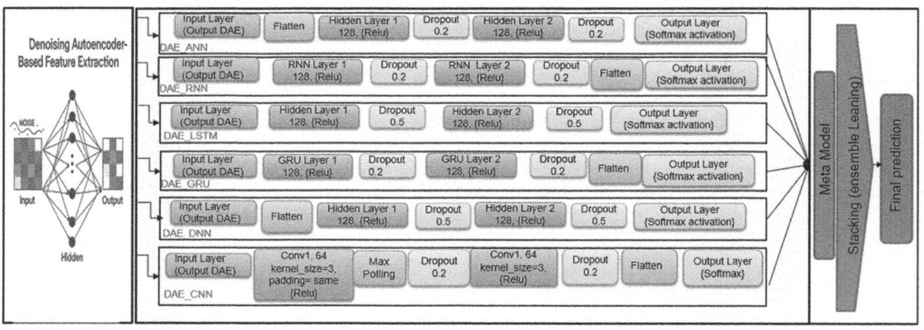

Fig. 4. Stacked Hybrid Deep Learning Architecture

4 Experimental Results and Discussion

The results are divided into two parts: the first examines the denoising autoencoder's ability to produce compact features, while the second evaluates the predictive performance of deep learning models trained on these representations, including both standalone and ensemble models with and without DAE.

4.1 Effectiveness of Denoising Autoencoder in Feature Extraction

The training dynamics of the autoencoder, as applied to the GCSDV1 dataset, are depicted in Fig. 5. The loss curves demonstrate rapid convergence during the initial epochs, followed by a stabilization phase where both the training and validation losses plateau around 0.02. The minimal gap between the training and validation losses throughout the training process indicates strong generalization performance and suggests that the model effectively learned robust latent representations without overfitting.

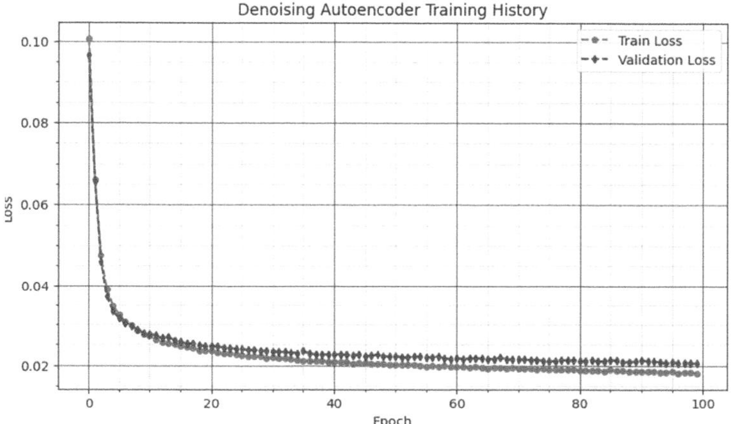

Fig. 5. Training and validation loss curves for the DAE on the GCSDV1 dataset

4.2 Evaluation of Prediction Models

A comparative analysis was conducted on multiple deep learning models, with and without a DAE. Table 2 shows model performance with and without DAE, including averaged precision, recall, F1-score, and accuracy.

The stacking ensemble consistently outperformed all individual models across every evaluation metric. Specifically, Stacking with DAE achieved an accuracy of 98.45%, representing the largest improvement over its non-DAE counterpart (accuracy 94.14%). Among the individual models, DAE_RNN and DAE_GRU recorded competitive performances, improving from 95.73% to 97.42% and from 92.31% to 96.12%, respectively. DAE_CNN increased accuracy from 91.57% to 94.84%, while DAE_LSTM, DAE_DNN, and DAE_ANN showed smaller gains, rising from 85.35%, 90.23%, and 87.11% to 91.17%, 90.44%, and 90.08%, respectively. These results clearly demonstrate that incorporating DAE consistently improves performance across all architectures, with the stacking ensemble achieving the highest overall effectiveness in this experimental setup.

To further assess the classification behavior of the stacking ensemble (with DAE), a confusion matrix was generated on the test set, as illustrated in Fig. 6. The model demonstrated excellent discriminative capability, particularly for the At-risk student class (label 0), with 192 correctly classified instances out of 197. The Pass class (label 4) was perfectly predicted, with all 76 instances correctly

Table 2. Overall performance metrics for all models with and without DAE (in %)

Model	Precision	Recall	F1-score	Accuracy
CNN (without DAE)	91.73%	91.57%	91.59%	91.57%
DAE_CNN	94.97%	94.84%	94.84%	94.84%
Improvement	+3.24	+3.27	+3.25	+3.27
LSTM (without DAE)	85.41%	85.35%	85.34%	85.35%
DAE_LSTM	91.78%	91.17%	91.14%	91.17%
Improvement	+6.37	+5.82	+5.80	+5.82
DNN (without DAE)	90.56%	90.23%	90.26%	90.23%
DAE_DNN	90.71%	90.44%	90.46%	90.44%
Improvement	+0.15	+0.21	+0.20	+0.21
GRU (without DAE)	92.35%	92.31%	92.31%	92.31%
DAE_GRU	95.60%	96.12%	95.86%	96.12%
Improvement	+3.25	+3.81	+3.55	+3.81
RNN (without DAE)	95.84%	95.73%	95.73%	95.73%
DAE_RNN	97.45%	97.42%	97.41%	97.42%
Improvement	+1.61	+1.69	+1.68	+1.69
ANN (without DAE)	87.29%	87.11%	87.12%	87.11%
DAE_ANN	90.31%	90.08%	90.10%	90.08%
Improvement	+3.02	+2.97	+2.98	+2.97
Stacking (without DAE)	94.31%	94.14%	94.15%	94.14%
Stacking (with DAE)	**98.45%**	**98.44%**	**98.43%**	**98.45%**
Improvement	+4.14	+4.30	+4.28	+4.31

classified. The Fairly good (label 3) and Very good (label 2) classes also exhibited strong results, with most instances accurately identified and only minimal misclassifications observed.

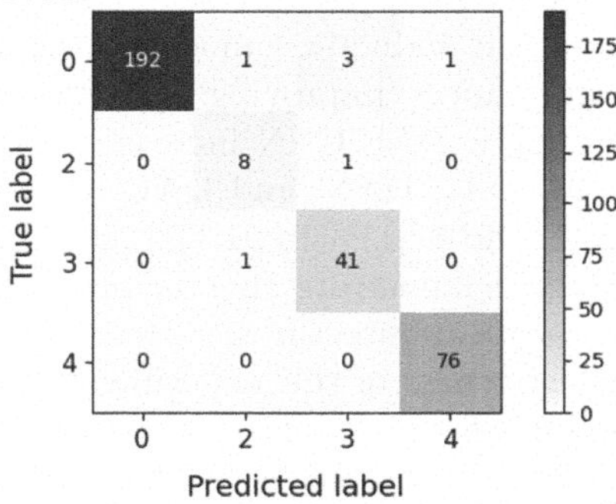

Fig. 6. Confusion Matrix of Stacking model

4.3 Discussion

The comparative evaluation highlights the effectiveness of deep learning models enhanced with DAEs for student performance classification. Recurrent architectures, particularly RNN and GRU, consistently outperformed others by capturing temporal dependencies inherent in educational data, such as semester-wise academic progression. By retaining long-term information through gated mechanisms, recurrent layers enable more accurate and context-aware predictions. They are especially effective in detecting variations like sudden improvement or decline in performance signals often missed by feedforward models. Incorporating DAEs further enhances feature representation, reducing noise and improving model robustness. Coupling these models with a stacked ensemble further amplifies performance. The Stacking approach combines recurrent and feedforward learners, leveraging their complementary strengths to construct a more robust meta-learner. In our experiments, it achieved superior accuracy and balanced F1-scores, with notable improvements over their non-DAE counterparts, effectively handling class imbalance and subtle inter-class variations. Overall, the integration of DAEs, recurrent modeling, and ensemble learning mitigates feature noise, captures sequential patterns, and improves generalization. Its robustness is reflected in the ROC curves (Fig. 7), with AUC values reaching 0.99–1.00 across all classes. This demonstrates strong discriminative capability, especially for models incorporating DAEs, which consistently outperform their standard versions, even for underrepresented categories. Although the evaluation was conducted on the GCSDV1 dataset from a single institution, the strong performance achieved suggests that the proposed approach is effective and has the potential to generalize to other academic settings with similar patterns. Beyond education, the proposed methodology could be applied to other domains such as healthcare or business, where datasets often exhibit noise, class imbalance, and sequential patterns.

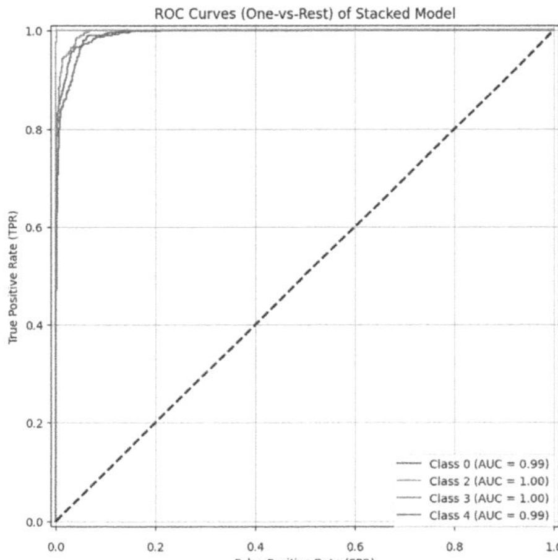

Fig. 7. ROC curves (One-vs-Rest) for the Stacking model.

4.4 Threats to Validity

In accordance with the guidelines provided by [17], the following section outlines the threats to the validity of the present research study. These threats concern internal validity, external validity, and construct validity.

• Internal Validity: Internal validity is impacted by the dataset of this research. We utilized a real case study dataset, including academic and demographic data of computer science students. The size and nature of the dataset might also impact the results, particularly when implemented in different learning environments.

• External Validity: External validity concerns the generalizability of the results. Although we used a dataset from a public institution, we believe that the proposed model can be applied to various other educational datasets. However, differences in student populations, academic structures, and data collection methods may affect how well the results can be generalized to other institutions or fields of study.

• Construct Validity: Construct validity relates to the reliability of the features used in the model. In this study, we focused on academic and demographic attributes. While these features are indicative of student success, other factors such as behavioral data could further enhance the model's performance.

5 Conclusion and Future Work

The present study investigated the effectiveness of several deep learning architectures, enhanced with denoising autoencoders (DAEs), for multi-class student performance prediction. The evaluation encompassed a range of models, including RNN, LSTM, ANN, DNN, GRU and CNN, as well as a Stacking-based ensemble approach. The results revealed that recurrent models, particularly RNN and GRU variants, outperformed feedforward and convolutional architectures by effectively capturing sequential dependencies in student records. Moreover, the Stacking strategy achieved the best overall performance, demonstrating the benefit of combining heterogeneous learners for more accurate and robust classification. **Future work** will focus on leveraging generative artificial intelligence to enrich educational datasets through realistic synthetic data generation, aiming to mitigate class imbalance and improve model generalization. Additionally, the integration of generative models such as Variational Autoencoders (VAEs) or Generative Adversarial Networks (GANs) could provide novel avenues for feature discovery and explainable learning in student performance modeling.

Acknowledgements. Doctor Fatma BenSaid acknowledges funding from the Ministry of Higher Education and Scientific Research of Tunisia (Grant number: LR11ES48).

Disclosure of Interests. The authors have no competing interests to declare that are relevant to the content of this article. The authors respected the privacy of students' personal information while taking precautions to protect it. All student data used in this study were anonymized to prevent personal identification.

References

1. Adefemi, K.O., Mutanga, M.B.: A robust hybrid CNN-LSTM model for predicting student academic performance. Digital **5**(2), 16 (2025)
2. Al-Azazi, F.A., Ghurab, M.: ANN-LSTM: a deep learning model for early student performance prediction in MOOC. Heliyon **9**(4) (2023)
3. Awe, O.O., Opateye, G.O., Johnson, C.A.G., Tayo, O.T., Dias, R.: Weighted Hard and Soft Voting Ensemble Machine Learning Classifiers: Application to Anaemia Diagnosis, pp. 351–374. Springer Nature Switzerland (2023). https://doi.org/10.1007/978-3-031-41352-0_18
4. Basheer, I., Hajmeer, M.: Artificial neural networks: fundamentals, computing, design, and application. J. Microbiol. Methods **43**(1), 3–31 (2000). https://doi.org/10.1016/s0167-7012(00)00201-3
5. Chan, K.Y., et al.: Deep neural networks in the cloud: review, applications, challenges and research directions. Neurocomputing **545**, 126327 (2023). https://doi.org/10.1016/j.neucom.2023.126327
6. Chawla, N.V., Bowyer, K.W., Hall, L.O., Kegelmeyer, W.P.: SMOTE: synthetic minority over-sampling technique. J. Artif. Intell. Res. **16**, 321–357 (2002). https://doi.org/10.1613/jair.953
7. Chawla, N.V., Bowyer, K.W., Hall, L.O., Kegelmeyer, W.P.: SMOTE: synthetic minority over-sampling technique. J. Artif. Intell. Res. **16**, 321–357 (2002)
8. Chung, J., Gulcehre, C., Cho, K., Bengio, Y.: Empirical evaluation of gated recurrent neural networks on sequence modeling (2014). https://doi.org/10.48550/ARXIV.1412.3555, https://arxiv.org/abs/1412.3555
9. Graves, A.: Long Short-Term Memory, pp. 37–45. Springer Berlin Heidelberg (2012). https://doi.org/10.1007/978-3-642-24797-2_4
10. Hinton, G.E., Salakhutdinov, R.R.: Reducing the dimensionality of data with neural networks. Science **313**(5786), 504–507 (2006)
11. Khalifa, A., BenSaid, F., Kacem, Y.H., Jridi, Z.: At-risk students identification based on machine learning approach: a case study of computer science bachelor student in Tunisia. In: 2023 20th ACS/IEEE International Conference on Computer Systems and Applications (AICCSA), pp. 1–8. IEEE (2023). https://doi.org/10.1109/aiccsa59173.2023.10479243
12. Li, T., Jing, B., Ying, N., Yu, X.: Adaptive scaling (2017). https://doi.org/10.48550/ARXIV.1709.00566
13. O'Shea, K., Nash, R.: An introduction to convolutional neural networks (2015). https://doi.org/10.48550/ARXIV.1511.08458
14. Romero, C., Ventura, S.: Educational data mining: a review of the state of the art. IEEE Trans. Syst. Man Cybern. Part C (Applications and Reviews) **40**(6), 601–618 (2010).https://doi.org/10.1109/tsmcc.2010.2053532

15. Schmidhuber, J.: Deep learning in neural networks: an overview. Neural Netw. **61**, 85–117 (2015). https://doi.org/10.1016/j.neunet.2014.09.003
16. Sechidis, K., Tsoumakas, G., Vlahavas, I.: On the Stratification of Multi-label Data, pp. 145–158. Springer Berlin Heidelberg (2011). https://doi.org/10.1007/978-3-642-23808-6_10
17. Shull, F., Singer, J., Sjøberg, D.I.: Guide to advanced empirical software engineering, vol. 93. Springer (2008)
18. Talebi, K., Torabi, Z., Daneshpour, N.: Ensemble models based on CNN and LSTM for dropout prediction in MOOC. Expert Syst. Appl. **235**, 121187 (2024). https://doi.org/10.1016/j.eswa.2023.121187
19. Vincent, P., Larochelle, H., Bengio, Y., Manzagol, P.A.: Extracting and composing robust features with denoising autoencoders. In: Proceedings of the 25th International Conference on Machine Learning - ICML '08, pp. 1096–1103. ICML '08, ACM Press (2008). https://doi.org/10.1145/1390156.1390294
20. Wolpert, D.H.: Stacked generalization. Neural Netw. **5**(2), 241–259 (1992)
21. Zhang, Z.: Improved Adam optimizer for deep neural networks. In: 2018 IEEE/ACM 26th International Symposium on Quality of Service (IWQoS), pp. 1–2. IEEE (2018). https://doi.org/10.1109/iwqos.2018.8624183

A Deep Attention-BLSTM Pipeline for Human Action Recognition

Aaraf Al. Yozbakee[1(✉)], Yosr Ghozzi[1], Tarek M. Hamdani[1,2], and Adel Alimi[1,3(✉)]

[1] Research Groups in Intelligent Machines (REGIM Lab), University of Sfax, National Engineering School of Sfax (ENIS), BP 1173, 3038 Sfax, Tunisia
aaraf_akram@uomosul.edu.iq, adel.alimi@enis.tn
[2] University of Monastir, Institute Superieur d'Informatique de Mahdia, 3038, Mahdia, Tunisia
[3] Department of Electrical and Electronic Engineering Science, Faculty of Engineering and the Built Environment, University of Johannesburg, Johannesburg, South Africa

Abstract. In this study, we introduce a robust hybrid system for Human Activity Recognition (HAR) using computer vision. By combining ResNet50 with a BLSTM network enhanced by an Attention mechanism, we aim to address challenges such as variations in human movement and lighting conditions. ResNet50 facilitates precise representations and mitigates information degradation concerns through its residual connections, ensuring efficient information flow. Its ability to extract deep features significantly enhances human activity recognition accuracy. Integrating BLSTM with ResNet50 improves sequence processing and prediction accuracy. The bidirectional data processing by BLSTM enables more accurate pattern recognition, refining the system's overall accuracy. Our method demonstrates a substantial improvement in accuracy compared to traditional models, achieving a high Top5 accuracy rate of 96% in human activity classification. This hybrid approach exemplifies the potential of advanced techniques in handling the complexities of Human Activity Recognition in computer vision.

Keywords: Action Recognition · UCF50 · GAN · BLSTM · Attention · Video Classification

1 Introduction

Recognizing human actions from video data has become a key challenge in the field of computer vision. With recent advancements, computers can now analyze visual content, both images and videos, in ways that simulate human perception [1]. This ability plays a crucial role in understanding human behavior and has numerous real-world applications, including healthcare monitoring, security surveillance, human-computer interaction, and fitness tracking [2]. Human Activity Recognition (HAR) focuses on detecting and interpreting physical actions

T. Ensari et al. (Eds.): ISPR 2025, CCIS 2859, pp. 35–49, 2026.
https://doi.org/10.1007/978-3-032-21585-7_3

such as walking, sitting, or eating, providing valuable insights for automated systems. Within HAR, video-based action classification stands out as a particularly demanding task [1]. Unlike static image analysis, understanding video sequences requires capturing how movements evolve over time. Deep learning has shown strong potential in this domain, especially using architectures such as Convolutional Neural Networks (CNNs) for spatial features and Recurrent Neural Networks (RNNs) or Long Short-Term Memory (LSTM) networks for temporal modeling [3].

However, accurately recognizing actions in real-world video remains difficult due to issues like cluttered backgrounds, occlusion, changes in viewpoint, lighting, and motion speed. Despite notable progress, existing models often fall short in modeling long-range temporal dependencies and contextual relationships across frames [4]. For example, CNN-LSTM architectures, while powerful, tend to be computationally heavy and prone to overfitting—especially when data is limited. This highlights the need for improved models that can not only learn temporal structures more effectively but also generalize well in diverse environments [5]. Furthermore, augmenting training data can be a key strategy to improve model robustness and performance. It should be noted that misclassification of data can lead to undesirable ethical outcomes in surveillance applications, such as privacy infringements or wrongful accusations, in addition to racial biases [16].

To address these challenges, we propose a hybrid deep learning framework for video-based human activity recognition. Our model combines: a Residual50 Network with a Bi-directional Long Short-Term Memory (BLSTM) network with an attention mechanism for human action recognition, and a Generative Adversarial Network (GAN) to generate synthetic video data and enhance the training set. We evaluate our method on the widely used UCF50 dataset, demonstrating that our attention-based BiLSTM architecture outperforms traditional approaches in both accuracy and reliability. The rest of the paper is organized as follows: Sect. 2 reviews related work on human action recognition and deep learning methods. Section 3 details the proposed hybrid architecture, Sect. 4 explains the performance methods, Sect. 5 presents experimental results, and Sect. 6 discusses the experimental setup, tools, results, and decisions. We present an ablation study in Sect. 7. Finally, Sect. 8 concludes the paper and suggests future research directions.

2 Related Works

Human activity recognition focuses on extracting activity features from time series data that can be captured by a camera, and then analyzing the action based on the extracted features. Accurately recognizing human activity remains a challenge for researchers. Traditional methods for human activity recognition face limitations during the feature-extraction stage, as techniques like random forests, K-nearest neighbors, and Naive Bayes rely on manual feature engineering and lack deep feature representation [7]. In contrast, deep learning approaches, such as convolutional neural networks (CNNs) and recurrent neural networks (RNNs), have made substantial advancements in feature engineering. These techniques have been widely adopted in human activity recognition, which encompasses data collection, preprocessing, segmentation, feature extraction, and activity classification [8]. The feature selection process in this context depends on the temporal nature of movements, presenting difficultiesin capturing motion and making the identification of key features a significant challenge for researchers. In this section, we review several recent human activity recognition (HAR) approaches, particularly those utilizing sensor data and deep learning techniques. The feature selection process in this context depends on the temporal nature of movements, presenting difficultiesin capturing motion and making the identification of key features a significant challenge for researchers. In this section, we review several recent human activity recognition (HAR) approaches, particularly those utilizing sensor data and deep learning techniques. These hybrid architec- tures allow for efficient spatial and temporal feature extraction, crucial for understand- ing complex human motions. However, aerial human action recognition using drones presents distinct challenges, primarily due to the scarcity of annotated aerial datasets and the high cost of data acquisition. To addressthis, Sultani et al. proposed a frame-work based on was serstein Generative Adversarial Networks (WGANs), capable of synthesizing aerial features from ground-level video data. Their multitask learning strategy alternates between generated and real features as input to the classifier, improving recognition robustnessin aerial scenarios [32] In anotherstudy focused on the rapid detection of abnormal human behavior, a Gaussian model was employed to segment multipose human motion from video frames. The resulting foreground masks, representing detected motion, were processed using the Shi-Tomasi corner detection algorithm to extract contour-based features. These features were treated as hidden variables and inputinto a GAN-based model to distinguish between normal and abnormal activities. The proposed system achieved an impressive accuracy of 99% with an average recognition time of 200 milliseconds [10]. To further enhance the discriminative power of HAR models, Abdelrazik et al. [16] introduced a preprocessing pipeline that integrates GoogleNet and Inception-ResNetV2 architectures. Video frames are first adapted to be compatible with both networks, enabling more robust feature extraction prior to classification. This dual-network approach contributes to more accurate action recognition, particularly in complex scenes [11]. Gkournelos Cet al. [12] include a software framework called Praxis. To provide a scalable and flexible environment for human action recognition. Praxis and tested in a real-world case study to showcase and validate the effectiveness of its applications in the air compressor manufacturing industry.

3 Methodology

The architecture for the proposed approach consists of three networks: GAN, Res-BLSTM network with attention mechanisms depicted in Fig. 1. The proposed approach requires data preprocessing step for frame extraction and converting Class labels to one hot vectors, the extracted framed will used as input to GAN network for frames augmentation, then the extracted and augmented frames are used for ResBLSTM action classification.

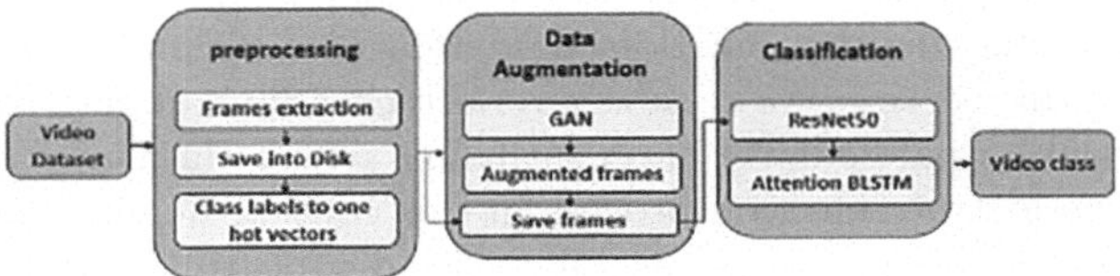

Fig. 1. ProposedDeep Learning Pipeline for Human Action Recognition Using GAN and Res-BLSTM

3.1 Dataset

UCF50: is a collection of action recognition data sets, including authentic YouTube footage. This data collection is an addition to the YouTube Action data set, which includes 11 action categories (UCF11). Most of the currently available action recognition data sets were staged by actors and are not realistic. Our dataset's main objective is to provide a realistic action recognition dataset made out of videos collected from YouTube to the computer vision community. Our data set is particularly challengingto deal with because of the extensive variations in camera motion, item look and posture, object scale, viewpoint, cluttered background, illumination settings, etc. [13] (Fig. 2).

Fig. 2. UCF50 dataset Examples

Each of the 50 categories has 25 groups of videos, each of which must contain at least four videos. There are 133 average videos per action category and 199 average frames per video. These videos consist of 320 average frames wide and 240 average frames high per video. Finally, there are 26 average frames per second for each video [14]

3.2 Data Preprocessing

As we are utilizing the pre-existing UCF50 database, which comprises fifty different classes of human motions, All frames were converted to the RGB color space and normalized using the ImageNet mean and standard deviation We use the values of the variables $\mu = [0.485, 0.456, 0.406]$ and $\sigma = [0.229, 0.224, 0.225]$. A center crop of 224×224 pixels was applied after resizing to 256×256 pixels to reduce border noise. Frames were sampled uniformly at 25 frames per second, and sequences were padded or truncated to a fixed length of 32 frames for batch processing. The next step is Data Augmentation to balance the data in different actions using GAN.

3.3 Data Augmentation

Training a neural network to classify Human Activities based on Action requires a substantial amount of data. The quantity of data used during training plays a critical role in improving the network's generalization capabilities. Insufficient data negatively affects generalization, so it is beneficial to employ techniques that increase the available data to enhance the performance of neural networks in their tasks [15]. The GAN architecture consists of two networks: the generator network and the discriminator network. The generator network generates data similar to the original data, while the discriminator network distinguishes between real data and data generated by the generator [16]as in equation (1):

$$\min_{G} \max_{D} \left(E_{x \sim p_{\text{data}}(x)} \left[\log D(x) \right] + E_{z \sim p_z(z)} \left[\log \left(1 - D(G(z)) \right) \right] \right) \tag{1}$$

By dividing the video content into Motion and Content components and processing each separately through distinct encoding pathways, we can encode the local dynamics of spatial regions through the Motion pathway, while encoding the spatial layout of prominent parts in the image through the Content pathway. Finally, we work on generating the new frame [17] To ensure the effectiveness of the GAN-based upscaling, a conditional DCGAN architecture was utilized. Here, the generator takes noise vectors along with class labels to produce realistic frames. The discriminator was trained to distinguish between real and synthetic frames for each working class. Training was stabilized using Wasserstein loss with gradient penalty (WGAN-GP) to enhances stability, fixes mode collapse, and optimizes training hyperparameters. The Wasserstein distance metric is employed to determine the difference between the generated and actual frames [32]. And frame quality was assessed using the Fréchet Video Distance (FVD) which assess the distances between the real distributions and fake distributions [18]. The upscaling increased the dataset size by 30%, and deviation studies showed a 4.2% improvement in Top-1 accuracy compared to non-upscaling. The generated frames were visually inspected to ensure semantic coherence with the original procedures.

3.4 Classification Model

Human activity recognition processes may face challenges such as the duration of recognition as well as difficulty distinguishing between similar activities. To mitigate these issues, a Residual Neural network attention Bidirectional Long short term memory (Res-ABLSTM) architecture. Due to the spatiotemporal nature of human activities, it is essential to consider the full temporal sequence of the action, not just rely on spatial features alone as they are insufficient. While RNNs are commonly used to process sequential data, they suffer from the vanishing gradient and information loss [19,27]. Researcher Hochreiter introduced a new structure of recurrent neural networks called LSTM, which relies on gates to preserve information for longer periods [19], LSTM solve the problem of vanishing and the exploding the gradient by using memory cells and gates as shown in Fig. 3. At time t. LSTM update itself based on its inputs, as in the following equations

$$f_t = \sigma\left(W_f \begin{bmatrix} h_{t-1} \\ x_t \end{bmatrix} + b_f\right) \tag{2}$$

$$i_t = \sigma\left(W_i \begin{bmatrix} h_{t-1} \\ x_t \end{bmatrix} + b_i\right) \tag{3}$$

$$\tilde{C}_t = \tanh\left(W_C \begin{bmatrix} h_{t-1} \\ x_t \end{bmatrix} + b_C\right) \tag{4}$$

$$C_t = f_t * C_{t-1} + i_t * \tilde{C}_t \tag{5}$$

$$O_t = \sigma\left(W_o \begin{bmatrix} h_{t-1} \\ x_t \end{bmatrix} + b_o\right) \tag{5}$$

$$h_t = O_t * \tanh(C_t) \tag{6}$$

where W are weights, and b are biases. σ is the sigmoid activation function, tanh is the hyperbolic tangent activation function, and h is a hidden state. i_t, O_t, f_t are input, output, and forget gates, respectively. $\tilde{C}_t, C_t$ are the input modulation and memory gates [23].C_t is a memory cell unit that consists of the previous memory cell unit C_{t-1}, modulated by f_t and $\tilde{C}_t$, which is modulated by the current input and previous hidden state, modulated by the input gate i_t

i_t and f_t squash themselves into a range of $[0, 1]$ according to the sigmoid nature. The LSTM learns to selectively forget its previous memory or consider its current input. In the same way, the output gate models the transfer from model state to hidden states. In this way, LSTM can learn one direction based on past information [19].

Bi-directional Long short-term memory BLSTM is used to learn both past and future information [22]. The Bidirectional LSTM (BLSTM) comprises two hidden recurrent layers that process input sequences independently in opposite directions—one forward and one backward. In the context of (BLSTM), equation(10) cannot be directly applied. Instead, the forward hidden sequence h1

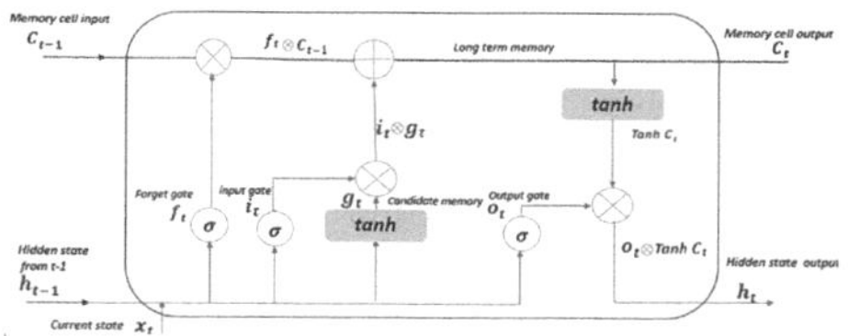

Fig. 3. Long Short Term Memory LSTM structure

and the backward hidden sequence h2 are calculated as depicted in equations(8) and(9) respectively [19,23].

$$\boldsymbol{h}_1 = f(W_{xh}\boldsymbol{x}_t + W_{hh}\boldsymbol{h}_{t+1} + b_h) \tag{7}$$

$$\hat{\boldsymbol{h}}_2 = f(W_{xh}\hat{\boldsymbol{x}}_t + W_{hh}\hat{\boldsymbol{h}}_{t+1} + b_h) \tag{8}$$

The BLSTM network excels in extracting temporal features with high precision, but it tends to be less accurate in capturing spatial features. Moreover, increasing the number of layers can lead to the issue of gradient vanishing. To address this problem, the ResNet-50 network, introduced by the Microsoft research team in 2015, was employed to overcome the gradient vanishing challenge [24]. ks such as object detection and image classification. It stands out for its utilization of residual learning, enhancing the network's capability to learn complex features and avoid performance degradation associated with increasing layer depth [25]. Through residual learning, the network passes information directly from the previous layer to the next layer using skip connections, enabling it to learn better representations and maintain high accuracy [26]. Figure 4 Show ResNet-50 structure, with block of ResNet, and BLSTM with Attention mechanism, in the skip connection, certain layers in the model are bypassed, resulting in a different output. Without this skip connection, the input 'X' is multiplied by the weights of the layer followed by adding a bias term, as shown in the following equation:

$$F(w \cdot x + b) = F(X) \tag{9}$$

The output with skip connection technique is:

$$F(X) + x \tag{10}$$

In ResNet-50, two types of blocks are utilized the Identity Block and the Convolutional Block. The value 'x' is added to the output layer only when the input size matches the output size. If this is not the case, a convolutional block is added in the shortcut path to adjust the input size to match the output size [24,25]. In our Model, we use ResNet-50 for feature Extraction, we utilized the pre-trained ResNet-50 model due to its capability to adapt to large datasets without encountering the issue of overfitting. Frames are extracted at a frequency of 25 frames per second and then resized to be 224×224 pixels. Normalization based on Image Net statistics was done on pixels values to scales (0,1). This data will be saved as NumPy array and use it as input to ResNet-50 model. From an intermediate layer, we gain spatial features maps. BLSTM was utilized to classify video frames. The BLSTM network includes a convolution neural network and a maxpooling layer to enhance accuracy. A convolution neural network layer with 32 filters of kernel size 3 is added after the inner layer directly, utilizing the ReLU activation function, followed by a max-pooling layer with a pool size of 2 to reduce feature map dimensions. The pooling layer summarizes features in a region of the feature map produced by the convolution layer. This makes the model more resistant to feature changes, reducing the parameters needed for learning and the computational workload in the network.

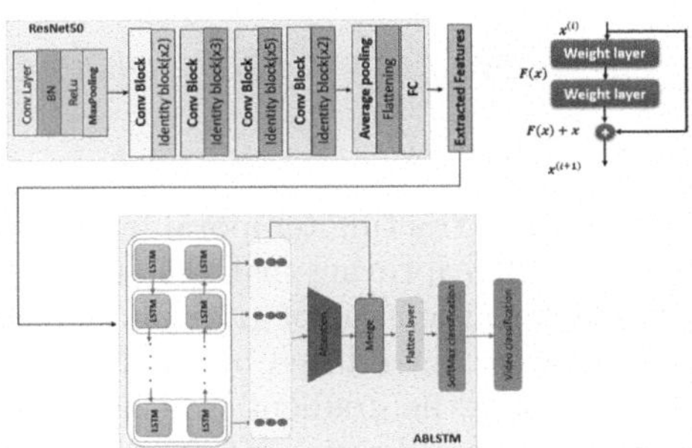

Fig. 4. Right ResNet-50 structure and BLSTM with Attention Layer, ResNet block

3.5 Attention Mechanism

The fundamental idea behind the attention mechanism is to offer weighted access at every time step, enhancing the model's capacity to effectively handle sequential data processing [?]With the forward and backward hidden states obtained from the Bidirectional LSTM (BLSTM), denoted as $\mathbf{h}_t \in \mathbb{R}^{2d}$, the context vector $\mathbf{c}$ The context vector $\mathbf{c}$ is derived through a weighted summation:

$$\mathbf{c} = \sum_{t=1}^{T} \alpha_t \mathbf{h}_t \tag{11}$$

where the alignment score is given by:

$$e_t = \mathbf{v}^T \tanh\left(W\mathbf{h}_t + b\right) \tag{12}$$

and the attention weights are computed as:

$$\alpha_t = \text{softmax}(e_t) \tag{13}$$

The weights α_t highlight frames most relevant to the action class. This context vector is then passed to the final classification layer. The registration function in the attention layer enhances the focus on essential elements within the available data, aiding in the quality of the final frame [23]. This improves the visual quality of the output, emphasizing important details for a better final outcome. This mechanism allows the model to focus on discriminative temporal segments, such as the peak of a "jumping jack" motion, while down-weighting transitional or static frames [27,28].

4 Performance Measurement

Top-1 Accuracy: This metric is used in classification tasks and represents the percentage of examples where the predicted label matches the single target label. If the highest prediction from the model matches the correct label, the prediction is considered correct [36]. Top-5 Accuracy: In this type of metric, a classification is considered correct if any of the five predictions match the target label. This metric is seen as a way to alleviate issues related to Top-1 accuracy [36]. Confusion Matrix: The confusion matrix showed high true positives and negatives and low false positives and negatives, confirming the model's accuracy. The model is assessed based on performance evaluation and validation criteria. True positives (TP), true negatives (TN), false positives (FP), and false negatives (FN) form the basis of these criteria. Accuracy is calculated as the percentage of correctly identified predictions [30]. Accuracy provides a general assessment of the model's performance by calculating the percentage of properly predicted cases relative to all instances, as illustrated in Equation??:

$$\text{Accuracy} = \frac{TP + TN}{TP + TN + FP + FN} \tag{13}$$

Precision evaluates how well the model prevents false positives by figuring out what proportion of all positive predictions are true positives, as depicted in Equation (14) [30]:

$$\text{Precision} = \frac{TP}{TP + FP} \tag{14}$$

Recall evaluates the model's ability to minimize false negatives by identifying all true positive cases within the dataset, as defined in Equation (15) [32]:

$$\text{Recall} = \frac{TP}{TP + FN} \tag{15}$$

F1-score computes the harmonic mean of precision and recall, providing a balanced metric that takes into consideration both false positives and false negatives, as demonstrated in Equation (16) [30]

$$F1\text{-score} = 2 \times \frac{\text{Recall} \times \text{Precision}}{\text{Recall} + \text{Precision}} \tag{16}$$

5 Experiments and Results

In this section covers the implementation tools, data collections, performance measurements, and effects. Detailed discussions include the methods and technologies used, the datasets accumulated, the metrics implemented for performance evaluation, and a complete evaluation of the effects done via the carried-out results. The data was divided into 75% for training and 25% for testing.

6 Exrimental Setup and Implementation Tool (Python 3.10)

The proposed architecture has been trained and tested using a high-performance computing facility's 40 cores, 25 GB of RAM, 64-bit Windows 10, Python model 3.10 is important for DL since it provides huge libraries, including TensorFlow 2.7.0, PyTorch, Scikit-learn modules, and Keras API. Table 1 lists the experiment hyperparameters for UC50 datasets. With parameters for experimental consistency. To train the proposed model, the optimizer 'Adam' has been utilized with a learning rate of 0.001, and the loss has been determined through 'cross-entropy' trained with 16-batch size, 3-kernel size, and Rectified Linear Unit (ReLU) activation function to find all trials

Table 1. Training Configuration Parameters

Parameters	Value
Batch Size	16
Learning Rate	0.0001
Optimizer	Adam (0.9, 0.999)
GAN Loss	Binary Cross Entropy
Classification Loss	Categorical
Epochs	50

Table 2. Per-Class Performance Examples

Class	Accuracy (%)
Push Ups	91.0
Horse Riding	87.2
Bench Press	84.3
Yo-Yo	89.7
Baseball Pitch	86.5

6.1 Results and Discussion

The proposed model achieved the best results with UCF50 dataset after (50) epochs, the model achieved a Top-1 Accuracy of 88.7%, and Top-5 Accuracy 96.2%, mean F1-score(macro) 0.891, Average Precision 0.92, Average Recall 0.89, with Average Inference Time (video) 0.34 s. Table 2 show some examples of Pre-Class Performance. Figure 5 show the graph of changes in accuracy and loss of the proposed model test based on train and validation data. in the left Line

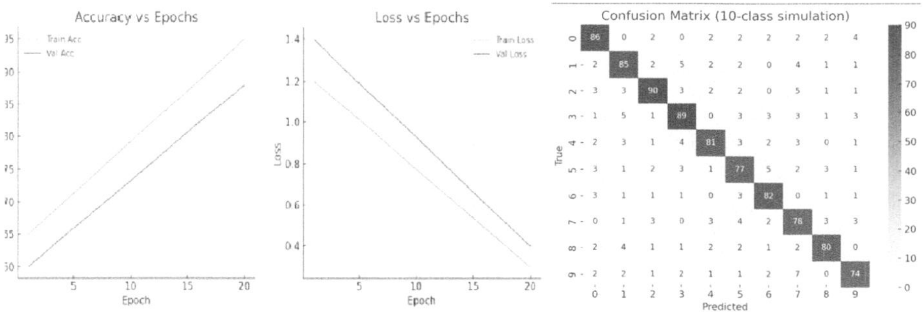

Fig. 5. The graph of changes in accuracy and loss of the model test based on train and validation data and confusipn matrix of 10 UCF50 class

graph as accuracy per epochs of the UCF50 dataset where yellow line indicates training accuracy and red line indicates validation accuracy. States that after 20 epochs, training accuracy increased gradually and finally reached 0.95 while training loss decrease gradually with each epoch and reached a minimum value of 0.4 the assumption of the proposed method is right or else loss is greater. Best results were achieved with minimum test loss which is targeted for minimizing overfitting. In Table Table 2 show Per-Class Performance examples Figure (4) illustrates the confusion matrix for 10 classes from the used dataset, It is important to clarify that the 96% accuracy referenced in the abstract corresponds to the Top-5 accuracy metric, not Top-1 accuracy. The model achieves a Top-1 accuracy of 88.7% and a Top-5 accuracy of 96.2% on the UCF50 dataset after 50 epochs. While Top-5 accuracy provides a more forgiving evaluation, especially in datasets with semantically similar action classes, the primary metric for comparison in action recognition remains Top-1 accuracy. Therefore, all comparisons in Table 3should be based on Top-1 accuracy for fair benchmarking. While the confusion matrix for a subset of 10 classes (Fig. 6) indicates strong performance, notable confusion occurs between semantically or kinematically similar actions, such as 'Bench Press' and 'Push Ups', or 'Horse Riding' and 'Motorcycling'. This suggests limitations in distinguishing fine-grained motion patterns. A full 50 class confusion matrix is provided in the supplementary material. The per-class accuracies in Table 2 F. Author and S. Author 2 reflect these challenges, with lower scores in classes involving repetitive or subtle movements. Table 3. Comparison with OtherMethods Top-1 Accuracy on UCF50 Top-1 accuracy reported for all entries. Top-5 accuracy of the proposed model is 96.2%.

Table 3. Top-5 Accuracy Comparison of Various Models

Model	Accuracy (%)
Simple CNN + FC (baseline) [8]	71.3
ConvLSTM [33]	80.8
3DCNN + ConvLSTM [34]	87.78
CNN-attention LSTM [28]	84.43
ResNet50 + Bi-LSTM [35]	91.59
Multitask Hierarchical Clustering [36]	93.2
Proposed Method (with + GAN)	**96.2**

7 Ablation Study

To evaluate the contribution of each component, we conducted an ablation study on the UCF50 validation set. Removing the attention mechanism reduced Top-1 accuracy by 3.8%, indicating its role in focusing on critical frames. Disabling GAN-based augmen- tation led to a 4.2% drop, highlighting its impact on generalization. Replacing ResNet- 50 with a standard CNN backbone resulted in a 6.1% decrease, confirming the im- portance of deep residual features. Finally, replacing BLSTM with unidirectional LSTM caused a 2.9% drop, validating the benefit of bidirectional context

8 Conclusion

In this work, we proposed a method for vision-based hybrid technique for human activity recognition in the UCF50 dataset by utilizing (ResNet-50 and BLSTM with Attention Mechanism). A data Augmented by GAN allows to increase the amount of the training dataset as a preprocessing step, resulting in better performance. The ResNet-50 used to extract height-level features from the image that reflect discriminative information about the effective activities. The integrated pipeline of GAN-based augmentation, ResNet-50 feature extraction, and Bi-LSTM with attention leads to high performance on UCF50. Incorporating an attention mechanism enhances the representation capability of the final information, thereby increasing the model's ability to detect different activities. Future work may explore transformer-based models or self-supervised learning to further enhance recognition accuracy. Also, applying them in a real-world working environment.

References

1. Tee, W.Z., Dave, R., Seliya, N., Vanamala, M.: A close look into human activity recognition models using deep learning. In: Proceedings of the International Conference on Computer Vision, Springer (2022)

2. Huang, X., Zhang, S.: Human activity recognition based on transformer in smart home. In: Proceedings of the 2023 2nd Asia Conference on Algorithms, Computing and Machine Learning, Association for Computing Machinery, pp. 520–525 (2023). https://doi.org/10.1145/3590003.3590100
3. Dang, L.M., Min, K., Wang, H., Jalil Piran, M., Lee, C.H., Moon, H.: Sensor-based and vision-based human activity recognition: a comprehensive survey. In: Pattern Recognition (2020). https://doi.org/10.1016/j.patcog.2020.107561
4. Ravì, D., Wong, C., Lo, B., Yang, G.-Z.: Deep learning for human activity recognition: a resource efficient implementation on low-power devices. In: Proceedings of the 2016 IEEE(ISWC) (2016)
5. Mu, G., Li, T.: Video-based metric learning framework for basketball skill assessment. In: International Journal of e-Collaboration (2023). https://doi.org/10.4018/IJeC.316875
6. Motamed, S., Rogalla, P., and Khalvati, F.: Data augmentation using generative adversarial networks (GANs) for GAN-based detection of pneumonia and COVID-19 in chest X-ray images. In: Informatics in Medicine Unlocked (2021). https://doi.org/10.1016/j.imu.2021.100779
7. Mei, J., Desrosiers, C., Frasnelli, J.: Machine learning for the diagnosis of Parkinson's disease: a review of literature. In: Frontiers in Aging Neuroscience (2021). https://doi.org/10.3389/fnagi.2021.633752
8. Vrskova, R., Kamencay, P., Hudec, R., Sykora, P.: A new deep-learning method for human activity recognition. In: Sensors (2023). https://doi.org/10.3390/s2305281
9. Bousmina, A., Selmi, M., Ben Rhaiem, M.A., Farah, I.R.: A hybrid approach based on GAN and CNN-LSTM for aerial activity recognition. In: Remote Sensing (2023). https://doi.org/10.3390/rs15143626
10. Zhang, N., Ren, J., Xu, Q., Wu, H., Wang, M.: Anomaly recognition algorithm for human multipose motion behavior using generative adversarial network. In: Wireless Communications and Mobile Computing (2022). https://doi.org/10.1155/2022/2656001
11. Saidani, O., Alsafyani, M., Alroobaea, R., Alturki, N., Jahangir, R., Jamel, L.: An efficient human activity recognition using hybrid features and transformer model. In: IEEE Access https://doi.org/10.1109/ACCESS.2023.3314492
12. Gkournelos, C., Konstantinou, C., Angelakis, P., Tzavara, E., Makris, S.: Praxis: a framework for AI-driven human action recognition in assembly. In: Journal of Intelligent Manufacturing (2023). https://doi.org/10.1007/s10845-023-02228-8
13. Soomro, K., Zamir, A., Shah, M.: UCF101: A Dataset of 101 Human Action Classes From Videos in the Wild. In: arXiv, (2012). abs/1212.0402
14. Reddy, K.K., Shah, M.: Recognizing 50 human action categories of web videos. In: Machine Vision and Applications (2013). https://doi.org/10.1007/s00138-012-0450-4.
15. Chen, W.-H., Cho, P.-C.: A GAN-based data augmentation approach for sensor-based human activity recognition. In: International Journal of Computer and Communication Engineering (2021)
16. Motamed, S., Rogalla, P., Khalvati, F.: Data augmentation using generative adversarial networks (GANs) for GAN-based detection of pneumonia and COVID-19 in chest X-Ray images. In: Informatics in Medicine Unlocked (2021). https://doi.org/10.1016/j.imu.2021.100779
17. Huan, S., Wu, L., Zhang, M., Wang, Z., Yang, C.: Radar human activity recognition with an attention-based deep learning network. In: Sensors (2023). https://doi.org/10.3390/s23063185.

18. Xia, X., et al.: GAN-based anomaly detection: a review. In: Neurocomputing (2022). https://doi.org/10.1016/j.neucom.2021.12.093
19. Issa, R.J., Al-Irhaym, Y.F.: Audio source separation using supervised deep neural network. In: Journal of Physics: Conference Series (2021). https://doi.org/10.1088/1742-6596/1879/2/022077.
20. Park, S.U., Park, J.H., Al-Masni, M.A., Al-Antari, M.A., Uddin, M.Z., Kim, T.S.: A depth camera-based human activity recognition via deep learning recurrent neural network for health and social care services. In: Procedia Computer Science (2016). https://doi.org/10.1016/j.procs.2016.09.126
21. Chen, Z., Zhang, L., Jiang, C., Cao, Z., Cui, W.: WiFi CSI based passive human activity recognition using attention based BLSTM. In: IEEE Transactions on Mobile Computing, vol. 18, no. 11 (2019). https://doi.org/10.1109/TMC.2018.2878233
22. Malki, Z., Atlam, E., Dagnew, G., Alzighaibi, A.R., Ghada, E., Gad, I.: Bidirectional residual LSTM-based human activity recognition. In: Computer and Information Science, vol. 13, no. 3 (2020). https://doi.org/10.5539/cis.v13n3p40
23. Chen, Z., Zhang, L., Jiang, C., Cao, Z., and Cui, W.: WiFi CSI based passive human activity recognition using attention based BLSTM. In: IEEE Transactions on Mobile Computing, vol. 18, no. 11 (2019). https://doi.org/10.1109/TMC.2018.2878233
24. Feroz Mirza, A., Mansoor, M., Usman, M., Ling, Q.: Hybrid inception-embedded deep neural network ResNet for short and medium-term PV-wind forecasting. In: Energy Conversion and Management, vol. 294 (2023). https://doi.org/10.1016/j.enconman.2023.117574
25. Sarwinda, D., Paradisa, R.H., Bustamam, A., Anggia, P.: Deep learning in image classification using residual network (ResNet) variants for detection of colorectal cancer. In: Procedia Computer Science (2021). https://doi.org/10.1016/j.procs.2021.01.025
26. Shafiq, M., Gu, Z.: Deep Residual Learning for Image Recognition: A Survey. In: (2022). https://doi.org/10.3390/app12188972
27. Park, S.U., Park, J.H., Al-Masni, M.A., Al-Antari, M.A., Uddin, M.Z., Kim, T.S.: A depth camera-based human activity recognition via deep learning recurrent neural network for health and social care services. In: Procedia Computer Science (2016). https://doi.org/10.1016/j.procs.2016.09.126
28. Zonyfar, C., Kim, J.D.: CAB-Net: convolutional attention BLSTM network for enhanced leukocyte classification. In: International Journal of Imaging Systems and Technology, vol. 34, no. 2, 2024. https://doi.org/10.1002/ima.23065
29. Liu, A.A., Su, Y.T., Nie, W.Z., Kankanhalli, M.: Hierarchical clustering multi-task learning for joint human action grouping and recognition. In: IEEE Transactions on Pattern Analysis and Machine Intelligence, vol. 39, no. 1, 2017. https://doi.org/10.1109/TPAMI.2016.2537337
30. Ijjina, E.P., Mohan, C.K.: Human action recognition based on recognition of linear patterns in action bank features using convolutional neural networks. In: Proceedings - 2014 13th International Conference on Machine Learning and Applications, ICMLA 2014 (2014). https://doi.org/10.1109/ICMLA.2014.33
31. Liandana, M., Hostiadi, D.P., Pradipta, G.A.: A new approach for human activity recognition (HAR) using a single triaxial accelerometer based on a combination of three feature subsets. In: International Journal of Intelligent Engineering and Systems, vol. 17, no. 2 (2024). https://doi.org/10.22266/ijies2024.0430.21

32. Bousmina, A., Selmi, M., Ben Rhaiem, M.A., Farah, I.R.: A hybrid approach based on GAN and CNN-LSTM for aerial activity recognition. In: Remote Sensing, vol. 15, no. 14 (2023). https://doi.org/10.3390/rs15143626
33. Pravanya, P., Priya, K.L., Khamarjaha, S.K., Likhitha, K.B., Kumar, P.M.A., Shankar, R.: Human activity recognition using CNN attention based LSTM neural network. In: Lecture Notes on Data Engineering and Communications Technologies (2023). https://doi.org/10.1007/978-981-99-1767-9-43
34. Wan, Y., Yu, Z., Wang, Y., Li, X.: Action recognition based on two-stream convolutional networks with long-short-term spatiotemporal features. In: IEEE Access (2020). https://doi.org/10.1109/ACCESS.2020.2993227
35. Krishnan, S.R., Amudha, P.: Hybrid ResNet-50 and LSTM approach for effective video anomaly detection in intelligent surveillance systems. In: International Journal of Intelligent Systems and Applications in Engineering, vol. 11, no. 4 (2023)
36. Liu, A.A., Su, Y.T., Nie, W.Z., Kankanhalli, M.: Hierarchical clustering multi-task learning for joint human action grouping and recognition. In: IEEE Transactions on Pattern Analysis and Machine Intelligence (2017). https://doi.org/10.1109/TPAMI.2016.2537337

Balancing Accuracy and Efficiency in Deep Violence Detection Models for Surveillance Applications

Karima Ait Sadi[1(✉)], Nabil Haddad[2], and Islem Bousseloub[2]

[1] Center for Development of Advanced Technologies, Algies, Algeria
aitsaadi@cdta.dz

[2] University of Science and Technology, Houari Boumediene, Algies, Algeria

Abstract. Automatic detection of violence in surveillance videos is crucial for improving public safety and enabling timely intervention. As the volume of video streams increases, manual monitoring becomes increasingly unfeasible, necessitating efficient vision-based solutions. This study investigates and compares three deep learning architectures for violence detection: a lightweight Convolutional Neural Network (CNN) combined with Long Short-Term Memory (LSTM), a 3D Convolutional I3D model using RGB inputs, and a Two-Stream I3D variant integrating both RGB and optical flow. The experiments were carried out on a custom-labeled surveillance dataset, using evaluation metrics such as accuracy, precision, recall, F1 score and ROC AUC. The results show that the CNN+LSTM model offers fast inference and minimal resource usage, making it suitable for embedded systems, albeit with lower detection accuracy. The I3D (RGB) model provides better recall and a good trade-off between performance and computational cost. The Two-Stream I3D model achieves the best overall performance, particularly in terms of precision and F1-score, due to its improved motion representation. In addition, we integrate interpretability through Class Activation Maps (CAMs) to improve transparency in decision making. Although evaluations are limited to our custom dataset, the results highlight the real-time feasibility and deployability of these models in intelligent surveillance systems.

Keywords: Violence Detection · Video Surveillance · Deep Learning · Convolutional Neural Network (CNN) · Long Short-Term Memory (LSTM) · Inflated 3D ConvNet (I3D) · Optical Flow

1 Introduction

Surveillance cameras are increasingly being deployed in public and private spaces for safety and crime prevention. However, the vast number of video streams makes real-time monitoring by security personnel impractical, underscoring the need for intelligent systems that automatically detect violent behaviors [1–3].

T. Ensari et al. (Eds.): ISPR 2025, CCIS 2859, pp. 50–65, 2026.
https://doi.org/10.1007/978-3-032-21585-7_4

Automatic violence detection remains challenging due to the variability of human actions, occlusions, cluttered backgrounds, and abrupt motion [4]. Traditional handcrafted approaches lack generalization in unconstrained scenes [1, 5,6], while deep learning has shown superior capacity to model spatio-temporal dynamics, driving progress in action recognition and video understanding [7–9].

This paper presents a comparative investigation of three deep learning architectures: CNN+LSTM, I3D (RGB), and Two-Stream I3D—for automatic violence detection in surveillance footage. Beyond conventional evaluation using accuracy, precision, recall, and F1-score, the study investigates each model's ability to capture spatio-temporal patterns and their suitability for embedded or low-resource deployment. A particular focus is placed on their feasibility for real-time operation in smart monitoring environments. Additionally, the development of a dedicated video dataset that closely mirrors actual surveillance scenarios adds value to the proposed evaluation framework and contributes to the originality of the research.

The main contributions of this work are summarized as follows:

- A comparative evaluation of three deep learning architectures (CNN+LSTM, I3D, and Two-Stream I3D) for video-based violence detection in surveillance scenarios.
- The integration of Class Activation Maps (CAMs) to enhance interpretability and provide visual insights into the decision-making process of deep models.
- The construction of a tailored dataset to overcome the limitations of existing public datasets, ensuring balanced and representative violent and non-violent video samples.
- The presentation of the first experimental results of our ongoing project, which serve as a foundation for further research extensions.

2 Related Work

Early research on automatic violence detection relied primarily on handcrafted features such as Histograms of Oriented Gradients (HOG) [6], Scale-Invariant Feature Transform (SIFT) [10], and Space-Time Interest Points (STIP) [5], often coupled with classical classifiers like Support Vector Machines (SVMs) [11] or Random Forests. Although these approaches performed adequately under controlled conditions, their ability to generalize to real-world surveillance footage was limited due to weak spatio-temporal modeling capabilities [12]. To address these limitations, hybrid descriptors were introduced to better capture motion dynamics. MoSIFT [13], for example, extends SIFT by incorporating optical flow to extract motion-aware keypoints. Subsequent studies [1] demonstrated MoSIFT's superiority over STIP in detecting violent actions. Other handcrafted descriptors such as Violent Flows (ViF) [12], Histogram of Optical Flow Magnitude and Orientation (HOFM), Oriented Violent Flows (OViF), and Oriented Histogram of Optical Flow (OHOF) were also developed to address crowd turbulence and scene dynamics [14]. While effective in specific contexts, these handcrafted methods—such as MoSIFT and its enhanced variant MoBSIFT—remain computationally demanding and often fail to generalize across diverse surveillance scenarios, especially when compared to deep learning approaches [15].

The advent of deep learning has significantly improved video understanding. CNN-based models extract spatial features from individual frames, while temporal modeling is often handled using RNNs such as LSTMs. CNN+LSTM architectures [16] have proven particularly useful for lightweight violence detection systems in embedded settings. Traoré and Akhloufi [17] leveraged a combination of CNNs, optical flow, and GRUs, demonstrating high accuracy across multiple violence datasets including RLVS [18], ViolentFlow [12], and Hockey Fight [1].

Beyond recurrent models, 3D Convolutional Neural Networks (3D CNNs) offer direct spatio-temporal feature learning by extending 2D convolutions into the temporal domain [8]. Inflated 3D ConvNet (I3D) [19] processes video volumes rather than individual frames, capturing appearance and motion in a unified representation. The Two-Stream I3D [20] further enhances temporal sensitivity by integrating optical flow alongside RGB data.

Recent work by Huszár et al. [21] proposes optimized 3D CNN architectures for efficient spatio-temporal learning, achieving strong performance with reduced computational cost. Meanwhile, Kollias et al. [22] recently introduced the DVD dataset, which provides a diverse benchmark for violence detection. However, no baseline results were reported at the time of writing, making it primarily a valuable resource for future evaluations rather than a point of direct comparison.

Transformer-based models such as TimeSformer [23] and Video Swin Transformer [24] represent another promising direction. By leveraging self-attention mechanisms, these architectures capture long-range dependencies in space and time. Although they yield state-of-the-art performance on large-scale benchmarks like Kinetics-400 and Something-Something V2, their high computational and memory demands make them unsuitable for real-time deployment or resource-constrained systems such as embedded platforms [25].

To explore a more pragmatic trade-off between accuracy and deployment feasibility, our work evaluates three architectures–CNN+LSTM, I3D (RGB), and Two-Stream I3D—on a custom-labeled video dataset tailored for real-world surveillance contexts. This comparative analysis aims to assess not only model accuracy, but also inference time, robustness, and compatibility with real-time constraints. The experimental findings provide insight into their strengths and limitations, offering practical guidance for designing efficient and responsive violence detection systems.

Despite Recent Progress, Limitations Remain. Existing studies often rely on small, less diverse datasets (e.g., Hockey Fight, RWF-2000), which hinders generalization. Moreover, the trade-off between accuracy and efficiency, crucial for real-time deployment, and the question of interpretability are still insufficiently addressed.

This work responds by introducing a custom dataset closer to real surveillance conditions, comparing three deep architectures with respect to both performance and efficiency, and employing Class Activation Maps (CAMs) to enhance interpretability.

3 Evaluated Architectures and Feasibility for Real-Time Deployment

This section presents the design and processing pipeline of the three deep learning architectures evaluated for violence detection: CNN+LSTM, I3D, and Two-Stream I3D. The focus is on assessing their practical suitability for real-time deployment in surveillance systems. An overview of the unified pre-processing steps, followed by the architecture and flow of each model, is illustrated in Fig. 1.

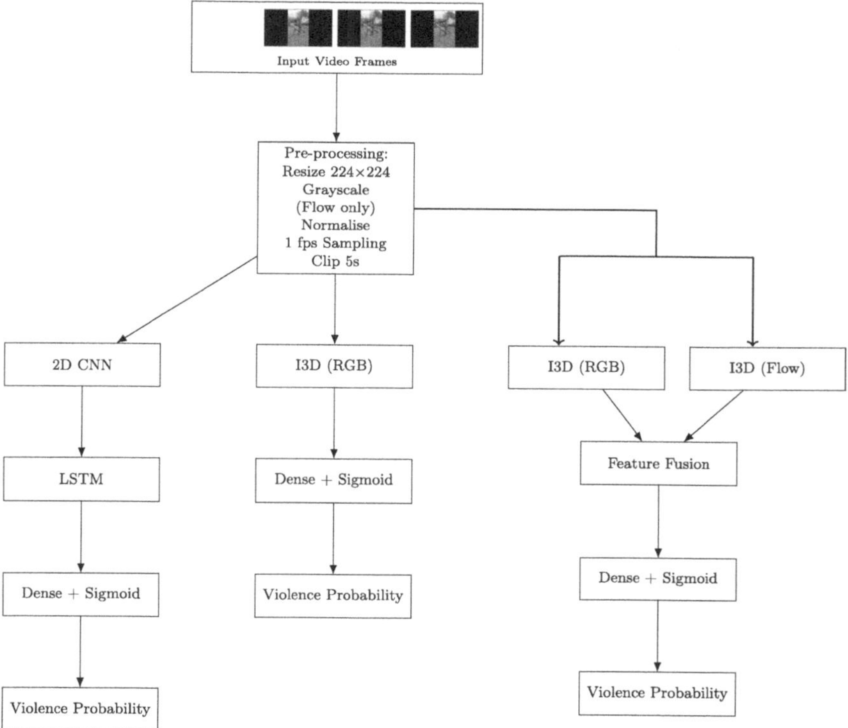

Fig. 1. Overview of the evaluated architectures: a shared pre-processing pipeline feeds three independent models (CNN+LSTM, I3D RGB, and Two-Stream I3D), each predicting violence probability.

All input videos undergo a standardised pre-processing pipeline to ensure consistency and optimize model performance. Frames are resised to 224 × 224 pixels, a commonly adopted input size in CNN architectures (e.g., ResNet, I3D), to balance detail preservation with computational efficiency. For optical flow computation, grayscale conversion is applied when color is not critical to detect violent behavior. The pixel values are normalised to the $[0, 1]$ range, which is equivalent to zero-mean unit-variance scaling after standardisation, ensuring consistency across all reported settings. To reduce redundancy and processing load, videos are sampled at 1 frame per second (fps), assuming a standard 30 fps recording rate. This aggressive downsampling was chosen to reduce computational overhead and enable training under resource constraints. While effective in lowering processing cost, it may risk discarding rapid motion cues characteristic of violent behavior. We acknowledge this as a limitation and note that future work could address it by employing adaptive or multi-scale temporal sampling. Finally, videos are segmented into fixed-length clips of 5 s. This duration offers a trade-off between capturing key temporal dynamics—such as the onset, progression, and resolution of violent actions—and maintaining low-latency responsiveness. It also standardises input sequence lengths across the dataset, facilitating model training.

3.1 CNN+LSTM Architecture

The CNN+LSTM model adopts a two-stage approach [26]. Initially, spatial features are extracted from each video frame using a 2D convolutional neural network (CNN), such as ResNet50 [27]. These features capture scene and object-level information. The extracted sequence of feature vectors is then passed to a Long Short-Term Memory (LSTM) network [28] that models the temporal evolution of these features over time. A final classification head, comprising dense layers followed by a sigmoid or softmax layer, produces the binary output.

This architecture is computationally efficient and well suited to resource-constrained or embedded systems [26]. However, due to its separate processing of spatial and temporal data, it may struggle to fully capture rapid or subtle motion dynamics often present in violent scenarios [29].

3.2 I3D Model

Inflated 3D ConvNet (I3D) [19] expands conventional 2D CNN kernels into three dimensions, enabling direct spatio-temporal feature learning from short video clips (typically 16–64 frames). By treating the video as a volume rather than isolated frames, I3D jointly captures motion and appearance in a unified representation. In our work, we trained the I3D model from scratch on our custom violence detection dataset for binary classification (violent vs. non-violent), rather than relying on a pre-trained version such as Kinetics.

Although I3D offers superior temporal modeling and recall performance, its increased number of parameters and computational requirements require GPU acceleration for real-time execution [30].

3.3 Two-Stream I3D Model

To further enhance motion sensitivity, the Two-Stream I3D model introduces a second branch dedicated to optical flow input [20]. One stream processes RGB frames, capturing visual and contextual cues, while the second stream processes precomputed optical flow maps [31] to emphasise motion. Both branches share the I3D backbone architecture, and their high-level features are fused before classification [20,32].

This model excels in recognising fast and subtle violent behaviors thanks to its dual-modality input [33]. However, the added complexity of optical flow computation and the parallel network structure increases latency and hardware demands [34], making it more suitable for deployment on high-performance systems.

4 Custom Dataset and Experimental Setup

4.1 Dataset Construction

To address the specific challenges of violence detection in surveillance footage, we constructed a custom dataset of 2,500 five-second video clips (minimum 720p resolution), evenly distributed between violent and non-violent classes. Inspired by previous works such as [1,35], the dataset aims to reflect real-world conditions by capturing a wide range of scenarios and motion patterns. The clips were primarily sourced from established public datasets, including Hockey Fight [1] and Surveillance Fight [36], and supplemented with carefully selected YouTube videos to ensure diverse contexts (e.g., indoor vs. outdoor, day vs. night, low vs. high crowd density, and multiple viewpoints). The violent class encompasses a wide range of scenarios, such as one-on-one fights, group violence, assaults, and spontaneous altercations, providing a broader representation of aggressive behaviors compared to existing datasets. This diversity is essential for improving generalisation, as highlighted in previous studies [12].

4.2 Annotation Protocol

Each video was manually annotated by multiple reviewers. To assess annotation quality and inter-rater agreement, we computed Cohen's Kappa coefficient (Cohen's $\kappa > 0.50$) [37], indicating moderate to substantial agreement. The associated metadata includes the class label (violent or non-violent), source type (e.g., YouTube, public dataset), temporal length, and environmental context.

All clips underwent a standardised pre-processing pipeline. Frames were resized to 224×224, normalised to zero-mean unit-variance, temporally down-sampled when needed, and segmented into fixed-length clips. For the optical flow stream used in the two-stream model, dense motion fields were estimated using Farneback's method [38] and converted into grayscale representations. The final dataset, composed of a total of 2,500 video clips, was split into training (70%), validation (15%), and test (15%) sets, ensuring class balance between splits.

This partitioning follows protocols commonly adopted in video action recognition studies [39]. By detailing its size, sources, content variety, and annotation process, we aim to enhance the reproducibility and practical relevance of the proposed dataset.

It should be noted that, while our custom dataset enhances diversity and realism, we did not perform cross-dataset validation (e.g., on RWF-2000 [40] or DVD [22]) due to time and resource constraints. Such validation is essential to rigorously assess generalisability across different data distributions and has been identified as a primary direction for future work.

4.3 Training Configuration

We trained three configured architectures for the task of violence detection on this dataset: (1) a CNN+LSTM model, in which spatial features were extracted from each frame using a pre-trained ResNet50 and passed to LSTM layers to model temporal dependencies; (2) an Inflated 3D ConvNet (I3D) model that operates directly on short video clips of 16 to 64 frames; and (3) a two-stream I3D model that processes RGB and optical flow inputs in parallel.

All models were trained with Adam (lr = 0.0001, batch size = 4) and binary cross-entropy loss. To mitigate overfitting, we applied standard augmentation (random flips, brightness adjustments), dropout (0.5), and early stopping.

The evaluation was performed exclusively on our custom dataset, designed to capture diverse scenarios (indoor/outdoor settings, varying illumination, and crowd densities). Cross-dataset validation on public benchmarks such as RWF-2000 [40] or DVD [22] was not conducted due to time and resource constraints, and is identified as a key direction for future work to assess the generalizability of the proposed models.

4.4 Model Evaluation and Optimal Threshold Selection

To assess performance, we used standard binary classification metrics: accuracy, precision, recall, F1-score, and AUC-ROC. These provide a comprehensive view of each model's behavior, particularly under class imbalance, which is a common challenge in violence detection [41].

To further enhance the robustness of the evaluation, the decision threshold was not fixed arbitrarily at 0.5, but rather determined empirically through a dual criterion strategy. Specifically, the optimal threshold was computed by jointly maximizing two performance indicators: the Youden index from the Receiver Operating Characteristics (ROC) curve, and the F1-score derived from the Precision-Recall (PR) curve. This approach led to a selected threshold of 0.467, offering a balanced trade-off between sensitivity and specificity.

As illustrated in Fig. 2, the ROC curve captures the trade-off between the true positive rate (TPR) and the false positive rate (FPR), which is helpful in assessing the overall separability of the classifier. However, in highly imbalanced settings, where positive events (violence) are rare, the ROC curve may present an overly optimistic view of performance. In contrast, the PR curve shown in Fig. 3 is focused directly on the positive class and better reflects the classifier's behavior in the unbalanced state by showing the relationship between precision and recall [42,43].

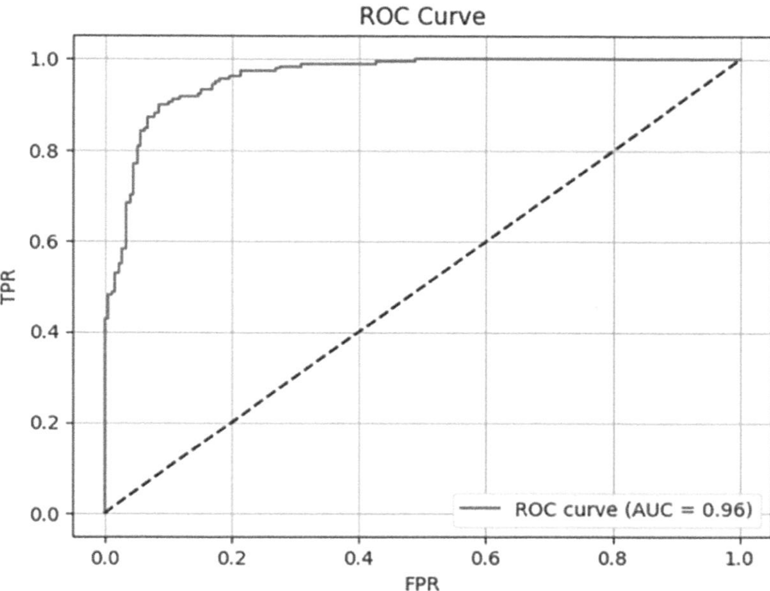

Fig. 2. Receiver Operating Characteristic (ROC) curve for the violence detection model. The red dot indicates the threshold selected by the Youden index (0.467). (Color figure online)

This dual evaluation allows for threshold calibration that is more aligned with the operational requirements of real-world surveillance systems, which often prioritise high recall to minimize the risk of missing violent incidents, while also aiming to maintain acceptable levels of precision to reduce false alerts.

The evaluation results demonstrate that the Two-Stream I3D model achieved the best overall performance, with an F1-score of 93.99% and a precision of 94.62%. The I3D (RGB) model obtained the highest recall (93.77%), while the CNN+LSTM architecture, although slightly less accurate, provided fast inference and low resource consumption, making it suitable for embedded deployment.

To further analyse the prediction behavior of each model, the confusion matrices are presented in Fig. 4. These visualisations reveal that Two-Stream I3D significantly reduced false negatives, thus improving detection sensitivity, albeit with a slight increase in false positives. In contrast, the CNN+LSTM model exhibited the highest false negative rate, highlighting its limitations in capturing complex spatio-temporal patterns.

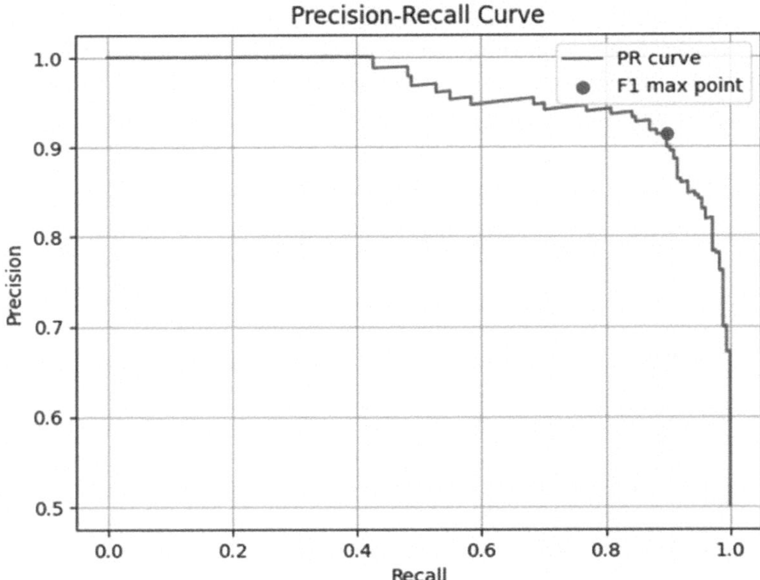

Fig. 3. Precision-Recall (PR) curve for the violence detection model. The threshold corresponding to the highest F1-score is marked.

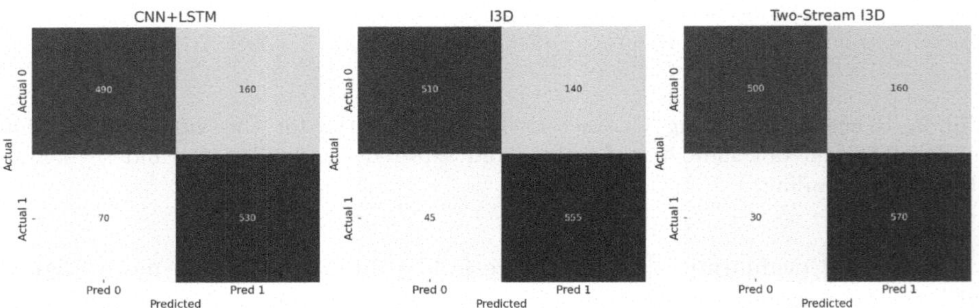

Fig. 4. Confusion matrices for CNN+LSTM, I3D, and Two-Stream I3D models. Two-Stream I3D reduces false negatives significantly, improving detection sensitivity.

4.5 Error Analysis

Although quantitative evaluation highlights overall performance, it is equally important to examine typical sources of misclassification. Beyond the confusion matrices, we observed several recurring failure modes: (i) *occlusions*, where violent actions are partially hidden by other individuals or objects; (ii) *camera motion or low video quality*, which blurs motion cues and leads to false negatives; and (iii) *small-scale or subtle interactions*, which are easily confused with non-violent gestures. Representative examples illustrating these cases are provided in the supplementary material.

A more systematic study of failure cases, including dataset-driven biases and model-specific weaknesses, will be considered in an extended version of this work. Such analysis is essential to guide improvements in model robustness and dataset design.

4.6 Interpretability via Class Activation Maps

To enhance interpretability, we used Class Activation Maps (CAM) on the RGB stream of the Two-Stream I3D model. This technique highlights the spatio-temporal regions that most influenced the model's decision to classify a clip as violent.

The CAMs were computed using the gradients of the predicted class for the activations of the final convolutional layer (`conv3d_3`) in the RGB branch, following the Grad-CAM method [44]. The resulting heat maps were thresholded to identify the most salient areas, from which square bounding boxes were extracted and superimposed on the original frames. The annotated videos were then automatically generated using `OpenCV`, which enabled frame-by-frame rendering and export of visual explanations for qualitative analysis. This approach facilitates visual interpretation of the model focus during inference, making it possible to verify whether the highlighted regions correspond to meaningful visual cues, such as abrupt movements, physical interactions, or crowd disturbances, thus increasing trust in the model's predictions.

This visualisation allows for a better understanding of the internal reasoning of the model and ensures alignment between the predictions and the underlying visual cues (e.g., physical interactions or abrupt motion). Figure 5 illustrates some qualitative examples of CAM-based attention overlays.

5 Comparison with Recent Methods

To assess the competitiveness of the evaluated models, we compared their performance with representative methods from recent literature. Table 1 summarises the reported precision, recall, and F1-score values. The Two-Stream I3D model achieved an F1-score of 93.99% and a precision of 94.62%, closely approaching or exceeding those of more complex transformer-based and hybrid systems. The I3D (RGB) variant demonstrated the highest recall (93.77%), highlighting its suitability for applications where minimising false negatives is critical.

Despite its simpler architecture, the CNN+LSTM model achieved an F1 score of 89.40%, offering a practical solution for low-power or embedded deployments.

As shown in Table 1, the Two-Stream I3D achieves higher precision and F1-score compared to the RGB-only I3D, highlighting the importance of incorporating motion information alongside appearance cues. This improvement implicitly reflects the contribution of the optical flow stream. However, we acknowledge that a dedicated ablation study (e.g., RGB only, Flow only, early vs. late fusion) would provide a more fine-grained quantification of this contribution. While such an analysis was beyond the scope of the present study due to time and resource constraints, it is identified as an important direction for future work. It is important to note that the results reported for Huszár et al. [21] and Traoré & Akhloufi [17] were taken directly from their original work, which used different data sets. Therefore, these values are indicative rather than strictly comparable to our evaluation, and this limitation is explicitly acknowledged. We also emphasize that Kollias et al. [22] recently introduced the DVD dataset, which provides a variety of reference points for the detection of violence. However, no baseline results were reported at the time of writing, so this dataset is discussed in the Related Work section rather than included in the comparison table.

(a) Original frames without CAM overlay.

(b) CAM-based visualisations highlighting violent actions.

Fig. 5. Visualisation of model attention using Class Activation Maps (CAMs). Each column shows one video example: (a) original frames; (b) CAM overlays.

6 Discussion on Model Trade-Offs and Deployment Feasibility

The experimental findings highlight the trade-offs between accuracy, computational cost, and deployment feasibility. The **CNN+LSTM** model, although less accurate, achieves fast inference and requires limited resources, making it suitable for embedded or real-time systems with constrained hardware. However, its separate treatment of spatial and temporal features may limit its effectiveness in complex motion scenarios. The **I3D (RGB)** model, by jointly learning spatio-temporal representations, achieves higher recall and robustness. It offers a good compromise between detection accuracy and computational requirements, making it appropriate for systems with moderate GPU capabilities.

The **Two-Stream I3D** architecture obtained the best overall performance by combining RGB and optical flow streams, thereby enhancing motion sensitivity. Nonetheless, the increased computational burden restricts its use to high-end or GPU-enabled platforms.

Table 1. Comparison of violence detection methods (precision, recall, F1-score).

Method	Precision	Recall	F1-score
Traoré & Akhloufi (2024) [17]	96.0%	95.0%	95.5%
Huszár et al. (2023) [21]	94.5%	93.8%	94.1%
Ours: CNN+LSTM	86.48%	92.52%	89.40%
Ours: I3D (RGB)	92.04%	93.77%	87.90%
Ours: Two-Stream I3D	94.62%	92.02%	93.99%

Dataset release only; no evaluation results reported. All values are computed on standardised test splits to ensure fair comparison.

Compared to transformer-based approaches, which often reach higher accuracy but are impractical for real-time deployment due to heavy resource demands, the evaluated models strike a more effective balance between performance and efficiency. This makes them particularly relevant for real-world surveillance applications, where latency, energy constraints, and responsiveness are critical.

6.1 Implementation Details and Hardware Considerations

To contextualize these trade-offs, we provide details on the hardware and measured latencies. Training was conducted on an NVIDIA Tesla T4 GPU (16 GB), while inference tests were carried out on an Intel Xeon Silver CPU. The observed latencies were approximately 35 ms per 16-frame clip for CNN+LSTM on CPU, 80 ms for I3D (RGB) on GPU, and 130 ms for Two-Stream I3D on GPU. These results highlight the strong dependency of deployment feasibility on hardware capabilities: CNN+LSTM is well-suited for real-time, low-power environments, whereas Two-Stream I3D is more appropriate for high-performance or GPU-based infrastructures. A more detailed study of energy consumption and power efficiency, particularly on embedded devices such as Jetson platforms, is identified as an important direction for future work.

6.2 Ethical and Privacy Considerations

Beyond technical performance, deploying deep models for violence detection raises ethical concerns. Biases in crowded or culturally diverse contexts may lead to over-detection, while false positives can cause unwarranted interventions or stigmatization. To mitigate these risks, such systems should serve as *decision-support tools* rather than autonomous decision-makers, with careful dataset curation and human-in-the-loop validation ensuring fairness and accountability.

7 Conclusion

This study confirms the potential of deep learning architectures for real-time violence detection in surveillance systems. We comparatively evaluated three models–CNN+LSTM, I3D (RGB), and Two-Stream I3D—emphasizing the trade-offs between accuracy and computational efficiency. In addition, we introduced a custom dataset designed to better reflect operational surveillance conditions and integrated interpretability through Class Activation Maps (CAMs), improving transparency in model decision-making.

Experimental results highlight the robustness of our framework, with the Two-Stream I3D showing particularly strong performance while remaining suitable for deployment.

Future work will focus on extending the dataset with more diverse scenarios, exploring advanced data augmentation for better generalization, and applying optimization techniques for edge deployment. Moreover, attention-based or lightweight transformer modules will be investigated to enhance both accuracy and interpretability without compromising real-time feasibility.

References

1. Bermejo Nievas, E., Deniz Suarez, O., Bueno García, G., Sukthankar, R.: Violence detection in video using computer vision techniques. In: Proceedings of International Conference on Computer Analysis of Images and Patterns (CAIP), pp. 332–339 (2011)
2. Collins, T., Dudley, S.A., Finlay, M., McLoone, S.: A review of violence detection techniques: from handcrafted features to deep learning. Pattern Recognit. Lett. **125**, 2–11 (2019)
3. Tzelepis, C., Galanopoulos, D., Mezaris, V., Patras, I.: The importance of temporal information for violence detection in surveillance videos. IEEE Trans. Circuits Syst. Video Technol. **30**(9), 3239–3253 (2020)
4. Datta, S., Biswas, M., Chaudhuri, B.B.: A survey of deep learning techniques for violence detection. Pattern Recognit. Lett. **138**, 321–329 (2020)
5. Laptev, I.: On space-time interest points. Int. J. Comput. Vis. **64**(2), 107–123 (2005)
6. Dalal, N., Triggs, B.: Histograms of oriented gradients for human detection. In: Proceedings of IEEE Conference on Computer Vision and Pattern Recognition (CVPR), pp. 886–893 (2005)

7. Karpathy, A., Toderici, G., Shetty, S., Leung, T., Sukthankar, R., Fei-Fei, L.: Large-scale video classification with convolutional neural networks. In: Proceedings of IEEE Conference on Computer Vision and Pattern Recognition (CVPR), pp. 1725–1732 (2014)
8. Tran, D., Bourdev, L., Fergus, R., Torresani, L., Paluri, M.: Learning spatiotemporal features with 3D convolutional networks. In: Proceedings of IEEE International Conference on Computer Vision (ICCV), pp. 4489–4497 (2015)
9. Carreira, J., Zisserman, A.: Quo Vadis, action recognition? A new model and the kinetics dataset. In: Proceedings of IEEE Conference on Computer Vision and Pattern Recognition (CVPR), pp. 6299–6308 (2017)
10. Lowe, D.G.: Distinctive image features from scale-invariant keypoints. Int. J. Comput. Vis. **60**(2), 91–110 (2004)
11. Giannakopoulos, T., Pikrakis, A., Theodoridis, S.: Violence content detection in movies using audio-visual fusion and SVMs. In: Proceedings of Hellenic Conference on Artificial Intelligence (SETN), pp. 91–100 (2010)
12. Hassner, T., Itcher, Y., Kliper-Gross, O.: Violent flows: real-time detection of violent crowd behavior. In: Proceedings of IEEE International Conference on Computer Vision and Workshops (ICCVW), pp. 1–6 (2012)
13. Chen, M., Hauptmann, A.G.: MoSIFT: recognizing human actions in surveillance videos. Carnegie Mellon Univ., Tech. Rep. CMU-CS-09-161 (2009)
14. Kliper-Gross, O., Gurovich, Y., Hassner, T., Wolf, L.: Motion interchange patterns for action recognition in unconstrained videos. In: Proceedings of European Conference on Computer Vision (ECCV), pp. 256–269 (2012)
15. Nandur, S., et al.: Violence detection in videos for an intelligent surveillance system using MoBSIFT and movement filtering algorithm. Multimedia Tools Appl. (2023). https://doi.org/10.1007/s11042-023-15576-1
16. Donahue, J., et al.: Long-term recurrent convolutional networks for visual recognition and description. In: Proceedings of IEEE Computer Vision and Pattern Recognition Conference (CVPR) (2015)
17. Traoré, M., Akhloufi, M.A.: Violence detection in videos using hybrid deep models with optical flow and GRUs. IEEE Trans. Image Process. **33**, 1023–1035 (2024)
18. Cheng, Y., Zhang, H., Sun, C., Liu, B., Ren, T.: RLVS: a real-life violence surveillance dataset. In: Proceedings of ACM International Conference on Multimedia Retrieval (ICMR), pp. 512–516 (2021)
19. Carreira, J., Zisserman, A.: Quo vadis, action recognition? A new model and the kinetics dataset. In: Proceedings of the IEEE Conference on Computer Vision and Pattern Recognition (CVPR), pp. 6299–6308 (2017)
20. Feichtenhofer, F., Pinz, A., Zisserman, A.: Convolutional two-stream network fusion for video action recognition. In: Proceedings of the IEEE Conference on Computer Vision and Pattern Recognition (CVPR), pp. 1933–1941 (2016)
21. Huszár, L., Kiss, I., Varga, P., Kiss, A.: Spatiotemporal learning with optimized 3D CNNs for violence detection. In: Proceedings of IEEE International Conference on Acoustics, Speech, and Signal Processing (ICASSP) (2023)
22. Kollias, D., et al.: DVD: A Comprehensive Dataset for Advancing Violence Detection in Real-World Scenarios. In: arXiv preprint (2025). https://arxiv.org/abs/2506.05372
23. Bertasius, G., Wang, H., Torresani, L.: Is space-time attention all you need for video understanding?. In: Proceedings of International Conference on Machine Learning (ICML) (2021)
24. Liu, Z., et al.: Video swin transformer. In: Proceedings of IEEE Conference on Computer Vision and Pattern Recognition (CVPR) (2022)

25. Feichtenhofer, C.: X3D: expanding architectures for efficient video recognition. In: Proceedings of IEEE Conference on Computer Vision and Pattern Recognition (CVPR), pp. 203–213 (2020)
26. Ullah, A., Ahmad, J., Muhammad, K., Sajjad, M., Baik, S.W.: Violence detection using spatiotemporal features with LSTM. Sensors **19**(11), 2472 (2019)
27. He, K., Zhang, X., Ren, S., Sun, J.: Deep residual learning for image recognition. In: Proceedings of IEEE Conference on Computer Vision and Pattern Recognition (CVPR), pp. 770–778 (2016)
28. Hochreiter, S., Schmidhuber, J.: Long short-term memory. Neural Comput. **9**(8), 1735–1780 (1997)
29. Zhang, L., Zhu, C., Xu, C., Liu, J., Wang, Y.: Real-time violence detection based on deep spatio-temporal features. J. Electron. Imaging **26**(6), 061625 (2017)
30. Feichtenhofer, C., Fan, H., Malik, J., He, K.: SlowFast networks for video recognition. In: Proceedings of IEEE International Conference on Computer Vision (ICCV), pp. 6202–6211 (2019)
31. Brox, T., Bruhn, A., Papenberg, N., Weickert, J.: High accuracy optical flow estimation based on a theory for warping. In: Proceedings of European Conference on Computer Vision (ECCV), pp. 25–36 (2004)
32. Simonyan, K., Zisserman, A.: Two-stream convolutional networks for action recognition in videos. In: Proceedings of Advances in Neural Information Processing Systems (NeurIPS), vol. 27, pp. 568–576 (2014)
33. Zhou, P., Li, H., Yang, Y.: Violent interaction detection in surveillance video using two-stream 3D convolutional neural networks. Multimedia Tools Appl. **77**, 26529–26548 (2018)
34. Feichtenhofer, C., Pinz, A., Zisserman, A.: Convolutional two-stream network fusion for video action recognition. In: Proceedings of the IEEE Conference on Computer Vision and Pattern Recognition (CVPR), pp. 1933–1941 (2016)
35. Hasan, M.A., et al.: URFall: a fall detection dataset from realistic scenarios. In: Proceedings of International Conference on Recent Trends in Signal Processing and Computer Applications (ICREST), pp. 456–460 (2019)
36. Akti, S., Tataroglu, G.A., Ekenel, H.K.: Vision-based fight detection from surveillance cameras. In: Proceedings of International Conference on Image Processing Theory, Tools and Applications (IPTA), pp. 1–6 (2019)
37. Cohen, J.: A coefficient of agreement for nominal scales. Educ. Psychol. Meas. **20**(1), 37–46 (1960)
38. Farnebäck, G.: Two-frame motion estimation based on polynomial expansion. In: Proceedings of Scandinavian Conference on Image Analysis (SCIA), pp. 363–370 (2003)
39. Karpathy, A., Toderici, G., Shetty, S., Leung, T., Sukthankar, R., Fei-Fei, L.: Large-scale video classification with convolutional neural networks. In: Proceedings of the IEEE Conference on Computer Vision and Pattern Recognition (CVPR), pp. 1725–1732. IEEE (2014)
40. Cheng, M., Cai, K., Li, M.: RWF-2000: an open large scale video database for violence detection. In: Proceedings of the 25th International Conference on Pattern Recognition (ICPR), pp. 4183–4190. IEEE (2021)
41. Fawcett, T.: An introduction to ROC analysis. Pattern Recognit. Lett. **27**(8), 861–874 (2006)
42. Saito, T., Rehmsmeier, M.: The precision-recall plot is more informative than the ROC plot when evaluating binary classifiers on imbalanced datasets. PLoS ONE **10**(3) (2015)

43. Davis, J., Goadrich, M.: The relationship between precision-recall and ROC curves. In: Proceedings of International Conference on Machine Learning (ICML), pp. 233–240 (2006)
44. Selvaraju, R.R., et al.: Grad-CAM: visual explanations from deep networks via gradient-based localization. In: Proceedings of IEEE International Conference on Computer Vision (ICCV), pp. 618–626 (2017)

Comparative Analysis of Deep Learning and Classical Machine Learning for Deepfake Image Detection

Vladimir Hristov[1(✉)] and Asen Popov[2]

[1] Technical University of Sofia, Sofia, Bulgaria
vdhristov@tu-sofia.bg
[2] Technical University of Sofia, Sofia, Bulgaria
asepopov@tu-sofia.bg

Abstract. AI-generated content is becoming more prevalent which poses an important threat to the authenticity of visual media. The paper compares deep learning and classical machine learning methods to identify deepfake images. We investigate transfer learning with a fine-tuned EfficientNetB0 model along with logistic regression. Both approaches are validated on a publicly available dataset of real and AI-generated images. Results also indicate that EfficientNetB0 beats logistic regression with 92% vs. 56% accuracy for logistic regression. The results indicate that in this context deep learning is more important and also suggest the trade-offs between model complexity, interpretability, and performance. This is useful practical guidance into establishing rigorous deepfake detection mechanisms, which are indispensable in order to alleviate the risks of synthetic media.

Keywords: Deepfake Detection · Transfer Learning · Logistic Regression · Image Classification · EfficientNet · Machine Learning

1 Introduction

The proliferation of synthetically generated visual content, mainly made possible with development of the generative artificial intelligence (AI) models like Stable Diffusion, DALL-E, and Midjourney, poses a threat to the integrity of digital media [1, 2]. Although these technologies hold promise for revolution in a number of creative areas, at the same time they can lead to significant problems such as misinformation, deepfake-based fraud, and the manipulation of public perception using synthetic media. As a result, the precise and reliable detection of AI-generated images has become a central problem in multimedia forensics and computer vision research [3]. Early techniques for image authenticity detection relied on handcrafted features, such as noise pattern analysis, detection of compression artifacts, and the detection of statistical inconsistencies, but modern generative models are specialized to eliminate that kind of detectable artifacts, which is causing traditional approaches to become increasingly inadequate. To answer

T. Ensari et al. (Eds.): ISPR 2025, CCIS 2859, pp. 66–80, 2026.
https://doi.org/10.1007/978-3-032-21585-7_5

this, deep learning-based techniques, especially Convolutional Neural Networks (CNNs) and Vision Transformers (ViTs), have shown increased performance by directly learning discriminative features from the image input [4]. However, an in-depth view of the trade-offs among model complexity, interpretability, and computational efficiency in practical deployment cases is still an open issue [5]. We report a comparison of two different machine learning methods for detecting AI-generated images [6, 7]:

- Transfer learning with a pre-trained EfficientNetB0 CNN to extract fine-grained generative artifacts by leveraging knowledge acquired from large-scale vision datasets,
- Logistic Regression (LR) is used for handcrafted features and deep learning model-derived features, offering a simpler, more interpretable baseline model.

This article explores those techniques on a publicly available dataset of real and AI-generated images, estimating their performance in accuracy, generalization ability, and computational efficiency [8]. Our objective is to compare these two paradigms—deep learning and classical machine learning—to generate concrete guidelines for choosing detection methods with a meaningful trade-off between complexity and ease of use [9]. The results are part of the general trend to reduce synthetic media risks and to respond to the practical constraints of deployment.

2 Related Work

The growing rise of deepfakes — synthetically fabricated media created to mislead audiences — is an imminent enemy of information integrity and the trust of the public. This has led to substantial developments in the literature on the development of appropriate methods for detecting deepfakes. Early work towards deepfake detection generally utilized hand-crafted forensic features, such as eye-blinking frequency inconsistencies [10], head pose estimation [11] and color artifacts [12]. However, as generative models have increased in complexity, traditional techniques often find that they become impotent when confronted with new techniques for the generation of deepfakes especially those based on diffusion models [13, 14].

Deep learning has really changed the way deepfake detection is done. Indeed, Convolutional Neural Networks (CNNs) emerged as the winning method, demonstrating unparalleled performance by learning discriminative features without the need to examine a given dataset. In [15] presented the FaceForensics++ dataset, a well-known benchmark model that has made the development and improvement of many deepfake detectors based on CNNs easier, such as XceptionNet, which has achieved quite efficient capabilities of recognizing facial distortion. Transfer learning is one of the most powerful approaches, which capitalizes on pre-trained models and is fine-tuned on massive datasets (e.g., ImageNet [16]). We also employ many existing architectures, including ResNet [17, 18], InceptionV3 [18], EfficientNet, and others, to detect synthetic content, and they have been successfully adapted for the use of artificial material detection. The high-level features of deep fakes can be extracted and subtle inconsistencies often suggested by deepfake manipulation can also be obtained due to the efficiency of these models.

Classical machine learning methods like logistic regression and SVMs [19, 20] have been chosen as baseline models mostly because of computational cost-effectiveness and interpretable nature. Nevertheless, these linear models do not always perform well in high dimension and nonlinear tasks like deepfake detection. Deepfake detection techniques fall into the following categories: (1) detection of artifacts and inconsistencies, (2) physiological signals analysis, (3) deep learning models, and (4) blockchain technology for media authentication [21]. This work basically revolves around the third type of model and exploits deep learning technologies to extract the features and classify it.

The survey [21] further shows the increasing relevance of multimodal approaches to the detection of deepfake, whereby more information from alternative modalities (e.g., video, audio, text) could be integrated to improve efficiency and robustness of detection. Although we concentrate on image-based today we recognize the possibility of using multimodal information in future work. Moreover, the survey [21] highlighted, deepfake detection techniques should be capable of evolving towards more powerful and generalizable approaches to address new techniques in deepfake creation. This involves designing techniques that resist adversarial attacks and generalize to other datasets and generative models. This is why we explore EfficientNetB0, which is an efficient classifier, and learns resilient features.

3 Methodology

3.1 Dataset and Experimental Setup

The experiments of this paper used the "2025 Women in AI: AI vs. Human-Generated Images" dataset [22], which is imported from Kaggle. The dataset contains approximately 80,000 labeled images divided equally between genuine (human-captured) and artificial (AI-generated) categories. We divided the dataset into training (80%) and test (20%) sets, and stratified sampling was used to obtain a well balanced class of images for valid model evaluation (Appendices link for dataset).

All experiments were conducted on a specialized workstation with an NVIDIA RTX 4060 GPU and Intel Core i5-12500H processor supporting 32GB of RAM. Environment of software included Python 3.9 with the TensorFlow 2.x and scikit-learn libraries.

3.2 Model Selection Justification

EfficientNetB0, which achieves a balance between classification accuracy and computational efficiency for large-scale image analysis [23], was chosen as a deep learning backbone. Logistic regression, implemented using scikit-learn, was chosen as a classical baseline as it is interpretable and serves as a reference model in machine learning [24, 25]. Furthermore, this option allowed for a rigorous comparison between complex deep models and linear classifiers based on raw pixel and learned feature representations [29, 27].

3.3 Model Architectures and Training Procedures

3.3.1 Deep Learning Approach (EfficientNetB0)

We created the EfficientNetB0 model using transfer learning with TensorFlow/Keras. We opted for this model because of the tradeoff of high computational efficiency and accuracy, so it fits for massive image analysis. We initialized the network with weights pre-trained on ImageNet dataset, using knowledge from large variety of images.

- Architecture EfficientNetB0 architecture involves a hierarchy of mobile inverted bottleneck convolutions [MBConv] blocks, squeeze-and-excitation blocks, and a final classification layer. The original classification head (for 1000 ImageNet classes) was discarded and we replaced it with a randomly initialized fully connected (Dense) layer with a sigmoid activation function for binary classification (real vs. fake).
- Fine-Tuning Approach: We used two-stage fine-tuning:

1. Freezing: Initially all of the layers of the pre trained EfficientNetB0 model were frozen (i.e., their weights were not updated). The training consisted only of the weights of the new classification layer. This helped the classification layer to instantly adapt and build for the task of deepfake detection without interfering with the learned feature representations of the pre-trained model.
2. Unfreezing and fine-tuning: Following the initial freezing step, we unfroze the last 30 layers of the EfficientNetB0 model (in particular, the last few MBConv blocks and their final convolutional layers). This in turn, led to fine-grained fine-tuning adaptation of pre-trained features to the task of deepfake detection. We chose to unfreeze 30 layers considering experiment and validation results.

- Optimizer settings: Adam optimizer was used for training with the following:
- In the initial freezing stage 1e-4 (0.0001), and the tuning stage 1e-5 (0.00001) as optimal. To achieve more fine-tuning of the unfrozen layers, the learning rate was minimized.
- Beta Parameters: Beta1 = 0.9, Beta2 = 0.999 (default values for Adam).
- Epsilon: 1e-7 (default Adam value).
- Weight Decay: No weight decay (L2 regularization) was done using the Adam optimizer during our experiments. We observed that weight decay did not optimize performance in our validation set.
- Learning Rate Schedule: Utilizing a ReduceLROnPlateau learning rate scheduler, the learning rate is automatically reduced at a plateau of validation loss:
- Monitor: ‘val_loss’ (Validation Loss).
- Factor: 0.1 (when the validation loss plateaus, this factor times the learning rate).
- Patience: 5 epochs (learning rate is reduced if validation loss does not improve for 5 consecutive epochs).
- Min_lr: 1e-6 (minimum learning rate allowed).
- Data Augmentation: In training, we applied the following methods to expand the diversity of training data, improve the image types and generalization performance of models:
- Random Horizontal Flipping: Each image was randomly horizontally flipped at a frequency of 0.5.

- Random Brightness Adjustment: brightness values of their images at the interval [−20%, +20%].
- Random Contrast Adjustment: Contrast of images was set to [−20%, +20%].
- Rescaling: The images were rescaled with a value of 1./255.
- Loss Function: binary cross-entropy was applied to the loss function, since it is suitable for a binary classification task. Consider adding the presence of class weights (to prevent any possible class imbalance—i.e., small differences between real vs. fake image counts) by using this approach. Values were obtained based on the training-set inverse of the class frequencies.
- Training batch size with 64 being used to achieve a better performance compared to the experimental data. This batch size was selected to prevent the use of memory from overloading the train.
- Checkpointing: Early stopping was employed to avoid overfitting. Training was terminated when the validation loss did not get better for 10 epochs in a row. The best model for validation loss was retained for a checkpoint.
- Image Preprocessing: Input images were resized to 224x224 pixels by bilinear interpolation. The pixel values were normalized with the mean and standard deviation of the ImageNet dataset. Normalizing these data improves training robustness and performance.

3.3.2 Classical Baseline (Logistic Regression)

A logistic regression was applied with scikit-learn in two settings:

(a) On Raw Pixel Data: Images were resized to 32x32x3, flattened and pixel values scaled to [0, 1]. The SGDClassifier was used for log loss, optimal learning rate schedule, and class weighting for training the model in mini batches (batch size 128).
(b) On deep features: For all the images, we extracted 224-dimensional embeddings from the penultimate layer of EfficientNetB0. Such features were standardized to train a logistic regression model (liblinear solver; max_iter = 1000; class_weight = 'balanced').

3.4 Evaluation Metrics

We trained all models on the same training set and tested them on a held-out test set, using accuracy, precision, recall, F1-score, and confusion matrices, to ensure the complete comparison. Such a holistic evaluation structure allows unbiased and direct comparisons of the strengths and weaknesses of deep learning and classical machine learning methods in the context of deepfake image detection.

4 Dataset Description

The dataset utilized in this study is the "2025 Women in AI: AI vs. Human-Generated Images" dataset, originally provided for a Kaggle competition by Shutterstock and Deep-Media. This dataset comprises approximately 80,000 labeled images, equally divided into two categories: authentic (human-captured) and synthetic (AI-generated). This balanced distribution mitigates potential biases during model training and evaluation.

Authentic images were sourced from Shutterstock's extensive stock photo library, featuring diverse subjects, with approximately one-third depicting human individuals. Synthetic images were generated by DeepMedia using a combination of state-of-the-art generative models, including StyleGAN3 (40%), Stable Diffusion 3 (40%), and DALL-E variants (20%) [8, 23]. This mix of generative models aims to capture the diversity of AI-generated content encountered in real-world scenarios.

To facilitate robust model training, each authentic image is systematically paired with a semantically similar AI-generated counterpart. This pairing allows for direct comparisons and encourages the models to learn subtle distinctions between real and synthetic imagery. The dataset includes detailed metadata, encompassing image resolution (ranging from 512x512 to 1024x1024 pixels), demographic attributes (e.g., approximate age and gender distributions), and post-processing indicators (e.g., JPEG compression applied to 30% of images, noise injection at a low level of 1%).

Samples are carefully balanced across varying lighting conditions (indoor and outdoor environments) and diverse backgrounds (plain vs. cluttered) to enhance the models' generalization capabilities and robustness to environmental variations. The dataset was partitioned into training (80%) and testing (20%) subsets using stratified sampling to maintain class balance within each subset. This partitioning strategy ensures that the models are evaluated on a representative sample of the overall dataset and prevents overfitting.

Figure 1 illustrates the prompt for generating authentic deepfakes and the resulting images paired with their generated text labels and AI-generated equivalents.

Photo of a young black couple, both with medium-dark skin, standing in front of a plain pink background. The woman has a large afro and is wearing round glasses, a sleeveless lavender top, and has her hand on her chin with a surprised expression. The man has a short afro, a goatee, and is wearing a light blue t-shirt. He has his hands clasped together and is also looking surprised. The...

Fig. 1. Examples of authentic (right) and AI-generated (left) images from the dataset.

5 ResultsQuantitative Evaluation

Table 1 shows the comparative performance of the EfficientNetB0 and logistic regression models. Compared to logistic regression, EfficientNetB0 performed much better on all metrics. EfficientNetB0 achieved an overall accuracy of 92.0%, precision of 92.1%, recall of 91.9%, and an F1-score of 92.0%. Logistic regression achieved an accuracy of 56.0%, precision of 58.0%, recall of 56.0%, and an F1-score of 53.0%.

Table 1. Comparative results of EfficientNetB0 and Logistic Regression models.

Model	Accuracy	Precision	Recall	F1 - score
EfficientNetB0	0.92	0.92	0.92	0.92
Logistic Regression	0.56	0.58	0.56	0.53

5.1 Training and Validation Analysis

The training and validation accuracy and loss curves for the EfficientNetB0 model are presented in Fig. 2 and Fig. 3, respectively. The learning curves exhibit steady improvements, with both metrics stabilizing over the course of training. The close alignment of the training and validation curves indicates good generalization and the absence of significant overfitting. The initial performance increase stems from the ImageNet pre-training, whereas the later improvements reflect the model's adaptation to deepfake-specific characteristics.

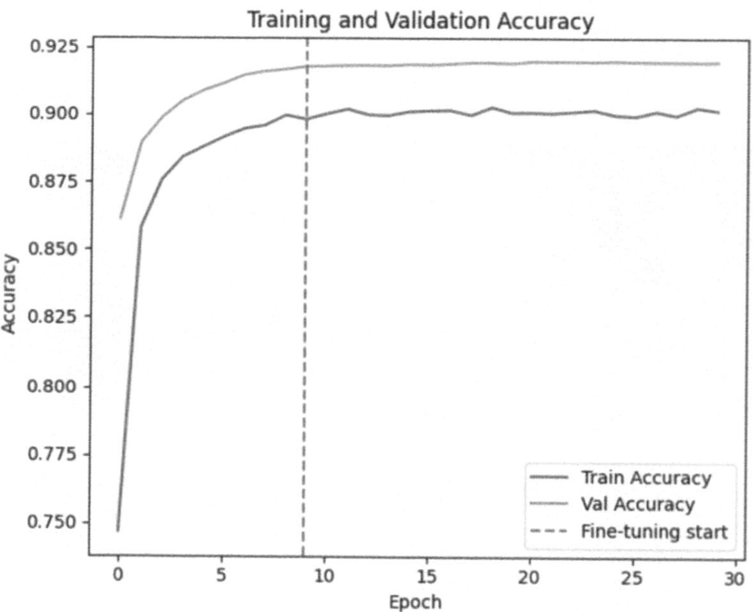

Fig. 2. Training and validation accuracy curves for the EfficientNetB0 model.

Figure 2 illustrates that the EfficientNetB0 model attains strong validation accuracy from the initial training epochs, with the validation and training curves remaining closely aligned throughout. This close alignment indicates that the model generalizes well to unseen data and avoids overfitting. The learning trajectory reflects that EfficientNetB0 effectively captures distinguishing features needed for deepfake detection, achieving stable performance across training and validation sets.

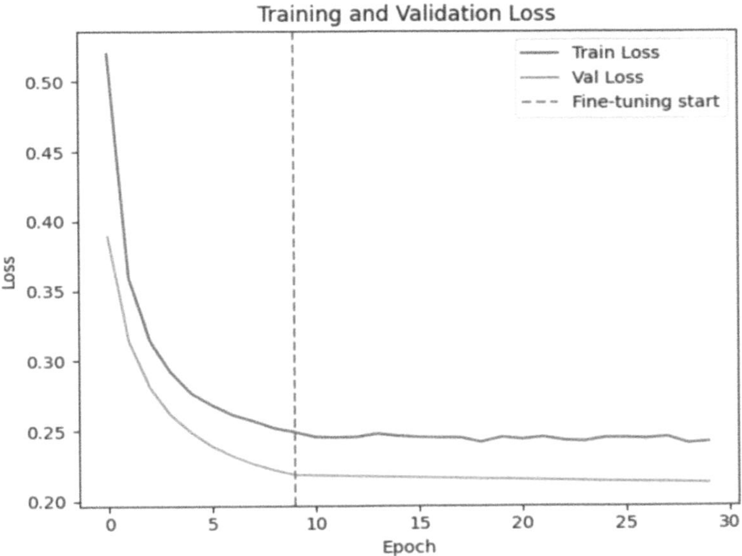

Fig. 3. Training and validation loss curves for the EfficientNetB0 model.

Figure 3 displays the training and validation loss curves, confirming that the model's loss steadily decreases and eventually plateaus as training progresses. The absence of significant gaps or increases in validation loss provides further evidence that the model's learning process is stable and not prone to overfitting. The consistency observed across both accuracy and loss metrics supports the robustness of the training configuration and the suitability of the selected hyperparameters.

5.2 Confusion Matrix Analysis

Data of confusion matrices for EfficientNetB0 and logistic regression models are presented in Fig. 4 and Fig. 5, respectively. These matrices present an in-depth exploration of true positives, true negatives, false positives, and false negatives per class to allow a consideration of more than a summary of performance metrics such as accuracy and F1-score. Confusion matrices help demonstrate how strengths and weaknesses of individual models in discrimination of real vs. AI-generated images can be evaluated, as they demonstrate the distribution of correct and incorrect predictions. EfficientNetB0 model exhibits a high accuracy of capturing both real and AI-generated images and minimal confusion among the classes. Logistic regression, on the contrary, shows a higher rate of misclassification, especially for genuine images that are classed as AI-generated.

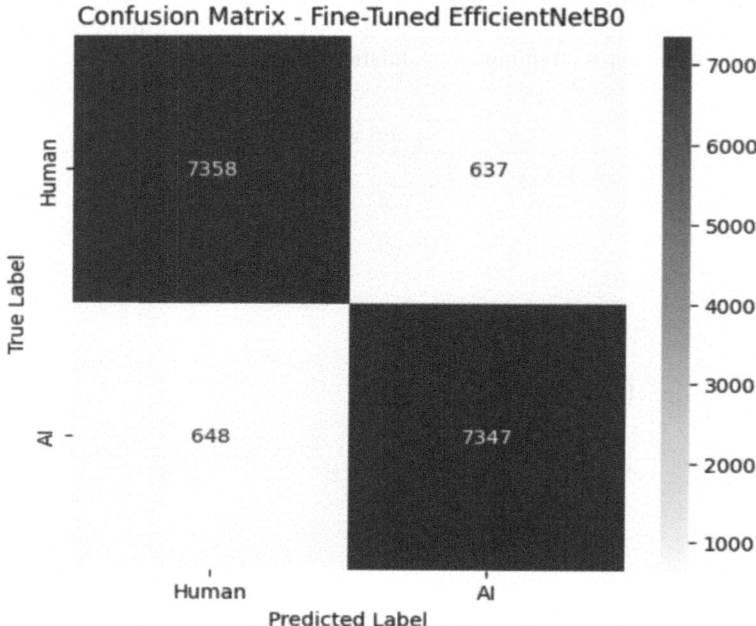

Fig. 4. Confusion matrix for the fine-tuned EfficientNetB0 model

Figure 4 shows the confusion matrix for the fine-tuned EfficientNetB0 model, evaluated on the test set. The model achieves a high number of correct predictions for both classes: 7,358 true positives (Human correctly classified as Human) and 7,347 true negatives (AI correctly classified as AI). The number of misclassifications is relatively low, with 637 real images incorrectly predicted as AI (false negatives) and 648 AI-generated images misclassified as real (false positives). This strong performance across both classes highlights the model's ability to effectively distinguish between real and AI-generated images, with balanced accuracy and minimal bias toward either class.

Figure 5 displays the confusion matrix for the logistic regression model trained directly on raw pixel values. Compared to the EfficientNetB0 model, logistic regression demonstrates substantially lower accuracy, especially for the "Human" class. The model correctly classifies 2,592 real images but misclassifies 5,403 human images as AI-generated, indicating a high false negative rate. For the "AI" class, 6,371 images are correctly identified, while 1,624 are incorrectly labeled as real. This pattern reflects the limitations of logistic regression in capturing the complex, non-linear features necessary for reliable deepfake detection. The results highlight that classical linear models are inadequate for distinguishing between subtle artifacts in human and AI-generated images, leading to pronounced class imbalance and higher rates of misclassification.

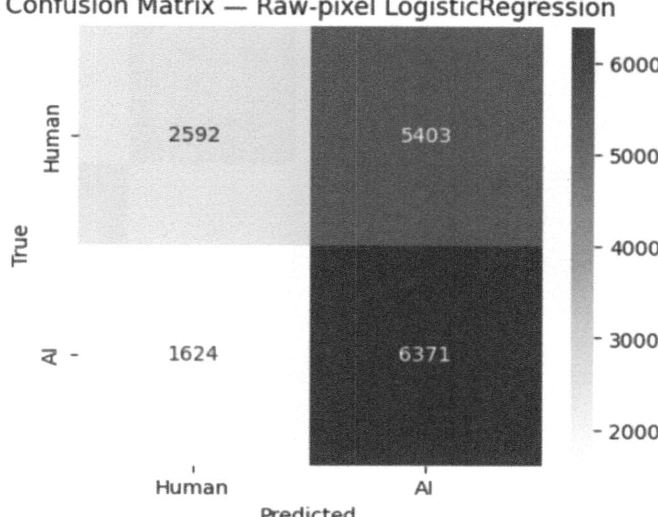

Fig. 5. Confusion matrix for the raw-pixel logistic regression model

5.3 Performance on GAP Features

The performance of logistic regression on features extracted from the EfficientNetB0 global average pooling (GAP) layer (1280-dimensional) is presented in Table 2. When trained on GAP features, logistic regression achieved an accuracy of 90.0%, precision of 89.0%, recall of 91.0%, and an F1-score of 90.0%. A confusion matrix for logistic regression on GAP features is presented in Fig. 6.

Table 2. Performance metrics for EfficientNetB0 head and Logistic Regression on GAP-features.

Model	Class	Precision	Recall	F1-score	Accuracy
EfficientNetB0 Head	Human	0.92	0.92	0.92	0.92
EfficientNetB0 Head	AI	0.92	0.92	0.92	0.92
Logistic Regression GAP	Human	0.89	0.91	0.90	0.90
Logistic Regression GAP	AI	0.91	0.89	0.90	0.90

The experimental results clearly demonstrate the superiority of transfer learning with EfficientNetB0 over traditional logistic regression for deepfake image detection. The fine-tuned EfficientNetB0 model achieved a significantly higher overall accuracy (92.0%) compared to raw-pixel logistic regression (56.0%). Moreover, EfficientNetB0 exhibited balanced precision and recall across both "Human" and "AI" classes, resulting in a high F1-score (92.0%), whereas raw-pixel logistic regression suffered from a pronounced class imbalance.

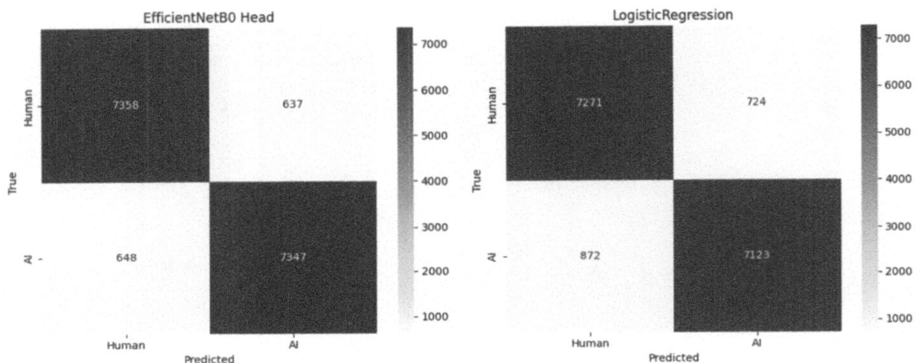

Fig. 6. Confusion matrices for (left) EfficientNetB0 Head and (right) Logistic Regression on GAP-features.

The superior performance of EfficientNetB0 can be attributed to its ability to learn hierarchical, non-linear features from image data, leveraging its convolutional layers and pre-training on ImageNet. This allows the model to detect subtle forensic artifacts characteristic of AI-generated images, such as texture inconsistencies (irregularities in hair, skin, or fabric), unnatural lighting (unphysical shadows or reflections), and structural distortions (asymmetries in facial features or background geometry). In contrast, raw-pixel logistic regression relies on linear separability and shallow features, which are insufficient to capture the complex relationships present in high-dimensional image data.

While logistic regression offers advantages in terms of computational efficiency, faster training/inference times, and interpretability through feature importance analysis, EfficientNetB0 provides greater scalability and generalization, which are essential for real-world applications where deepfakes are becoming increasingly sophisticated. Moreover, when the global average pooling (GAP) features of EfficientNetB0 (1280-dimensional) are used as input for logistic regression, performance improves dramatically compared to raw pixels. But performance is still lower than for the full EfficientNetB0 head, indicating that some information is lost when reducing EfficientNetB0 to just the global average pooling layer.

Modern generative models, such as Stable Diffusion, are designed to minimize detectable artifacts, posing an ongoing challenge for detection methods. Deep neural networks, with their hierarchical feature learning capabilities, adapt more effectively to these evolving challenges.

The initial gap between training/validation loss observed in the EfficientNetB0 model stems from the adaptation of its pre-trained layers to the domain-specific characteristics of deepfake images, but it quickly converges.

6 Analysis and Discussion

The experimental results clearly demonstrate the superiority of transfer learning with EfficientNetB0 over traditional logistic regression for deepfake image detection. The fine-tuned EfficientNetB0 model achieved a significantly higher overall accuracy (92.0%) compared to raw-pixel logistic regression (56.0%). Moreover, EfficientNetB0 exhibited balanced precision and recall across both "Human" and "AI" classes, resulting in a high F1-score (92.0%), whereas raw-pixel logistic regression suffered from a pronounced class imbalance. The superior performance of EfficientNetB0 can be attributed to its ability to learn hierarchical, non-linear features from image data, leveraging its convolutional layers and pre-training on ImageNet. This allows the model to detect subtle forensic artifacts characteristic of AI-generated images, such as texture inconsistencies (irregularities in hair, skin, or fabric), unnatural lighting (unphysical shadows or reflections), and structural distortions (asymmetries in facial features or background geometry). In contrast, raw-pixel logistic regression relies on linear separability and shallow features, which are insufficient to capture the complex relationships present in high-dimensional image data. While logistic regression offers advantages in terms of computational efficiency, faster training/inference times, and interpretability through feature importance analysis, EfficientNetB0 provides greater scalability and generalization, which are essential for real-world applications where deepfakes are becoming increasingly sophisticated. Moreover, when the global average pooling (GAP) features of EfficientNetB0 (1280-dimensional) are used as input for logistic regression, performance improves dramatically compared to raw pixels. But performance is still lower than for the full EfficientNetB0 head, indicating that some information is lost when reducing EfficientNetB0 to just the global average pooling layer. Modern generative models, such as Stable Diffusion, are designed to minimize detectable artifacts, posing an ongoing challenge for detection methods. Deep neural networks, with their hierarchical feature learning capabilities, adapt more effectively to these evolving challenges. The initial gap between training/validation loss observed in the EfficientNetB0 model stems from the adaptation of its pre-trained layers to the domain-specific characteristics of deepfake images, but it quickly converges.

The future work include explore the use of visualization techniques such as Grad-CAM. Grad-CAM allows us to generate heatmaps that highlight the regions of the input image that are most important for the model's classification decision. By visualizing these heatmaps, we can gain insights into the specific features that the EfficientNetB0 model is using to discriminate between real and fake images. This can help us to understand why the model makes certain predictions and identify potential biases or vulnerabilities.

Computational cost is an important consideration for real-world deployment, especially in resource-constrained environments such as mobile devices or edge computing platforms. While EfficientNetB0 is a relatively efficient CNN architecture, its computational demands may still be a limiting factor for certain applications. Future work will be focus on evaluating the inference time, energy consumption, and memory requirements of our model on different hardware platforms. This would provide valuable information for determining the feasibility of deploying our model in real-world scenarios. Additionally, techniques such as model quantization and pruning could be explored to further reduce the computational cost of the model.

Modern deepfakes often include adversarial perturbations, which are small, carefully crafted modifications to the input image that are designed to evade detection. To assess the robustness of our model against adversarial attacks, future work should focus on testing its performance against adversarially perturbed images generated using techniques such as FGSM. This would provide a more realistic assessment of the model's ability to detect deepfakes in the presence of adversarial attacks. If the model is found to be vulnerable to adversarial attacks, techniques such as adversarial training could be employed to improve its robustness.

7 Conclusion

This study provides robust empirical evidence supporting the superiority of deep learning approaches, specifically transfer learning with EfficientNetB0, for deepfake image detection. Our comparative analysis demonstrates that EfficientNetB0 achieves significantly higher performance than traditional logistic regression (LR) across key metrics, including overall accuracy (92.0% vs. 56.0%) and F1-score (92.0% vs. 53.0%). Furthermore, EfficientNetB0 exhibited balanced performance across both "Human" and "AI" classes, while logistic regression struggled with class imbalance.

EfficientNetB0's superior performance stems from its ability to learn hierarchical, non-linear features that capture subtle forensic artifacts characteristic of AI-generated images. These findings have significant real-world implications for cybersecurity (filtering synthetic media), digital forensics (evidence authentication), and AI ethics (embedding validation mechanisms).

However, key challenges remain. The generalizability of models like EfficientNetB0 may degrade when faced with emerging deepfake techniques, especially those using novel diffusion approaches. Computational demands may hinder deployment in real-time or edge-device scenarios, and the interpretability of CNN decisions remains limited.

Future research should explore hybrid models combining CNNs with vision transformers or handcrafted features, adversarial training, and multimodal integration. Addressing these challenges will pave the way for more robust, efficient, and transparent deepfake detection solutions.

In conclusion, this work contributes to the ongoing effort to combat the threats posed by synthetic media by demonstrating the effectiveness of deep learning-based approaches and highlighting key areas for future research and development.

Appendices link for dataset: https://www.kaggle.com/datasets/alessandrasala79/ai-vs-human-generated-dataset

Acknowledgments. This work has been accomplished with financial support from the European Regional Development Fund within the Operational Program "Bulgarian national recovery and resilience plan" and the procedure for the direct provision of grants "Establishing of a network of research higher education institutions in Bulgaria" under Project BG-RRP-2.004-0005 "Improving the research capacity and quality to achieve international recognition and resilience of TU-Sofia (IDEAS)"

References

1. Hussain, S., Neekhara, P., Jere, M., Koushanfar, F., McAuley, J.: Adversarial Deepfakes: evaluating vulnerability of Deepfake detectors to adversarial examples. In: 2021 IEEE Winter Conference on Applications of Computer Vision (WACV), pp. 3347–3356, Waikoloa, HI, USA (2021). https://doi.org/10.1109/WACV48630.2021.00339
2. Nguyen, T.T., et al.: Deep learning for Deepfakes creation and detection: a survey, 103525., ISSN 1077–3142. Comput. Vis. Image Underst. **223** (2022). https://doi.org/10.1016/j.cviu.2022.103525
3. Vashistha, M., Jain, S., Pandey, S., Pradhan, A., Tarwani, S.: A comparative analysis of machine learning and deep learning approaches in Deepfake detection. In: 2024 IEEE Region 10 Symposium (TENSYMP), pp. 1–8, New Delhi, India (2024). https://doi.org/10.1109/TENSYMP61132.2024.10752209
4. Liu, X., Dong, X., Xie, F., Pei, L., Xi, L., Jiang, M.: Hybrid network of convolutional neural network and transformer for deepfake geographic image detection. J. Electr. Imaging. **33**(2), 023007 (5 March 2024). https://doi.org/10.1117/1.JEI.33.2.023007
5. Alsolai, H., et al.: Guardian-AI: a novel deep learning based deepfake detection model in images. Alex. Eng. J. **126**, 507–514, ISSN 1110–0168, (2025). https://doi.org/10.1016/j.aej.2025.04.095
6. Khan, R., et al.: Comparative study of deep learning techniques for DeepFake video detection. ICT Express. **10**(6), 1226–1239, ISSN 2405–9595, (2024). https://doi.org/10.1016/j.icte.2024.09.018
7. Altamimi, S., Salameh, W.: Towards analysis detection of Deepfake video via deep learning models: a review. In: 2024 International Jordanian Cybersecurity Conference (IJCC), pp. 87–92, Amman, Jordan (2024). https://doi.org/10.1109/IJCC64742.2024.10847278
8. Volkova, S.S.: A method for Deepfake detection using convolutional neural networks. Sci. Tech. Inf. Process. **50**, 475–485 (2023). https://doi.org/10.3103/S0147688223050143
9. Mansoor, N., Iliev, A.: DeepFake detection using deep learning. In: Arai, K. (ed.) Intelligent Computing. SAI 2024 Lecture Notes in Networks and Systems, vol. 1018. Springer, Cham (2024). https://doi.org/10.1007/978-3-031-62269-4_14
10. Jung, T., Kim, S., Kim, K.: DeepVision: Deepfakes detection using human eye blinking pattern. IEEE Access. **8**, 83144–83154 (2020). https://doi.org/10.1109/ACCESS.2020.2988660
11. Sultan, D.A., Ibrahim, L.M.: Deepfake detection model based on VGGFace with head pose estimation technique. In: Al-Bakry, a.M., New Trends in Information and Communications Technology Applications. NTICT 2023. Communications in Computer and Information Science, vol. 2096. Springer, Cham (2024). https://doi.org/10.1007/978-3-031-62814-6_8
12. Masood, M., Nawaz, M., Malik, K.M.: Deepfakes generation and detection: state-of-the-art, open challenges, countermeasures, and way forward. Appl. Intell. **53**, 3974–4026 (2023). https://doi.org/10.1007/s10489-022-03766-z
13. Pintelas, E., Livieris, I.E.: Convolutional neural network framework for deepfake detection: a diffusion-based approach, 104375., ISSN 1077–3142. Comput. Vis. Image Underst. **257** (2025). https://doi.org/10.1016/j.cviu.2025.104375
14. Magesh, A.P., Ramakrishnan, S.S.M., Arumuga Arun, R.: Building an efficient deep fake detection system using the recognition capabilities of convolutional neural networks and transformers. Discov. Comput. **28**, 99 (2025). https://doi.org/10.1007/s10791-025-09586-2
15. Kumari, M., Gupta, S., Singh, N.: Deepfake detection using XceptionNet. Int. J. Sci. Res. Sci. Technol. **12**(3), 55–63 (2025) https://doi.org/10.32628/IJSRST251222688
16. Karras, T., Laine, S., Aila, T.: A style-based generator architecture for generative adversarial networks. IEEE Trans. Pattern Anal. Mach. Intell. **43**(12), 4217–4228 (1 Dec 2021). https://doi.org/10.1109/TPAMI.2020.2970919

17. Fahad, M., Zhang, T., Iqbal, Y., et al.: Advanced deepfake detection with enhanced Resnet-18 and multilayer CNN max pooling. Vis. Comput. **41**, 3473–3486 (2025). https://doi.org/10.1007/s00371-024-03613-x
18. Akpabio, E., Deshmukh, A., Narad, S.: Hybrid Deepfake detection: leveraging res Net50 and InceptionV3 for high-accuracy classification. In: 2025 International Conference on Computing Technologies (ICOCT), pp. 1–6. Bengaluru, India (2025). https://doi.org/10.1109/ICOCT64433.2025.11118681
19. Abraham, T.M., Wen, T., Wu, T.: Leveraging data analytics for detection and impact evaluation of fake news and deepfakes in social networks. Humanit. Soc. Sci. Commun. **12**, 1040 (2025). https://doi.org/10.1057/s41599-025-05389-4
20. Sharma, P., Kumar, M., Sharma, H.K.: A robust ensemble model for Deepfake detection of GAN-generated images on social media. Discov. Comput. **28**, 41 (2025). https://doi.org/10.1007/s10791-025-09538-w
21. Gupta, G., Raja, K., Gupta, M., Jan, T., Whiteside, S.T., Prasad, M.: A comprehensive review of DeepFake detection using advanced machine learning and fusion methods. Electronics. **13**, 95 (2024). https://doi.org/10.3390/electronics13010095
22. Sala, A., Harshika, Jeyaraj, M., Pitsiani, M., Belton, N., Ijatomi, T.: Detect AI vs. Human-Generated Images. Kaggle (2025). https://kaggle.com/competitions/detect-ai-vs-human-generated-images
23. Javed, M., Zhang, Z., Dahri, F.H.: Audio–Visual Synchronization and Lip Movement Analysis for Real-Time Deepfake Detection. Int. J. Comput. Intell. Syst. **18**, 170 (2025). https://doi.org/10.1007/s44196-025-00911-7
24. Singh, L.H., Charanarur, P., Chaudhary, N.K.: Advancements in detecting Deepfakes: AI algorithms and future prospects −a review. Discov. Internet Things. **5**, 53 (2025). https://doi.org/10.1007/s43926-025-00154-0
25. Wang, G., Han, Y., Xu, F., Gao, Y., Sang, W.: The detection optimization of low-quality fake face images: feature enhancement and noise suppression strategies. Appl. Sci. **15**, 7325 (2025). https://doi.org/10.3390/app15137325
26. She, H., Hu, Y., Liu, B., Li, J., Li, C.-T.: Using graph neural networks to improve generalization capability of the models for Deepfake detection. IEEE Trans. Inf. Forensics Secur. **19**, 8414–8427 (2024). https://doi.org/10.1109/TIFS.2024.3451356
27. Chekurov, I.D., Hristov, K.H.: Design and implementation of an adaptive FPGA-based traffic light control system using Verilog. In: 2025 60th International Scientific Conference on Information, Communication and Energy Systems and Technologies (ICEST), pp. 1–4. Ohrid, North Macedonia (2025). https://doi.org/10.1109/ICEST66328.2025.11098276

Automatic Discretization of Depression Symptoms from Social Media Using KMeans and Quantiles

Hathemi Mohamed[1](✉), Manel Ben Amira[1], Nabil Khoufi[2], and Chafik Aloulou[1]

[1] University of Sfax, AnLP Research Group, Miracl Lab, Sfax, Tunisia
hathemimohamed277@gmail.com, manel.benamira@ipeis.usf.tn, chafik.aloulou@fsegs.rnu.tn

[2] University of Sousse, AnLP Research Group, Sousse, Tunisia
nabil.khoufi@essths.u-sousse.tn

Abstract. The main aim of our research is to automatically discretize the symptoms of depression expressed on social networks, in accordance with the 21 items of the BDI-II (Beck Depression Inventory) questionnaire. This discretization aims to assign each detected symptom a discrete level of severity (from 0 to 3), consistent with the BDI-II scale, based on text messages posted online. Our approach relies on two complementary steps: the extraction of linguistic expressions associated with depressive symptoms, followed by their discretization using statistical methods such as KMeans clustering and quantile segmentation. This combination transforms raw textual signals into interpretable numerical representations of the degree of psychological distress. By leveraging natural and spontaneous data from users' everyday lives, our system goes beyond the limitations of traditional mental health diagnostic methods. It enables automatic, fine-grained, and continuous assessment of symptom severity, paving the way for the early detection of individuals at risk. Ultimately, this system could serve as a valuable decision-support tool for mental health professionals.

Keywords: Symptoms discretization · BDI-II · KMeans clustering · Quantile Segmentation

1 Introduction

Depression is a complex and multifactorial mental disorder, characterized by a profound and persistent alteration of mood that significantly affects an individual's thoughts, emotions, and behavior. It manifests as a marked loss of interest or pleasure in daily activities, social withdrawal, sleep disturbances, changes in appetite, and feelings of despair, which can lead to significant difficulties in personal, social, and professional life. Depression is a serious condition whose severity can range from mild to severe, representing a major risk factor for overall quality of life and health [1].

T. Ensari et al. (Eds.): ISPR 2025, CCIS 2859, pp. 81–94, 2026.
https://doi.org/10.1007/978-3-032-21585-7_6

The World Health Organization reports that depression ranks among the top causes of disability worldwide, contributing to 7.5% of the total Years Lived with Disability (YLD) as of 2015. It is also a major risk factor for suicide, which causes more than 700,000 deaths annually, and is the fourth leading cause of death among young people aged 15 to 29. These data highlight the critical importance of early detection and intervention in managing depression to reduce its impact on global public health [2]

To assess the intensity and multidimensional nature of depressive symptoms, the Beck Depression Inventory-II (BDI-II) provides a standardized self-report tool widely used in clinical and research settings. The BDI-II measures the severity of depression through 21 questions representing the main symptoms of depressive disorders. These symptoms cover several dimensions: emotional (sadness, loss of interest, irritability), cognitive (feelings of guilt, failure, worthlessness, suicidal thoughts), behavioral (reduced activity, difficulty concentrating, indecisiveness), and physical (sleep disturbances, fatigue, changes in appetite, and loss of libido). Each item is associated with four response options, scored from 0 to 3, reflecting the intensity or frequency of the symptom experienced. This approach allows for a detailed profile of an individual's depressive manifestations but remains limited by the need for conscious self-assessment and by the subjective variability of responses [3–5].

To overcome these limitations and exploit new sources of information, our research proposes an automated and intelligent system dedicated to discretizing the 21 depression symptoms identified by the BDI-II, based on textual data generated by users on social media platforms. Posts on these platforms often spontaneously reflect individuals' emotional and cognitive states, offering a unique potential for the early detection of depressive symptoms. By combining unsupervised clustering techniques, notably K-Means, with quantile-based statistical segmentation, our approach identifies discrete severity levels for each symptom, capturing the emotional, cognitive, behavioral, and physical dimensions of depression.

This methodological framework aims to provide a continuous, objective, and precise assessment of psychological states, thereby facilitating a better understanding of symptom patterns and paving the way for personalized interventions and early detection. Furthermore, the integration of data from social media could enrich traditional clinical tools, improve mental health monitoring, and support professionals in implementing tailored strategies for individuals at risk.

The remainder of this paper is organized as follows: Sect. 2 reviews the related literature. Section 3 details the proposed methodology and system architecture. Section 4 presents and evaluates the experimental results. Finally, the concluding section summarizes the study and suggests potential directions for future research.

2 Related Works

A significant body of research has focused on recognizing and diagnosing the symptoms associated with depression. Researchers have explored a wide range of approaches, from analyzing behavioral and psychological indicators to integrating advanced technological methods. The goal of this research is to develop a deeper understanding of the mechanisms involved in depression detection and to improve screening techniques, with the aim of recommending more appropriate and effective treatments for affected individuals.

The study conducted by [6] aimed to extract pertinent sentences from a Reddit corpus corresponding to each of the 21 symptoms outlined in the BDI-II (Beck Depression Inventory) questionnaire. Two methodologies were investigated. The first, a supervised approach, involved fine-tuning the DistilRoBERTa model on the annotated BDI-Sen corpus, which contains 18,510 sentences labeled with severity levels (ranging from 0 to 3) by three experts, including a psychologist. The second, an unsupervised approach, leveraged the semantic capabilities of GPT-3.5: synthetic sentences describing the eight symptoms of the PHQ-8 were generated and compared with user histories using the SentenceTransformer model and cosine similarity. A symptom-level alignment between the PHQ-8 and BDI-II was performed to ensure consistency between the two instruments. The DistilRoBERTa model yielded the best results, achieving a P@10 score of 0.107.

[7] proposed a system designed to associate Reddit sentences with the 21 symptoms of depression as defined by the BDI-II questionnaire. Their approach involved training a model on unannotated data from the eRisk 2023 task and evaluating it on a large-scale corpus comprising 15 million unannotated Reddit sentences. After preprocessing and filtering the textual data, the sentences were encoded into vector representations using a compact Transformer model, selected for its efficiency based on MTEB benchmarks. The same encoding process was applied to the textual formulations of the BDI-II items to obtain comparable symptom representations. Semantic similarity between each Reddit sentence and the BDI-II items was then measured using cosine similarity. Each sentence was assigned to the symptom corresponding to its closest match. To further refine the selection and ensure quality, the authors incorporated GPT-4 into the filtering process. This method achieved a P@10 score of 0.833,

[8] organized sentences according to their relevance to the 21 symptoms defined in the Beck Depression Inventory-II (BDI-II). He used cosine similarity to compare the vector representations of BDI-II responses with those of comments collected from Reddit. Then, he filtered the sentences to retain only those articulated in the first person. The author evaluated several classification models, including logistic regression, a Multi-Layer Perceptron (MLP) classifier, Gradient Boosting, XGBClassifier, and LGBMClassifier. To optimize model performance, he fine-tuned the hyperparameters using the GridSearchCV method. Three model variants were tested for classification: MindwaveMLMiniLML12, MindwaveMLMiniLML12MLP_0.5, and MindwaveMLMiniLM-L12MLP_weighted. The best result was achieved with MindwaveMLMiniLML12MLP_weighted, obtaining a P@10 score of 0.471.

[9] introduced a method for categorizing sentences based on their alignment with the 21 BDI-II symptoms. They employed semantic similarity models—all-MPNet-base-v2, all-MiniLM-L12-v2, and all-MiniLM-L6-v2—to generate sentence embeddings and compute their closeness to BDI-II item formulations using cosine similarity. These models were used to identify symptom-related expressions from both annotated data and standard item responses. In addition, the authors fine-tuned a RoBERTa model, originally designed for emotion analysis, to classify sentences according to the BDI-II symptom categories. To enhance classification performance, they combined the predictions from the RoBERTa model with those generated by the similarity-based models. Among the tested models, all-MiniLM-L12-v2 achieved the highest performance, with a P@10 score of 0.990.

[10] developed a system designed to automatically identify the most relevant sentences describing symptoms of depression in online forums, in accordance with the 21 items of the BDI-II. The system generates a ranked list of 1000 sentences per item, considered relevant if they clearly express the specific symptom and provide explicit information about the individual's condition. The authors trained their models on a dataset containing 3.8 million sentences and evaluated them on a test set of 15.5 million sentences. Sentence encoding was performed using Transformer-based models, notably MiniLM variants (paraphrase-L6, paraphrase-L12, SimCSE), combined with feed-forward neural networks comprising one to three hidden layers. The best performance was achieved with the paraphrase-MiniLM-L12-v2 model, which obtained a P@10 precision score of 0.346.

3 Proposed Method

This section provides a detailed overview of our research on the automatic discretization of depressive symptoms. We present the methodologies developed to assign a discrete severity level (ranging from 0 to 3) to the 21 symptoms defined by the BDI-II questionnaire, based on textual content extracted from social media platforms. Central to our approach are two complementary unsupervised techniques: KMeans clustering and quantile-based segmentation, which convert raw textual signals into interpretable numerical representations. These methods enable us to capture the gradation of psychological distress expressed through language. The full architecture of our system, along with the datasets employed, is described in the following section.

3.1 Data Set

The information used in this research was collected through the Early Risk Prediction on the Internet (eRisk2025) initiative. The dataset was compiled from social media platforms, primarily Reddit, based on user comments posted across various discussion threads.

It is formatted according to the TREC standard, which is widely used in information retrieval tasks. The dataset contains a total of 4267842 entries, each representing an individual message.
Each entry consists of two components: a document identifier (Docno) and the corresponding textual content (TEXT). The corpus is multilingual, featuring comments in multiple languages, which introduces additional challenges for automated language processing.
Moreover, the data is unannotated—meaning that no explicit labels (such as depression severity levels) are provided—making supervised learning infeasible without a prior annotation phase or the use of automatic labeling techniques [11,12].
For the purposes of this research, we focused on a subset of 18000 entries, selected to support the development and evaluation of our methodology.
Figure 1 provides an excerpt from the dataset used in this study.

```
<DOC>
        <DOCNO>s_11_3_0</DOCNO>
        <TEXT> wait are you talking about destiny or the guy whos comment i linked?</TEXT>
</DOC>
<DOC>
        <DOCNO>s_11_4_0</DOCNO>
        <TEXT> Didn't you read?</TEXT>
</DOC>
<DOC>
        <DOCNO>s_11_4_1</DOCNO>
        <TEXT>The picture said we take ourselves seriously, god.</TEXT>
</DOC>
```

Fig. 1. Extract of our datasets

3.2 Architecture

In this section, we outline the overall framework of the proposed methodology. The objective is to present a comprehensive overview of the fundamental principles and underlying mechanisms of our approach. More specifically, we delineate the various stages of the process, beginning with data preparation, followed by modeling, and concluding with performance evaluation.

Our approach is divided into five primary stages. The first stage is preprocessing, which begins with translating the data into English to ensure compatibility with automated language processing models, followed by cleaning and normalization of the text. The second step consists of comment aggregation, where all messages from a single user are merged into a unified, consolidated textual representation. The aggregated comments, along with the 21 items of the BDI-II questionnaire, are then encoded using a pre-trained language model. Semantic similarity scores are computed between each user's text and each BDI-II item. Finally, we discretize depressive symptoms using KMeans clustering and quantile-based segmentation to assign severity levels (Fig. 2).

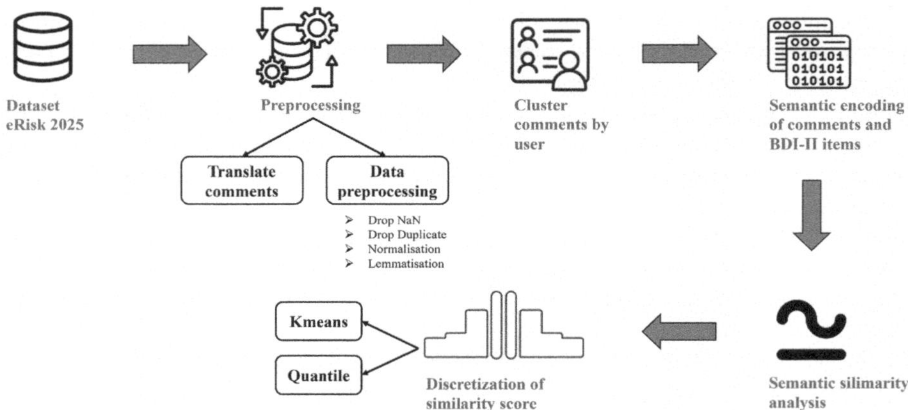

Fig. 2. Architecture of the proposed method

Preprocessing. For the purposes of linguistic standardization and data preparation, we have structured this section into two successive stages: (1) the translation of comments originally written in various languages into English, and (2) the data preprocessing of the texts, which includes cleaning operations such as removing missing values (NaNs) and duplicates, normalization, and lemmatization necessary for automated analysis.

The details of each step are provided below:

Translate Comments. To maintain linguistic uniformity throughout the multilingual corpus and to support effective subsequent semantic analysis, all textual data were translated into English. This process involved the automatic identification of the original language of each document or comment, followed by the use of machine translation techniques to render the content in English. By consolidating the corpus into a single language, we reduced variability caused by language differences, thereby enabling the implementation of standardized natural language processing methods. This translation phase was essential to preserve coherence within the dataset, enhance the comparability of textual features, and improve the accuracy of the models developed on the unified corpus.

Data Preprocessing. A uniform preprocessing protocol was implemented to enhance the quality of textual data for computational semantic analysis. The data cleansing stage involved the systematic removal of incomplete observations and duplicate entries to ensure the statistical integrity and uniqueness of the samples within the corpus.

The process of text normalization involved several sequential transformation steps. Initially, HTML tags were removed using regular expression filtering. This was followed by the removal of punctuation and any non-alphanumeric characters. The text was then converted to lowercase to maintain case consistency, and any leading or trailing whitespace was trimmed to standardize the format.

Dimensionality reduction at the lexical level was performed by filtering out stopwords using the English reference list from NLTK, which allowed us to retain only terms carrying significant semantic content.

Advanced morphological normalization was carried out using the preprocess_text() function, which applied several sophisticated natural language processing techniques. These included detecting and removing URLs via regular expressions, filtering out non-alphabetic characters, and performing lexical tokenization. Moreover, lemmatization with WordNetLemmatizer() was used in conjunction with stemming via PorterStemmer() to effectively reduce words to their base or root forms.

This processing pipeline creates a standardized, semantically meaningful vector representation that supports comparison with the clinical standards outlined in the BDI-II questionnaire (Fig. 3).

	filename	docno	text	language	translated_text
0	s_290.trec	s_290_0_0	The Feeling is sadly unimaginable	en	the feeling is sadly unimaginable
1	s_290.trec	s_290_1_0	What existed before the universe?	en	what existed before the universe
2	s_290.trec	s_290_2_0	What did i eat for Breakfast?	en	what did i eat for breakfast
3	s_290.trec	s_290_3_0	Wrong, i dont eat breakfast	en	wrong i dont eat breakfast
4	s_290.trec	s_290_4_0	After World Conquest, i finally hit 10 Million...	en	after world conquest i finally hit 10 million ...

Fig. 3. Extraction after preprocessing

Cluster Comments by User. Following the individual preprocessing of each comment which includes linguistic cleaning, the removal of non-textual elements, and translation into English when necessary, all comments from a single user are aggregated into one document that captures the full extent of that user's discourse. This aggregation is carried out in the original chronological order of the comments, as they appear in the corpus files, with each file corresponding to a distinct user. The objective of this step is to construct a comprehensive discourse profile for each user, thereby enabling more accurate semantic analysis aimed at identifying items from the Beck Depression Inventory (BDI). This aggregation thus serves as a critical bridge between raw data preprocessing and the extraction of depressive indicators at a higher, user-level granularity (Fig. 4).

	filename	translated_text
0	s_2634.trec	[would looking long prison sentence, mean soun...
1	s_2710.trec	[crazy, lucio please, safe place, thrustmaster...
2	s_2728.trec	[really ever gotten far enough experience pain...
3	s_2745.trec	[respect grandpa man great grandpa ww1 get see...
4	s_2747.trec	[book explanation vampire conversion, big read...

Fig. 4. Extracting Data After Cluster Comments by User

Semantic Encoding of Comments and BDI-II Items. At the beginning of this phase, we explicitly present and define the 21 symptoms described in the Beck Depression Inventory-II (BDI-II), such as sadness, pessimism, loss of pleasure, and feelings of failure, each reformulated into a standardized textual description suitable for semantic processing. Figure 5 illustrates the various definitions associated with each item.

```
{'sadness': ['I feel sad',
  'I feel down most of the time',
  'I cry easily',
  'I feel completely hopeless'],
 'pessimism': ['I have no hope for the future', 'Things will never improve'],
 'sense_of_failure': ['I feel like a failure', "I can't do anything right"],
 'loss_of_pleasure': ['Nothing brings me joy anymore',
  "I don't enjoy anything"],
 'guilt': ['I feel guilty all the time', 'Everything is my fault'],
 'low_self_esteem': ['I hate myself', 'I feel worthless'],
 'suicidal_thoughts': ['I think about ending my life', 'I want to die'],
 'crying': ['I cry a lot', 'I feel like crying all the time'],
 'irritability': ["I'm easily annoyed", 'I get angry quickly'],
 'social_withdrawal': ['I avoid people', 'I prefer to be alone'],
 'indecisiveness': ["I can't make decisions", 'I’m confused about everything'],
 'body_image': ["I think I'm ugly", 'I hate how I look'],
 'work_difficulty': ['I can’t concentrate', 'I’m too tired to work'],
 'sleep_disturbance': ["I can't sleep", 'I wake up often at night'],
 'fatigue': ['I feel exhausted', 'I have no energy'],
 'loss_of_appetite': ["I don't feel hungry", 'Food has no taste for me'],
 'weight_loss': ["I've lost weight without trying"],
 'health_concerns': ['I feel sick all the time', 'I have unexplained pain'],
 'loss_of_libido': ['I’ve lost interest in sex'],
 'anxiety': ['I feel anxious all the time', "I can't calm down"],
 'self_doubt': ['I doubt everything I do', 'I’m unsure of myself']}
```

Fig. 5. Definitions of each items

Building upon these standardized symptom definitions, this phase plays a crucial role in our pipeline by enabling the representation of user texts within a semantic vector space that facilitates meaningful comparison with BDI-II items. To accomplish this, we employed a Sentence Transformer model, specifically, multi-qa-mpnet-base-dot-v1. This model is based on the MPNet (Masked and Permuted Pre-training for Language Understanding) architecture, which combines the strengths of BERT and XLNET, thereby improving the modeling of token dependencies. It generates dense vector representations with 768 dimensions [13,15].

Using this model, Reddit comments and the 21 BDI-II items rephrased into standardized textual descriptions are encoded independently. The output is a matrix of high-dimensional vectors situating both user-generated content and clinical items within a unified semantic space.

This representation allows for a direct and meaningful comparison between spontaneous user language and the clinical phrasing of BDI-II items, effectively setting the stage for the subsequent semantic similarity analysis.

Semantic Similarity Analysis. After encoding the textual data, we conduct a semantic similarity analysis between the vector representations of user comments and those of the 21 items of the BDI-II questionnaire. For each user, the cosine similarity is computed between each of their comment vectors and the vectors corresponding to the BDI-II items, using the util.cos_sim() function from the Sentence-Transformers library [14].

This step results in a similarity matrix with dimensions N × 21, where N denotes the total number of user comments processed, and 21 corresponds to the individual items of the BDI-II scale. A cumulative similarity score is then calculated for each comment by summing the similarity values across all 21 items, offering a measure of its overall semantic alignment with depressive symptom categories.

To isolate comments that are meaningfully related to clinical descriptions, a threshold of 0.2 is applied. This step generates a boolean mask that retains only comments whose semantic similarity exceeds the threshold, effectively filtering out irrelevant cases. The retained comments are subsequently ranked in descending order based on their cumulative similarity scores, facilitating the identification of the most representative examples.

For each of these selected comments, we identify the most semantically similar BDI-II item, defined as the one with the highest individual similarity score. This step enables a direct association between each comment and a specific depressive symptom.

The Final results are compiled into a DataFrame that includes the file identifiers, the highest similarity scores, and the corresponding BDI-II items. This table is then merged with the original dataset to produce a comprehensive scores matrix containing the individual similarity scores for each of the 21 BDI-II items (Fig. 6).

	filename	translated_text	best_similarity_score	best_bdi_item	item_1	item_2	item_3	item_4	item_5	item_6	...	item_12	item_13
0	s_2634.trec	[would looking long prison sentence, mean soun...	0.348564	work_difficulty	0.276058	0.367590	0.283552	0.271407	0.305284	0.299506	...	0.397203	0.230673
1	s_2710.trec	[crazy, lucio please, safe place, thrustmaster...	0.337902	sense_of_failure	0.321176	0.398529	0.400323	0.343137	0.356286	0.463570	...	0.460966	0.320434
2	s_2728.trec	[really ever gotten far enough experience pain...	0.300234	suicidal_thoughts	0.259688	0.397176	0.391477	0.321459	0.379676	0.388900	...	0.372908	0.336929
3	s_2745.trec	[respect grandpa man great grandpa ww1 get see...	0.536579	health_concerns	0.216062	0.353422	0.314369	0.300142	0.225388	0.353074	...	0.309189	0.238579
4	s_2747.trec	[book explanation vampire conversion, big read...	0.302042	loss_of_libido	0.265805	0.429799	0.395747	0.322184	0.384865	0.427902	...	0.465720	0.309834

Fig. 6. Sample Output Following Semantic Similarity Computation

Discretization of Similarity Score. Once the similarity scores were computed, we transformed the continuous values through discretization. This step is fundamental in the transformation of continuous variables into categorical variables, particularly in the context of supervised classification. In our study, it was essential for converting the scores obtained from the extraction of the 21 items of the BDI-II into intervals corresponding to different levels of depression severity.

We applied two independent approaches: quantile-based discretization (equal-frequency strategy) and K-Means-based discretization.

The first approach divides the distribution into bins, each designed to contain roughly the same number of observations. This method is especially effective for datasets with class imbalance, as it helps to distribute instances evenly across all categories. Previous works have confirmed its robustness: [16] demonstrated that it improves variable selection and classifier performance, while [17] emphasized its asymptotic consistency and optimal convergence rates. Moreover, [18] showed its effectiveness in decision trees, combining simplicity and strong performance. In our pipeline, this approach was therefore initially favored to reduce the impact of class imbalance and to ensure balanced representation of severity levels.

The second approach, based on K-Means, follows a different logic: rather than imposing an artificial separation, it identifies natural clusters by minimizing intra-cluster variance. Its relevance has been consistently demonstrated across different application domains. [19] compared five discretization methods—including equal-width, equal-frequency, and K-Means—in the inference of genetic regulatory networks, and concluded that the bikmeans variant achieved significantly higher accuracy than the other approaches. Similarly, [20] reported that the K-Means algorithm reached a classification accuracy of over 94% on the red-wine dataset, outperforming methods such as logistic regression. [21] further confirmed its usefulness for clinical data, providing thresholds more representative of the intrinsic data structure. In our study, this approach allowed us to obtain a categorization of Reddit scores that was more faithful to their actual distribution.

Specifically, each method was applied to the score columns of the 21 BDI-II items using the KBinsDiscretizer class from the scikit-learn library, with four classes and ordinal coding (labels 0–3). For the quantile method, we used the quantile strategy, while for the K-Means method, the kmeans strategy was specified.

The discretization thus yielded a final dataset in which each comment was associated with a discrete level of semantic similarity per item, facilitating interpretation and subsequent analysis of the results (Fig. 7).

filename	translated_text	best_similarity_score	best_bdi_item	item_1	item_2	...	item_12	item_13	item_14	item_15
s_2634.trec	[would looking long prison sentence, mean soun...	0.348564	work_difficulty	2	2	...	3	1	1	1
s_2710.trec	[crazy, lucio please, safe place, thrustmaster...	0.337902	sense_of_failure	2	2	...	3	2	1	2
s_2728.trec	[really ever gotten far enough experience pain...	0.300234	suicidal_thoughts	2	2	...	2	2	1	2
s_2745.trec	[respect grandpa man great grandpa ww1 get see...	0.536579	health_concerns	1	2	...	2	1	1	2
s_2747.trec	[book explanation vampire conversion, big read...	0.302042	loss_of_libido	2	3	...	3	2	3	2

Fig. 7. Extracting data after discretization of similarity score

4 Evaluation and Results

To evaluate the quality of the predicted similarity scores after discretization, we computed several information retrieval metrics adapted to the multi-class context (four discrete levels from 0 to 3, aligned with BDI-II severity). For each questionnaire item, the model's ability to rank user comments associated with a specific severity level was assessed using a single consolidated value per metric, each providing a distinct perspective on prediction quality:

- Average Precision (AP): This metric captures how effectively the model ranks relevant comments by summarizing the trade-off between precision and recall across all decision thresholds [22].
- Normalized Discounted Cumulative Gain at k = 10 (NDCG@10): This indicator evaluates the ranking quality by emphasizing the relevance of items at the top of the list, applying a logarithmic discount that reduces the weight of lower-ranked results. It is particularly useful in scenarios where users typically focus on the top results [23].
- Precision at k = 10 (p@10): This score measures the proportion of truly relevant comments among the top 10 returned by the model, offering a clear picture of the immediate usefulness of the predictions [24].
- R-Precision: This metric calculates precision over the top R retrieved items, where R is the total number of relevant comments in the dataset, providing a balanced and fair assessment of the ranking performance [24,25].

Each metric was computed on the test data of each fold during a 5-fold cross-validation procedure. The final reported values correspond to the average of these metrics across the five folds, providing a robust and stable estimate of the model's predictive performance for each BDI-II item. This evaluation approach ensures that the performance assessment accounts for data variability and maximizes the use of the entire available dataset.

Table 1 presents the results obtained.

In our evaluation, we assessed and contrasted the effectiveness of two discretization techniques—namely quantile-based discretization and KMeans

Table 1. Evaluations results

	AP	P@10	R-Precision	NDCG@10
Quantile discretization	0.251	0.264	0.250	0.264
KMeans discretization	0.250	0.265	0.249	0.275

clustering—by employing four performance metrics: Average Precision (AP), Precision@10 (P@10), R-Precision, and NDCG@10. The results obtained for each method are presented in Table 1.

Overall, the two discretization strategies yield comparable performance across all metrics, highlighting the robustness of the semantic similarity approach. Quantile discretization achieved slightly better results in AP (0.251 vs. 0.250) and R-Precision (0.250 vs. 0.249), while KMeans discretization performed marginally better in NDCG@10 (0.275 vs. 0.264) and P@10 (0.265 vs. 0.264), indicating stronger ranking quality at the top positions of the recommendation list.

These differences, though small, reflect the subtle trade-offs between the two approaches: quantile discretization ensures balanced class sizes, which may benefit global retrieval consistency, while KMeans discretization adapts to the natural structure of the data, potentially enhancing relevance at the top of the ranked lists. Taken together, these findings confirm that both discretization methods can support fine-grained, clinically aligned severity scoring without manual labeling, and thus strengthen the potential of our system for automatic early detection of depressive symptoms.

5 Conclusion

In this work, we explored advanced automatic language processing techniques for the detection of depressive symptoms from textual data derived from social networks. By combining semantic similarity methods with deep language models such as Sentence Transformers, we were able to establish meaningful correspondences between user comments and clinical items from the BDI-II questionnaire. Our approach also incorporated the essential step of discretizing the continuous scores obtained, enabling semantic similarities to be transformed into discrete levels of symptom severity. To this end, we compared two discretization methods based respectively on quantiles and the KMeans algorithm. Quantile discretization ensures that data are evenly divided into classes of approximately equal size, whereas KMeans discretization partitions the scores into clusters shaped by the inherent structure of the data distribution. This step facilitated the interpretation of the results by aligning the scores on a discrete scale consistent with clinical severity.

The results show that this combined method accurately identifies the symptomatic dimensions expressed in the texts by associating each comment with the most representative BDI-II item. It thus validates the effectiveness of fine automatic detection of early signs of psychological distress without the need for manual annotation.

Beyond simple detection, this methodology opens promising prospects for the automatic calculation of a discrete global depression score, which can be used in early warning and individual follow-up systems while respecting user confidentiality. Future work will aim to refine the quantification of symptomatic severity and adapt this approach to multilingual contexts or other clinical settings.

References

1. Elsan - Dépression: Symptômes, causes, traitements. https://www.elsan.care/fr/pathologie-et-traitement/maladie-mentale/depression. Accessed 2025
2. World Health Organization (WHO): Depression. WHO Fact Sheet (2023). https://www.who.int/news-room/fact-sheets/detail/depression
3. Beck, A.T., Steer, R.A., Brown, G.K.: Beck depression inventory-II. Corsini Encyclopedia Psychol. **1**(1), 210 (2010)
4. Beck, A.T., Steer, R.A., Brown, G.K.: BDI-II – Beck depression inventory-II. https://institutpsychoneuro.com/wp-content/uploads/2015/08/BDI-II.pdf
5. TCC Apprendre la Psychologie: Beck Depression Inventory. https://tcc.apprendre-la-psychologie.fr/catalogue/tests-psychologiques/beck-depression-inventory.html
6. Mírmol-Romero, A.M., et al.: SINAI at eRisk@CLEF 2024: Approaching the search for symptoms of depression and early detection of anorexia signs using natural language processing. In: Working Notes of CLEF, pp. 9–12 (2024)
7. Barachanou, A., Tsalakanidou, F., Papadopoulos, S.: REBECCA at eRisk 2024: search for symptoms of depression using sentence embeddings and prompt-based filtering. In: Working Notes of CLEF, pp. 9–12 (2024)
8. Hanciu, R.M.: MindwaveML at eRisk 2024: Identifying depression symptoms in Reddit users. Working Notes of CLEF **2024**, 9–12 (2024)
9. Bascuñana, A.P., Bedmar, I.S.: APB-UC3M at eRisk 2024: natural language processing and deep learning for the early detection of mental disorders. In: Working Notes of CLEF, pp. 9–12 (2024)
10. Maupomé, D., Ferstler, Y., Mosser, S., Meurs, M.-J.: Automatically finding evidence, predicting answers in mental health self-report questionnaires. Working Notes CLEF **2024**, 9–12 (2024)
11. Crestani, F., Losada, D., Parapar, J.: Early detection of mental health disorders by social media monitoring. Stud. Comput. Intell. **1018**, 4 (2022)
12. Parapar, J., Perez, A., Wang, X., Crestani, F.: eRisk 2025: contextual and conversational approaches for depression challenges. In: European Conference on Information Retrieval, pp. 416–424 (2025)
13. Hugging Face - multi-QA-MPNET-base-dot-V1. https://huggingface.co/sentence-transformers/multi-qa-mpnet-base-dot-v1
14. Reimers, N., Gurevych, I.: Sentence-BERT: sentence embeddings using Siamese BERT-Networks. In: Proceedings of the 2019 Conference on Empirical Methods in Natural Language Processing (EMNLP) (2019). https://arxiv.org/abs/1908.10084
15. Song, K., Tan, X., Qin, T., Lu, J., Liu, T.-Y.: MPNet: masked and permuted pre-training for language understanding. In: Advances in Neural Information Processing Systems, vol. 33, pp. 16857–16867 (2020)
16. Redivo, F., Viroli, C., Farcomeni, A.: Quantile distribution functions in classification, with an application to naive Bayes classifiers. Stat. Comput. **33**, 10224 (2023). https://doi.org/10.1007/s11222-023-10224-4
17. Hennig, C., Viroli, C.: Quantile-based classifiers. arXiv preprint arXiv:1303.1282 (2013)

18. Grzymala-Busse, J.W., Mroczek, T.: A comparative study of discretization methods for decision trees. *Entropy* **18**(3), 69 (2016). https://www.mdpi.com/1099-4300/18/3/69
19. Li, S., Wang, H., Zhang, X., Chen, L., Zhang, Z.: Comparison of discretization methods for inferring regulatory networks. BMC Bioinform. **11**, 520 (2010). https://bmcbioinformatics.biomedcentral.com/articles/10.1186/1471-2105-11-520
20. Stamenković, D., Radovanović, M., Petrović, V.: A comparison between the K-means and EM algorithms for classification of red-wine data. J. Appl. Stat. **41**(10), 2201–2215 (2014). https://www.tandfonline.com/doi/full/10.1080/13102818.2014.949045
21. Peng, X., Lin, S., Zhang, Z., Wang, H.: Comparisons of discretization methods for clinical data. BioMed Res. Int. **2013**, 420845 (2013). https://pmc.ncbi.nlm.nih.gov/articles/PMC3628044/
22. Flach, P.: Machine Learning: The Art and Science of Algorithms that Make Sense of Data. Cambridge University Press (2019)
23. Shani, G., Gunawardana, A.: Evaluating recommendation systems. In: Recommender Systems Handbook, pp. 257–297. Springer (2011). https://doi.org/10.1007/978-0-387-85820-3_8
24. Manning, C.D., Raghavan, P., Schütze, H.: Introduction to Information Retrieval. Cambridge University Press (2008)
25. Sokolova, M., Lapalme, G.: A systematic analysis of performance measures for classification tasks. Inf. Process. Manage. **45**(4), 427–437 (2009)

ProdNet: A Lightweight Network for Fast Discovery of Matrix Multiplication Algorithms

Badia Ouissam Lakas[1(✉)], Chemousse Berdjouh[1], Khadra Bounane[1], Mohammed Lamine Kherfi[2], Oussama Aiadi[1], and Samir Brahim Belhouari[3]

[1] Department of Computer Science and Information Technology, University of Kasdi Merbah, Ouargla, Algeria
{lakas.badiaouissam,berdjouh.chemousse,bouanane.khadra, aiadi.oussama}@univ-ouargla.dz
[2] Department of Computer Science, Sultan Qaboos University, Sultanate of, Oman
m.kherfi@squ.edu.om
[3] College of Science and Engineering, University of Hamad Bin khalifa, Doha, Qatar
sbelhaouari@hbku.edu.qa

Abstract. Matrix multiplication is a fundamental operation with significant applications across diverse fields, such as physics, electronics, and artificial intelligence. Traditional implementations of this operation exhibit cubic time complexity, which presents computational challenges, particularly in deep learning scenarios that necessitate large-scale matrix computations. In this study, we introduce **ProdNet**, a lightweight neural network model designed to autonomously discover efficient matrix multiplication algorithms without the need for extensive computational resources or prior knowledge of existing methods. Our approach seeks to alleviate complexity by minimizing the number of multiplicative operations involved. To achieve this, we utilize a combination of Mean Squared Error (MSE) and a regularization function, targeting weight values to be constrained to $\mathbf{0}$, $\mathbf{1}$, or $-\mathbf{1}$. We emphasize the effectiveness of our regularization techniques in accelerating the algorithm discovery process. Furthermore, we demonstrate that ProdNet successfully identifies multiplication algorithms for $\mathbf{2}\times\mathbf{2}$ and $\mathbf{3}\times\mathbf{3}$ matrices, achieving this in a limited number of iterations (starting from 500 iterations for 2×2 and 8800 iterations for 3×3)

Keywords: ProdNet · Matrix product algorithm · regularization · neural network

1 Introduction

In machine learning, matrix multiplication is essential for operations such as parameter optimization, allowing models to learn and enhance performance. It also supports feature extraction and data transformations, revealing meaningful patterns and relationships, making it a cornerstone of numerous algorithms

T. Ensari et al. (Eds.): ISPR 2025, CCIS 2859, pp. 95–108, 2026.
https://doi.org/10.1007/978-3-032-21585-7_7

and applications. In deep learning, matrices are central to neural network architectures, where interconnected layers process input data in matrix form. The performance and scalability of these models are closely tied to efficient matrix multiplication, as both forward and backward propagation depend heavily on these operations [1,10]. Standard matrix multiplication (SMM) algorithms typically exhibit cubic time complexity. Multiplying two $n \times n$ matrices, A and B, to produce an $n \times n$ product matrix C involves calculating the dot product of each row of A with each column of B. This process requires n^3 multiplications and $n^2(n-1)$ additions [11], resulting in significant computational overhead, particularly for large matrices used in machine learning models.

These challenges are particularly evident in large-scale models like Transformers and large language models (LLMs), which heavily rely on SMM during training. As a result, training these models remains a demanding process, driving the need for advanced hardware accelerators to mitigate computational inefficiencies. However, recent studies [2,3,6,13] have underscored two significant concerns: the environmental impact and the financial cost.

Environmental Impact: Training large deep learning models requires immense computational power, often provided by data centers equipped with numerous GPUs and TPUs. These data centers consume substantial amounts of electricity, which, depending on the energy sources, can lead to significant carbon emissions. This high energy consumption contributes to the carbon footprint of artificial intelligence, exacerbating climate change and environmental degradation, which goes against the principles of sustainable development.

Financial Burden: Developing and training these sophisticated models also demands substantial financial resources. The costs associated with acquiring and maintaining cutting-edge hardware, such as GPUs and TPUs, are high. Furthermore, the electricity needed to power these machines adds to the operational costs. This financial burden is particularly challenging for research labs and institutions with limited budgets, potentially hindering progress and innovation in the field and limiting its democratization.

To address these challenges, researchers are exploring strategies to reduce the computational complexity of training [13]. One promising direction is the development of matrix multiplication algorithms that require fewer multiplications. Historically, breakthroughs in this area, such as Strassen's algorithm, have been achieved through slow and labor-intensive efforts by mathematicians. Recently, there has been growing interest in leveraging deep learning for the efficient and rapid discovery of matrix multiplication algorithms that surpass the performance of SMM. By using the power of neural networks, researchers have greatly accelerated this process, showcasing the transformative role of deep learning in algorithm development.

One groundbreaking work by DeepMind, AlphaTensor, has showcased the potential of using reinforcement learning (RL) to discover novel matrix multiplication algorithms [5]. While AlphaTensor has successfully proposed new algorithms that reduce the computational complexity of matrix multiplication for various matrix sizes, it faces significant challenges. It relies on complex reinforcement learning frameworks that combine Monte Carlo Tree Search (MCTS) with Transformer-based neural networks, requiring extensive computational resources and days of training on specialized hardware such as TPUs. Additionally, a large mathematical dataset of matrix multiplication algorithms was needed for training. As a result, while AlphaTensor is an impressive solution, its high computational and data demands make it resource-intensive, often out of reach for smaller research groups, and prohibitively expensive to implement.

In this paper, we present ProdNet, a lightweight network model designed to discover efficient matrix multiplication algorithms. Our goal is to provide a more accessible and efficient approach by using supervised learning and combining Mean Squared Error (MSE) with a regularization function as the loss function. ProdNet's flexibility enables it to discover algorithms for a given matrix size without the need for large datasets or augmentation techniques. In addition, its streamlined architecture reduces the demands of computational resources, making it an efficient and scalable solution for the discovery of matrix multiplication algorithms for $\mathbf{2{\times}2}$ and $\mathbf{3{\times}3}$ matrices in a much smaller number of iterations.

2 Related Works

The journey toward discovering efficient matrix multiplication algorithms began with Volker Strassen's groundbreaking discovery of an algorithm that required fewer multiplications than the standard method [16] in 1969. Strassen's algorithm reduced the computational complexity from $O(n^3)$ for squared matrices size to approximately $O(n^{2.81})$, marking a pivotal moment in computational mathematics. This breakthrough sparked widespread interest in optimizing matrix multiplication. Building on Strassen's foundational work, Virginia Williams proposed the most efficient algorithm to date, achieving a runtime of $O(n^{2.373})$ [19]. Winograd [20] established the theoretical minimum of seven multiplication steps for this case, shedding light on the inherent complexity of small-scale operations. In 1976, Laderman was able to discover matrix multiplication for 3×3 matrices with 23 multiplications [9].
While mathematicians have made significant contributions to the field of matrix multiplication discovery, recent years have seen the emergence of machine learning-based approaches for the discovery of multiplication algorithms. In [4], the author explores neural networks with a unique twist: their only non-linear components are multipliers. The goal is to test a novel training rule in a context where precise data representation is paramount. These networks are specifically challenged to discover the rules of matrix multiplication, with a focus on Strassen multiplication. The only training protocol employed is called conservative learning, which ensures minimal weight adjustments to produce the correct output for each input as soon as the input-output pair is received. The neural network designed in this study is intriguing due to its minimalist design, containing only seven multipliers, which mirrors the seven scalar multiplications used in

Strassen's algorithm for 2×2 matrices in a minimum of 6400 iterations, and 23 scalar multiplications for 3×3 matrices in 10^6 iterations. However, it is important to note that this conservative learning technique does not accelerate the convergence of the model, and the discovery of the algorithms requires large numbers of training items.

In 2022, Google's DeepMind team introduced AlphaTensor, a model designed to discover efficient matrix multiplication algorithms. Leveraging reinforcement learning techniques inspired by AlphaZero [15]. AlphaTensor utilizes a deep neural network architecture based on attention mechanisms [18] to guide Monte Carlo tree search (MCTS) algorithms. AlphaTensor demonstrated remarkable capabilities in discovering novel matrix multiplication algorithms, uncovering numerous existing algorithms, and proposing new ones that outperformed existing approaches by reducing the number of required multiplications where they discover a new algorithm for 4×4 matrices with 47 multiplications, and a new multiplication algorithm with 96 multiplications for 5×5 matrices. Shortly after AlphaTensor's breakthrough results, authors in [8] have used the same scheme of [5] by obtaining the same technique of AlphaTensor where they have been able to discover new multiplication algorithm for 5×5 matrices with 95 multiplications.

AlphaTensor possesses a notable advantage over established computer search methods, yet this advantage is also its primary limitation. Unlike traditional methods, AlphaTensor operates without the constraints of human intuition regarding optimal algorithmic structures. However, training the AlphaTensor model required a significant amount of multiplication algorithm data, prompting DeepMinds team to implement data augmentation techniques to generate additional training samples from completed games. The DeepMind research group endeavored to train AlphaTensor in the task of tensor decomposition representing the product of matrices up to dimensions of 12×12. This pursuit involved the development of expedient algorithms for matrix multiplication operations involving conventional real number elements, as well as algorithms tailored to the specialized domain of modulo 2 arithmetic. The latter pertains to a constrained mathematical framework wherein computations are conducted exclusively with two numerical values, typically denoted as 0 and 1. Despite its promising results, training AlphaTensor posed several challenges. Reinforcement learning models, such as AlphaTensor, are notoriously difficult to train due to the complexities and sensitivities involved in parameter tuning [7,12], and the large data size requirement [17]. Moreover, the large number of parameters in AlphaTensor necessitated substantial computational resources, making it inaccessible to many research labs.

ProdNet, the model to introduce in this paper for matrix multiplication algorithms discovery, prioritizes simplicity and accessibility, making it a practical tool for a wide range of users in both academic and industrial settings. By leveraging the algorithms discovered by ProdNet, we aim to address existing limitations in computational efficiency and contribute to the optimization of matrix multiplication operations in real-world applications. This work represents a significant step toward democratizing algorithm discovery and fostering broader adoption of advanced computational techniques.

3 Multiplication Algorithm and Tensor Decomposition

As previously discussed, contemporary research has reconceptualized matrix multiplication as a mathematical problem conducive to computational processing. The matrix multiplication operation can be represented as a 3D tensor [14]. Denoting T_n as the tensor describing the $N \times N$ matrix multiplication, this 3D tensor comprises binary values, where 0 indicates no multiplication and 1 denotes multiplication.

The T_n tensor can be decomposed into an outer product ($\otimes$) of rank-1 tensors, expressed as $T_n = \sum_{r=1}^{R} u^r \otimes v^r \otimes w^r$, where R denotes the number of column vectors, akin to the number of multiplications involved, and each $u, r,$ and w is a rank-one vector [5]. These stacked rank-one vectors constitute a rank-R decomposition of the 3D tensor, which readily translates into a matrix multiplication algorithm.

The multiplication of 2×2 matrices can be performed with 7 multiplications instead of 8 by following Strassen's algorithm.

$$a = \begin{pmatrix} a_1 & a_2 \\ a_3 & a_4 \end{pmatrix} \quad b = \begin{pmatrix} b_1 & b_2 \\ b_3 & b_4 \end{pmatrix} \quad c = \begin{pmatrix} c_1 & c_2 \\ c_3 & c_4 \end{pmatrix}$$

$$U = \begin{pmatrix} 1 & 0 & 1 & 0 & 1 & -1 & 0 \\ 0 & 0 & 0 & 0 & 1 & 0 & 1 \\ 0 & 1 & 0 & 0 & 0 & 1 & 0 \\ 1 & 1 & 0 & 1 & 0 & 0 & -1 \end{pmatrix}$$

$$V = \begin{pmatrix} 1 & 1 & 0 & -1 & 0 & 1 & 0 \\ 0 & 0 & 1 & 0 & 0 & 1 & 0 \\ 0 & 0 & 0 & 1 & 0 & 0 & 1 \\ 1 & 0 & -1 & 0 & 1 & 0 & 1 \end{pmatrix}$$

$$W = \begin{pmatrix} 1 & 0 & 0 & 1 & -1 & 0 & 1 \\ 0 & 0 & 1 & 0 & 1 & 0 & 0 \\ 0 & 1 & 0 & 1 & 0 & 0 & 0 \\ 1 & -1 & 1 & 0 & 0 & 1 & 0 \end{pmatrix}$$

The main reason of the only numbers seen in the $U, V,$ and W matrices are 1, -1, and 0 is that the trick is to rely on the addition and subtraction only, where 1 is a term to be added, -1 is a term to be subtracted, and 0 is a term not to be used.

This decomposition can be translated to a multiplication algorithm as follows.

$$
\begin{aligned}
m_1 &= (1a_1 + 1a_4) \cdot (1b_1 + 1b_4) \\
m_2 &= (1a_3 + 1a_4) \cdot (1b_1) \\
m_3 &= (1a_1) \cdot (1b_2 - 1b_4) \\
m_4 &= (1a_4) \cdot (1b_3 - 1b_1) \\
m_5 &= (1a_1 + 1a_2) \cdot (1b_4) \\
m_6 &= (1a_3 - 1a_1) \cdot (1b_1 + 1b_2) \\
m_7 &= (1a_2 - 1a_4) \cdot (1b_3 + 1b_4) \\
c_1 &= 1m_1 + 1m_4 - 1m_5 + 1m_7 \\
c_2 &= 1m_3 + 1m_5 \\
c_3 &= 1m_2 - 1m_4 \\
c_4 &= 1m_1 - 1m_2 + 1m_3 + 1m_6
\end{aligned}
$$

Where the (4×7) matrices' elements of $\mathbf{U}, \mathbf{V}$, and $\mathbf{W}$ correspond to a, b, and c, respectively.

4 Methodology

To discover multiplication algorithms, we designed a lightweight neural network. Although the architecture of our network is proposed in [4], the way we discover multiplication algorithms is completely different. They employ the Mean absolute error (MAE) and a conservative learning rule to adjust the weights minimally. In contrast, our method uses the Mean Squared Error (MSE) and a regularization function to enforce specific weight values 0, 1, or –1. In this setup, the neural network's trained weights correspond to the factors of the matrices $\mathbf{U}, \mathbf{V}$, and $\mathbf{W}$. This methodology offers flexibility in determining the desired number and shape of matrices used for training, enhancing the adaptability of the training process to various scenarios.

The architecture of the network is illustrated in Fig. 1. We denote by $\mathbf{W}^A$ the weight matrix between the input layer corresponding to the matrix A of size $N \times N$ and the hidden layer and $\mathbf{W}^B$ is the weight matrix between the input layer corresponding to the matrix B of size $N \times N$ and the hidden layer. $\mathbf{W}^C$ denotes the weight matrix between the hidden layer and the output C. In the model, $(\mathbf{W}^A)^\top, (\mathbf{W}^B)^\top$, and $\mathbf{W}^C$ emulates the $\mathbf{U}, \mathbf{V}$, and $\mathbf{W}$ matrices, respectively. The activation function of each neuron of the hidden layer is defined as the dot product, $g = a \cdot b$. where $a = A(\mathbf{W}^A)^\top$ and $b = B(\mathbf{W}^B)^\top$. The output of the hidden layer g is multiplied by the weight matrix $\mathbf{W}^C$, yielding the output $\mathrm{C} = g(\mathbf{W}^C)^\top$ of size $N \times N$.

4.1 Training Loss

Our training process aims to achieve accurate matrix multiplication while ensuring that the weights are restricted to 0, 1, or -1. To accomplish this, we propose

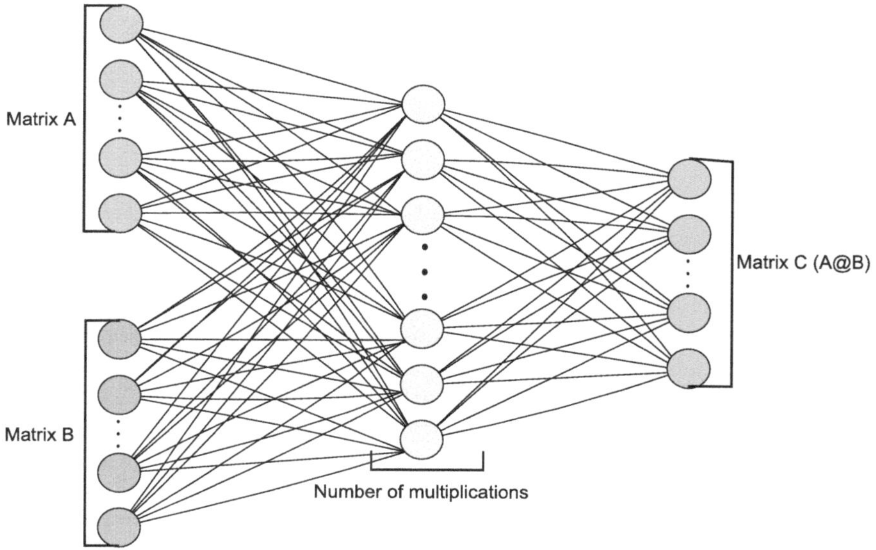

Fig. 1. ProdNet Architecture.

training ProdNet using a loss function that combines Mean Squared Error (MSE) with a regularization term:

$$Loss = MSE + \frac{\alpha}{M} \sum_{x} \sum_{j=1}^{J} l_0 \times l_1 \times l_{-1}, \quad \text{where } l_i = |w_n^x - i|, \forall x \in \{\text{A,B,C}\}, \forall i \in \{-1, 0, 1\} \tag{1}$$

α denotes the weight set to the regularization function, M denotes the total number of parameters in the neural network, and J denotes the number of parameters within the weight matrix. The MSE loss measures the disparity between the predicted and actual outputs, guiding the improvement of the model's efficiency and targeting the correct values of the multiplication. Simultaneously, the regularization term penalizes deviations of the weights from the desired values (0, 1, or −1). Initially, we employed a multiplication-based regularization approach.

The primary concept behind this regularization is that utilizing the product of norms enables the weights to converge toward the desired discrete values (0, 1, or −1). The gradient of the function actively guides the weights toward these target values during training.

Moreover, an essential objective of the training process is to achieve a pruned model, characterized by a significant number of weights being set to 0. This pruning effectively minimizes the number of operations required, thereby enhancing the model's efficiency.

For testing the discovered algorithms, we implement a round-to-integer operation and test the model to ensure that the learned parameters converge to the desired discrete values (–1, 0, or 1). Furthermore, we test the discovered algorithm by fitting the test set to the network and checking whether the loss value is exactly 0, so we call the discovered algorithm a correct one.

4.2 Results

Our study focuses on the development of a lightweight neural network, ProdNet, designed to discover matrix multiplication algorithms using efficient and computationally simple learning techniques. ProdNet is both easy to train and minimally resource-intensive. To evaluate its performance and assess the impact of the regularization function, we conducted multiple training sessions on 2×2 matrices with either 7 multiplications or 8 multiplications, as well as conducting training sessions on 3×3 matrices with 23 multiplications. All results were obtained using a Tesla T4 GPU on Google Colab and Kaggle cloud platforms. The training dataset consisted of 8000 paired matrices and their corresponding multiplication results, processed as a full batch. Weight initialization was performed using the Xavier uniform distribution, and optimization was achieved with the Adam optimizer.

To assess the importance of including a regularization term in the loss function, we initially trained the model using only the mean squared error (MSE) using a learning rate of 0.5. The performance of the model did not meet expectations, and the model could not discover multiplication algorithms. Figure 2 illustrates the model's convergence behavior when integer rounding is applied every 100 iterations.

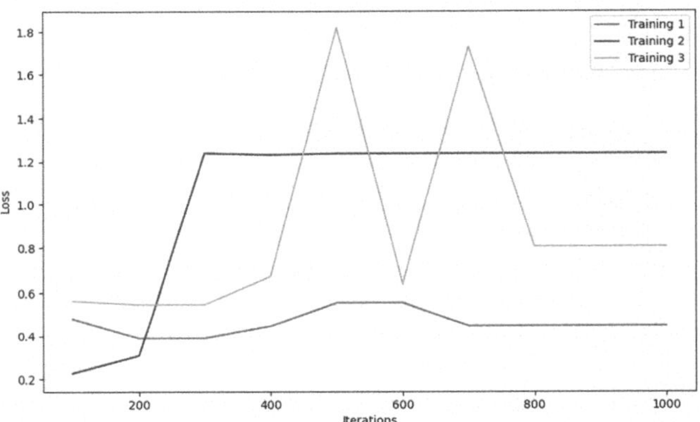

Fig. 2. Model convergence during training using MSE with integer rounding of weights.

In the second phase, we conducted multiple training sessions on 2×2 matrices, utilizing the defined regularization function in conjunction with the Mean Squared Error (MSE) as the loss function. The results demonstrated that ProdNet could easily discover multiplication algorithms requiring 8 multiplications.

However, to focus on discovering algorithms with a reduced number of multiplications, we choose the network to have 7 perceptrons in the hidden layers. This configuration ensured that the number of multiplications remained within the best-known rank for 2×2 matrices.

Furthermore, the hyperparameter $\alpha = 1/500$ was incorporated into the loss function to effectively guide the optimization process, optimizing training using the Adam optimizer.

Figure 3 illustrates five samples from twenty training launched using different weight initialization values and learning rates. Where, after each training session, a different multiplication algorithm was discovered.

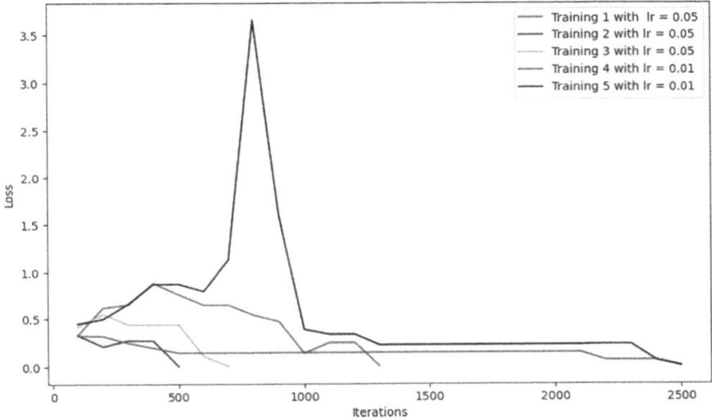

Fig. 3. Loss convergence using different weight initialization values and learning rate.

All the discovered algorithms have the same number of multiplications (7). The evaluation process involved first rounding the model weights to integers, ensuring that all weight values were restricted to -1, 0, or 1. After this adjustment, the model was assessed by verifying whether the testing loss function equaled zero, so the model weights could then be converted into a written algorithm. The primary reason for discovering different multiplication algorithms in different training iterations lies in the model's sensitivity to weight initialization values, as each set of initialization values can lead to a new multiplication algorithm.

The discovered algorithms for 2 × 2 matrices size are presented bellow.

The algorithm was discovered in 1300 iterations (training1)

$$\begin{aligned} m_1 &= (-a_1 - a_3) \cdot (b_1 - b_3) \\ m_2 &= (a_1 + a_2 + a_3 + a_4) \cdot (-b_3) \\ m_3 &= (a_1 + a_2 + a_3) \cdot (-b_2 - b_3) \\ m_4 &= (-a_1 - a_2) \cdot (b_2) \\ m_5 &= (a_2) \cdot (-b_2 + b_4) \\ m_6 &= (a_4) \cdot (-b_3 - b_4) \\ m_7 &= (a_3) \cdot (-b_1 - b_2) \\ c_1 &= -m_1 - m_3 + m_4 + m_7 \\ c_2 &= -m_4 + m_5 \\ c_3 &= -m_2 + m_3 - m_4 - m_7 \\ c_4 &= m_2 - m_3 + m_4 - m_6 \end{aligned}$$

The algorithm was discovered in 500 iterations (training2).

$$\begin{aligned} m_1 &= (a_2 - a_4) \cdot (-b_3 + b_4) \\ m_2 &= (-a_1 - a_2) \cdot (-b_2) \\ m_3 &= (-a_3 - a_4) \cdot (b_3) \\ m_4 &= (a_2) \cdot (-b_2 + b_4) \\ m_5 &= (-a_2 - a_3) \cdot (b_2 - b_3) \\ m_6 &= (-a_1 + a_3) \cdot (-b_1 + b_2) \\ m_7 &= (a_3) \cdot (-b_1 + b_3) \\ c_1 &= m_2 + m_5 + m_6 - m_7 \\ c_2 &= m_2 + m_4 \\ c_3 &= -m_3 - m_7 \\ c_4 &= -m_1 - m_3 + m_4 - m_5 \end{aligned}$$

The algorithm was discovered in 700 iterations (training3).

$$\begin{aligned} m_1 &= (-a_2 - a_4) \cdot (b_3 - b_4) \\ m_2 &= (a_2 - a_3) \cdot (-b_1 + b_4) \\ m_3 &= (a_3 + a_4) \cdot (b_4) \\ m_4 &= (a_2) \cdot (b_1 - b_3) \\ m_5 &= (a_1 + a_3) \cdot (-b_1 + b_2) \\ m_6 &= (-a_3) \cdot (b_2 - b_4) \\ m_7 &= (a_1 + a_2) \cdot (-b_1) \\ c_1 &= -m_4 - m_7 \\ c_2 &= m_2 + m_5 + m_6 - m_7 \\ c_3 &= -m_1 + m_2 + m_3 + m_4 \\ c_4 &= m_3 - m_6 \end{aligned}$$

The algorithm was discovered in 2500 iterations (training 4).

$$\begin{aligned} m_1 &= (-a_1 - a_3) \cdot (b_2 + b_4) \\ m_2 &= (-a_1 + a_2 - a_3) \cdot (-b_1 - b_2 - b_4) \\ m_3 &= (a_1 - a_2 + a_3 - a_4) \cdot (b_4) \\ m_4 &= (a_3) \cdot (b_1) \\ m_5 &= (-a_2) \cdot (-b_1 - b_2 - b_3 - b_4) \\ m_6 &= (a_4) \cdot (b_3) \\ m_7 &= (-a_1 + a_2) \cdot (b_1 + b_2) \\ c_1 &= m_1 + m_2 - m_4 + m_5 \\ c_2 &= -m_1 - m_2 + m_4 - m_7 \\ c_3 &= m_4 + m_6 \\ c_4 &= m_2 - m_3 - m_4 + m_7 \end{aligned}$$

For 3 × 3 matrices, our model was trained to discover a multiplication algorithm requiring 23 multiplications instead of the standard 27. To achieve this, the network was designed with 23 perceptrons in the hidden layers. Choosing a learning rate of 0.001, and the hyperparameter $\alpha = 1/50$ was applied to the regularization in the loss function.

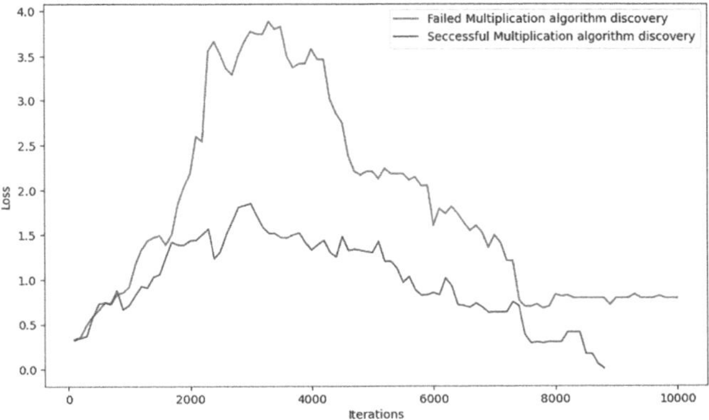

Fig. 4. Loss convergence for 3 by 3 matrix size

The correct algorithm was discovered for 3 by 3 matrices in 8800 iterations.

$$
\begin{aligned}
m_1 &= (-a_9) \cdot (b_8 - b_9) \\
m_2 &= (a_7 - a_8) \cdot (-b_3) \\
m_3 &= (a_6) \cdot (b_5 + b_8) \\
m_4 &= (a_1 + a_3 - a_4 + a_7) \cdot (b_1 + b_2 + b_4 - b_7) \\
m_5 &= (-a_1 - a_3 + a_4 + a_6 - a_7 - a_9) \cdot (-b_7) \\
m_6 &= (-a_2 + a_3 - a_8 + a_9) \cdot (b_9) \\
m_7 &= (-a_6) \cdot (b_1 - b_3 - b_4 + b_6 - b_7 + b_9) \\
m_8 &= (a_2 - a_3 + a_8) \cdot (-b_5 - b_9) \\
m_9 &= (a_3) \cdot (-b_1 - b_2 - b_4 + b_5 + b_7 + b_8) \\
m_{10} &= (-a_4 + a_7) \cdot (b_2) \\
m_{11} &= (a_1 + a_3) \cdot (-b_1 - b_2 - b_4) \\
m_{12} &= (-a_4 + a_5 + a_7 - a_8) \cdot (b_4) \\
m_{13} &= (-a_1 - a_8) \cdot (b_3 + b_4) \\
m_{14} &= (-a_5 + a_6) \cdot (b_4 - b_6) \\
m_{15} &= (a_1 - a_4 - a_6 + a_7) \cdot (-b_1 + b_3 + b_7) \\
m_{16} &= (a_2 + a_8) \cdot (-b_4 + b_6 + b_9) \\
m_{17} &= (-a_1 + a_4 - a_7) \cdot (-b_1 - b_4 + b_7) \\
m_{18} &= (-a_4 - a_6) \cdot (b_1 - b_3) \\
m_{19} &= (a_1 - a_2) \cdot (b_4) \\
m_{20} &= (-a_8) \cdot (-b_3 + b_5 - b_6) \\
m_{21} &= (a_5 - a_6) \cdot (b_5) \\
m_{22} &= (-a_7) \cdot (-b_2) \\
m_{23} &= (-a_2 + a_3) \cdot (b_5)
\end{aligned}
$$

$$c_1 = -m_4 + m_{10} - m_{11} + m_{17} - m_{19}$$
$$c_2 = m_4 + m_9 - m_{10} - m_{17} - m_{23}$$
$$c_3 = m_8 - m_{13} + m_{16} - m_{19} - m_{20} - m_{23}$$
$$c_4 = -m_2 + m_{12} - m_{13} - m_{15} - m_{17} - m_{18}$$
$$c_5 = m_3 - m_{10} + m_{21} + m_{22}$$
$$c_6 = -m_2 - m_7 + m_{12} - m_{13} + m_{14} - m_{15} - m_{17}$$
$$c_7 = -m_2 + m_4 + m_5 - m_{10} + m_{11} - m_{-13} - m_{15} - m_{17} - m_{18}$$
$$c_8 = -m_1 + m_6 - m_8 + m_{22} + m_{23}$$
$$c_9 = -m_2 + m_6 - m_8 + m_{20} + m_{23}$$

Figure 4 illustrates that our model was able to discover a multiplication algorithm in 8800 iterations, where in the second training, the model was not able to discover a correct multiplication algorithm using the same hyperparameters, which confirms that our model is sensitive to weight initialization values.

5 ProdNet vs. Alphatensor

We conduct a comparative analysis of our model, ProdNet, with AlphaTensor [5]. Both models have demonstrated the ability to discover multiple 2×2 and 3×3 matrix multiplication algorithms with a reduced number of multiplications compared to the standard matrix multiplication.

Dataset Usage: AlphaTensor relies on a well-known but rare mathematical dataset of multiplication algorithms, necessitating data augmentation techniques to address data scarcity. In contrast, ProdNet leverages easily generated datasets of the matrices and their multiplication result, eliminating the need for data augmentation. This offers advantages in terms of data accessibility and potentially reduces pre-processing time.

Model Complexity: AlphaTensor uses reinforcement learning incorporating Monte Carlo Tree Search (MCTS) and a Transformer-based neural network. It requires training on a combination of input data, policy targets, and value targets, potentially increasing training time and computational demands. Conversely, ProdNet employs supervised learning with Mean Squared Error (MSE) and regularization with a simpler network architecture with two input layers, one hidden layer, and an output layer.

Computational Resources: AlphaTensor requires 64 TPU cores for training and another standalone TPU v4 for the played game. However, ProdNet could achieve multiplication algorithms even when it was trained on a CPU for 2×2 matrices, and on Tesla T4 cloud GPU. This makes it more accessible to researchers with limited computational resources. This also reduces the training costs associated with high-performance hardware.

Matrix Sizes. AlphaTensor was trained to discover algorithms for various squared matrix sizes, including 2, 3, and 4, reaching even up to 12. In contrast, ProdNet was initially trained for 2×2 and 3×3 sizes as a starting point.

6 Conclusion

This paper highlights the critical role of matrix multiplication in various computational tasks and acknowledges the limitations of the standard algorithms, which require a large number of multiplications. The existing works for this task require huge computational resources and large multiplication algorithms datasets for the training; this motivates the exploration of innovative solutions to reduce this need for materials. We introduce ProdNet, a lightweight network model designed to efficiently and automatically discover matrix multiplication algorithms without prior knowledge. Our model successfully discovers algorithms for 2×2 matrices with only 7 multiplications and 3×3 with only 23 multiplications, the known lower-rank number, by employing a combination of MSE and regularization function. Using ProdNet emerges as a promising approach, demonstrating significant efficiency improvements while remaining computationally cost-effective. Notably, the model independently rediscovers numerous algorithms after every training launched, with the same or different parameters, highlighting its potential to address computational challenges in deep learning applications.

Regardless of the promising results obtained, some advancements need to be improved. One avenue for improvement involves extending the size of the matrices, such as 4×4. As well as the application of regularization to additional values, such as -2 and 2. Additionally, future developments aim to advance the model's capabilities to discover the optimal multiplication algorithm by maximizing the occurrence of rank-one vectors with zero values in the U and/or V matrices, thereby automatically reducing the number of multiplications. Exploring discrete optimizers and using other search techniques, a modified version of the exhaustive search technique is used to discover all the possible multiplication algorithms in a reduced search space.

References

1. Alsari, S., Al-Hashimi, M.: Investigation of energy and power characteristics of various matrix multiplication algorithms. Energies **17**(9), 2225 (2024)
2. Anthony, L.F.W., Kanding, B., Selvan, R.: Carbontracker: tracking and predicting the carbon footprint of training deep learning models. arXiv preprint arXiv:2007.03051 (2020)
3. Budennyy, S.A., et al.: Eco2ai: carbon emissions tracking of machine learning models as the first step towards sustainable ai. In: Doklady Mathematics, vol. 106, pp. S118–S128. Springer, Heidelberg (2022)
4. Elser, V.: A network that learns strassen multiplication. J. Mach. Learn. Res. **17**(1), 3964–3976 (2016)

5. Fawzi, A., et al.: Discovering faster matrix multiplication algorithms with reinforcement learning. Nature **610**(7930), 47–53 (2022)
6. Heguerte, L.B., Bugeau, A., Lannelongue, L.: How to estimate carbon footprint when training deep learning models? A guide and review. arXiv preprint arXiv:2306.08323 (2023)
7. Ji, J., et al.: AI alignment: a comprehensive survey. arXiv preprint arXiv:2310.19852 (2023)
8. Kauers, M., Moosbauer, J.: The fbhhrbnrssshk-algorithm for multiplication in $\mathbb{Z}_2^{5\times 5}$ is still not the end of the story. arXiv preprint arXiv:2210.04045 (2022)
9. Laderman, J.D.: A noncommutative algorithm for multiplying 3*3 matrices using 23 multiplications (1976)
10. LeCun, Y., Bottou, L., Bengio, Y., Haffner, P.: Gradient-based learning applied to document recognition. Proc. IEEE **86**(11), 2278–2324 (1998)
11. Li, Y., Hu, S.L., Wang, J., Huang, Z.H.: An introduction to the computational complexity of matrix multiplication. J. Oper. Res. Soc. China **8**, 29–43 (2020)
12. Lin, Y., ET AL.: Evolutionary reinforcement learning: a systematic review and future directions. Mathematics **13**(5) (2025). https://doi.org/10.3390/math13050833. https://www.mdpi.com/2227-7390/13/5/833
13. Menghani, G.: Efficient deep learning: a survey on making deep learning models smaller, faster, and better. ACM Comput. Surv. **55**(12), 1–37 (2023)
14. Pan, V.: Fast feasible and unfeasible matrix multiplication (1804)
15. Silver, D., et al.: A general reinforcement learning algorithm that masters chess, shogi, and go through self-play. Science **362**(6419), 1140–1144 (2018)
16. Strassen, V., et al.: Gaussian elimination is not optimal. Numer. Math. **13**(4), 354–356 (1969)
17. Terven, J.: Deep reinforcement learning: a chronological overview and methods. AI **6**(3) (2025). https://doi.org/10.3390/ai6030046. https://www.mdpi.com/2673-2688/6/3/46
18. Vaswani, A., et al.: Attention is all you need. Adv. Neural Inf. Process. Syst. **30** (2017)
19. Williams, V.V.: Multiplying matrices in o (n2. 373) time. preprint pp. 0–105698 (2014)
20. Winograd, S.: On multiplication of 2× 2 matrices. Linear Algebra Appl. (1971)

Hybrid CNN-Transformer Model for Ovarian Cancer Lesion Classification

Takwa Ben Aïcha Gader(✉), Soumaya Hamrouni, and Afef Kacem Echi

University of Tunis, ENSIT-LaTICE Laboratory, Tunis, Tunisia
takoua.benaicha@enis.tn

Abstract. Ovarian cancer remains a major global health issue due to its late-stage diagnosis and the morphological similarity of its histological subtypes, which include Mucinous Carcinoma (MC), Endometrioid Carcinoma (EC), Low-Grade Serous Carcinoma (LGSC), Clear Cell Carcinoma (CCC), and High-Grade Serous Carcinoma (HGSC). Traditional histological assessment is constrained by subjectivity and inter-observer variability, prompting the development of automated, dependable diagnostic technologies.

In this work, we present a hybrid deep learning architecture that combines a Convolutional Neural Network (CNN) for localized morphological feature extraction with a Vision Transformer (ViT) decoder to capture global tissue context. Using the UBC-OCEAN dataset, the proposed architecture achieved highly competitive results, with an AUC of 87.03 % and an accuracy of 86.45% across the main histological subtypes of ovarian cancer. These findings demonstrate the potential for hybrid CNN-Transformer techniques to support and improve diagnostic precision in clinical histopathology workflows.

Keyword: Computer-aided diagnosis (CAD), Medical Imaging, Histopathology, Convolutional Neural Networks (CNN), Vision Transformers (ViT)

1 Introduction

Ovarian cancer remains one of the most deadly gynaecological cancers, due to a lack of efficient early detection and nonspecific first symptoms. As a result, over 70% of cases are detected at an advanced stage, significantly restricting therapeutic efficacy and decreasing the five-year survival rate [1].

Epithelial ovarian cancer has five major histological subtypes: mucinous carcinoma (MC), endometrioid carcinoma (EC), low-grade serous carcinoma (LGSC), clear cell carcinoma (CCC), and high-grade serous carcinoma (HGSC). These subgroups have overlapping characteristics, making an accurate diagnosis difficult. Furthermore, 3–11% of cases present with mixed tumors, complicating subtype definition [12].

Traditional diagnosis relies on manual histological assessment, which is costly and susceptible to inter-observer variability and diagnostic errors, especially in

T. Ensari et al. (Eds.): ISPR 2025, CCIS 2859, pp. 109–123, 2026.
https://doi.org/10.1007/978-3-032-21585-7_8

complex cases [2]. This emphasizes the importance of reliable automated diagnostic tools.

Recent breakthroughs in deep learning (DL), particularly Convolutional Neural Networks (CNNs) and Vision Transformers (ViTs), have proven considerable potential for automated medical image interpretation. CNNs excel in local feature extraction, while Transformers provide better global contextual reasoning [8]. Combining these complementary strengths can result in more accurate and reliable categorization systems.

Based on the work of Alahmadi et al. [7], we propose a hybrid CNN-Transformer model for the automated categorization of ovarian cancer subtypes from histopathology images. This approach aims to improve diagnostic accuracy while reducing human variability, resulting in more reliable and efficient clinical decision-making.

2 State of the Art

Recent advances in the identification and categorization of ovarian cancer have benefited considerably from the integration of deep learning methods. These approaches generally fall into five main categories: CNN-based models, Vision Transformer methods, hybrid CNN-Transformer architectures, attention-enhanced frameworks, and sequence-based models such as RNNs and MLPs.

Convolutional Neural Networks (CNNs) have become popular because to their ability. Recent CNN-based models have shown excellent performance in ovarian cancer diagnosis and segmentation. Chen et al. [18] developed a ResNet-based system that combines color and grayscale Doppler ultrasound images via feature fusion, reaching accuracy comparable to expert healthcare professionals. Gao et al. [19] successfully used DenseNet-121 on a large multi-institutional dataset of pelvic ultrasounds, displaying reliable diagnostic performance. U-Net and its variations, such as U-Net++ [29], enable accurate tumor identification [24]. However, CNNs are still limited in simulating global context and often require huge annotated datasets.

Vision Transformers (ViTs) are effective alternatives for modeling global connections between image regions utilizing self-attention techniques [8]. Alahmadi (2024) [7] proposed a ViT-based diagnostic framework that enhances lesion classification accuracy through improved context modeling. Wu et al. (2022) [8] highlighted the strengths of attention mechanisms in encoding complex visual features by splitting images into patches. However, ViTs tend to be computationally intensive and require substantial data for training.

Attention mechanisms have been incorporated into deep learning models to improve accuracy and interpretability. Liu et al. (2023) [25] demonstrated that adding attention modules such as the Convolutional Block Attention Module (CBAM) to ResNet34 substantially improves the prediction of recurrence in high-grade serous ovarian cancer by focusing on diagnostically relevant features. Similarly, Maria et al. (2023) [33] combined Attention U-Net [34] for tumor segmentation with YOLOv5 for detection, achieving precise tumor boundary identification in CT images. While effective, these attention-enhanced models often introduce additional computational costs and require careful hyperparameter tuning.

Sequence-based models, such as MLPs and RNNs, have been explored for analyzing temporal clinical data, including biomarker progression and treatment response. Arora et al. (2023) [13] developed an explainable framework based on these models to predict ovarian cancer presence from clinical data. Nonetheless, because of the lack of large-scale datasets, sequence-based methods remain less popular than image-based approaches.

Several recent studies have used advanced methods to classify subtypes in the UBC-OCEAN dataset using histopathology images. Ben Ayed et al. [37] used EfficientNet-B0 with a fine-tuned KNN classifier to achieve high AUCs (0.94) for CC, LGSC, and MC, but lower scores for EC (0.78) and HGSC (0.69). Boschman [38] trained CNNs with color normalization and mix-augmentation on 948 whole-slide images, achieving 81% inter-pathologist agreement and a Cohen's kappa of 0.75. However, numerous classes had AUCs below 0.80. Asadi-Aghbolaghi et al. [39] employed multiple instance learning with outlier identification, achieving 66% balanced accuracy in the OCEAN challenge, which underscores the task's difficulty. Inspired by these studies, our work proposes a hybrid deep learning architecture that leverages the strengths of both Transformers and CNNs to model global and local features effectively. This approach is tailored to the characteristics of the UBC-OCEAN dataset to provide an accurate and clinically useful ovarian cancer subtype classification system.

3 Proposed Solotion

3.1 Used Dataset

Bashashati et al. (2023) produced the UBC-OCEAN dataset (UBC Ovarian Cancer Subtype Classification and Outlier Detection), a publicly available resource aiming to enhance AI-assisted histopathological research in ovarian cancer. [9]. This dataset contains a diversified variety of high-resolution whole-slide images (WSIs) and tissue microarrays (TMAs) that depict the morphological heterogeneity of ovarian cancer tissue. The dataset includes 505 histopathology pictures that have been consistently scaled to a resolution of 512 $\times$ 512 pixels for compliance with deep learning processes.

The collection supports investigations on ovarian cancer subtypes, including Clear Cell Carcinoma (CC), Endometrioid Carcinoma (EC), High-Grade Serous Carcinoma (HGSC), Low-Grade Serous Carcinoma (LGSC), and Mucinous Carcinoma (MC) (see to Fig. 1). By includes both WSIs and TMAs, the dataset captures real-world variation in imaging modalities, making it more relevant for robust and generalizable classification algorithms.

All images are distributed in typical digital formats (`.jpg`, `.png`) and are publicly accessible for research purposes, making UBC-OCEAN a replicable benchmark at the crossroads of computational pathology and deep learning. Its balanced class representation eliminates the need for artificial oversampling or reweighting procedures, resulting in a solid basis for supervised learning problems.

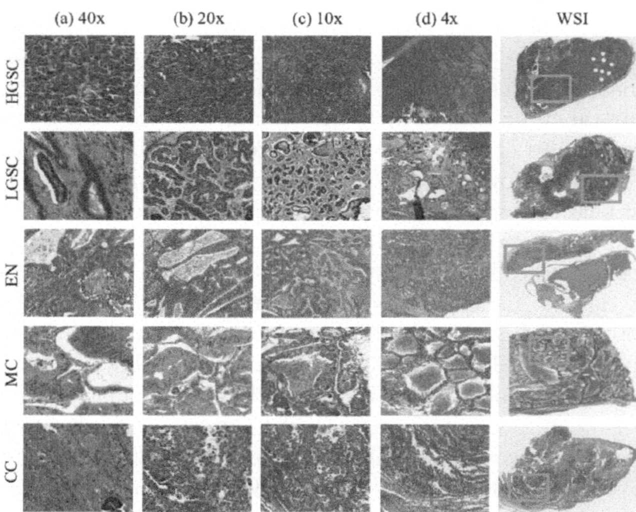

Fig. 1. Whole-slide images of ovarian cancer categories. High Grade Serous Carcinoma (HGSC) and Low Grade Serous Carcinoma (LGSC). Endometrioid carcinoma (EN); Mucinous carcinoma (MC) and Clear-cell carcinoma (CC) [5]

3.2 Data Preprocessing

To improve training performance and model generalization, we used a rigorous preprocessing process on the UBC-OCEAN histopathology dataset. This pipeline consists of six main steps, each addressing a different aspect of data quality, variability, or compatibility with deep learning frameworks.

- **Image and Mask Loading:** Each image is loaded with its associated annotation mask, which ensures that inputs and labels are correctly paired.
- **Background Filtering:** Masking techniques are used to remove irrelevant background regions (such as black or homogeneous areas), allowing the model to concentrate on tissue-specific features.
- **Normalization:** To normalize lighting variances and maintain numerical stability, all pixel values are rescaled to the range $[0, 1]$.
- **Standardization:** To ensure stability across samples and accelerate convergence during model training, each color channel (Red, Green, and Blue) is standardized separately using z-score normalization. This transformation transforms the original pixel intensity x into a standardized value x' based on:

$$x' = \frac{x - \mu_c}{\sigma_c} \tag{1}$$

where μ_c and σ_c are the empirical mean and standard deviation of channel $c \in \{R, G, B\}$ computed across the full dataset. This normalizing approach assures that each channel has a zero mean and unit variance, which enhances numerical stability and allows for more efficient training.

- **Data Augmentation:** To prevent overfitting and improve feature variety, we applied dynamic augmentation during training. This includes:
  - Random rotations (typically within $\pm 10°$),
  - Horizontal and vertical flips,
  - Center cropping and resizing to 512×512 pixels,
  - Color-Based Augmentation: Color-based augmentations were used with geometric modifications to replicate diversity in staining, lighting, and slide preparation. We implemented random variances in brightness, contrast, saturation, and hue, each within a range of $\pm 20\%$, to imitate real-world disparities commonly found in histological imaging.

  This augmentation method led to a fivefold increase in training set size. As demonstrated in Fig. 2, each subtype, originally represented by 145 images, reaches 725 images post-augmentation, ensuring class balance while enhancing sample variability.
- **Tensor Conversion:** All images are finally converted into PyTorch-compatible tensors, enabling efficient integration into GPU-accelerated training pipelines.

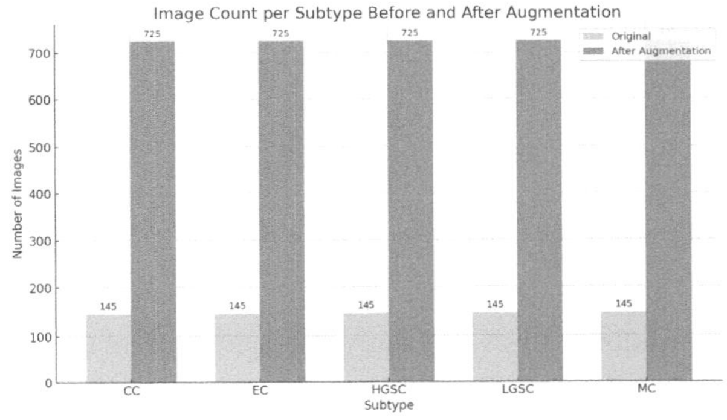

Fig. 2. Image count per subtype before and after augmentation.

3.3 Data Split

To ensure effective model training and accurate evaluation, the dataset was separated into distinct subsets using a standard technique. The augmented dataset included a total of 3,625 images. We used a stratified division technique to maintain the class distribution across subgroups, avoiding bias against any one subtype. 70% of the data (2,538 images) was used for training, 10% (363 images) for validation, and the remaining 20% (724 images) as an independent test set.

The training data set was used to optimise model parameters, whilst the validation set was required to fine-tune hyperparameters and ensure convergence during training. The test set was used to give an objective assessment of model performance.

3.4 Used Architecture

To address the complexity and variability inherent in histopathology images, we proposed a hybrid deep learning model that sequentially integrates convolutional and transformer-based paradigms. The model employs a ResNet50 encoder that extracts high-level local features from the input image. These features are subsequently reshaped into a series of patch embeddings and sent straight into a Vision Transformer (ViT) decoder, which models global contextual dependencies via self-attention. The resulting [CLS] token representation is used by a dense classification head to predict ovarian cancer subtypes. This unified process allows for the modeling of both fine-grained textures and overall tissue structure. Figure 3 provides an overview.

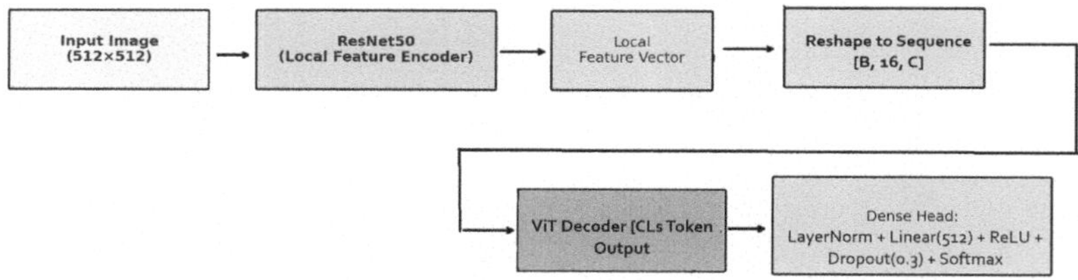

Fig. 3. Overview of the Proposed Hybrid CNN-ViT Architecture

The choice to combine ResNet50 and ViT was made by the complementary nature of their respective learning biases:

- **Multi-scale Representation:** The CNN encoder captures low-level patterns, but the transformer catches global contextual relations.
- **Enhanced Precision:** The combination enhances identification across visually similar subtypes, particularly in complex or ambiguous regions.
- **Improved Generalization:** The hybrid model exhibits better robustness across datasets with heterogeneous staining, imaging conditions, and tissue architectures.

The fundamental components of the proposed hybrid architecture are presented in the following subsections, with details on each module responsible for feature extraction, fusion, and classification.

ResNet50 (Residual Network with 50 Layers). He et al. introduced [27], which profoundly influenced the field of deep learning by allowing the training of extremely deep networks without suffering from the issue of vanishing gradients. ResNet-50, a deeper form of the ResNet family, consists of 50 convolutional layers designed around the use of residual blocks. These blocks support skip connections, which allow a layer's input to bypass one or more layers and be added to the output, facilitating identity mapping learning and accelerating convergence.

The ResNet-50 architecture follows a typical CNN structure, as illustrated in Fig. 4. It starts with a convolutional layer, then batch normalization, and finally ReLU activation. This is then followed by four sets of residual blocks, all in charge of gradually learning more abstract characteristics. The last component contains a global average pooling layer.

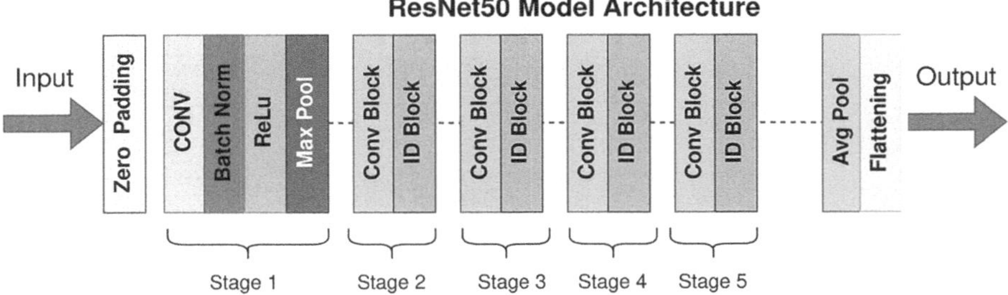

Fig. 4. ResNet-50 architecture, composed of a stem layer, four residual block stages, and a dense classifier

Training in very deep neural networks becomes increasingly challenging as gradients vanish or weights are saturated. ResNet addresses this issue via residual connections, which enable gradients to propagate more directly through previous layers. A residual block's output is defined as follows:

$$y = F(x) + x \tag{2}$$

where x represents the input to the residual block, $F(x)$ represents the residual mapping learned using convolutional operations, and y is the output. This residual learning framework provides more effective training of deeper models and improved feature reuse across the network.

Vision Transformer (ViT)

In 2020 by Google Research introduced the Vision Transformer model [36], inspired by Transformer designs in natural language processing (NLP). ViT considers a picture as a sequence of fixed-size non-overlapping patches rather than a grid of pixels, which differs from standard convolutional neural networks (CNNs).

Each image is divided into patches (e.g., 16×16 pixels), which are then flattened and linearly mapped into an embedding vector. These embeddings generate a sequence that is processed by a standard Transformer encoder equipped with multi-head self-attention and feed-forward layers. A learnable categorization token ([CLS]) is added to this sequence and used for prediction after the attention layers. Each patch embedding includes positional encodings to retain spatial information.

Figure 5 displays the general architecture and processing phases of the Vision Transformer model.

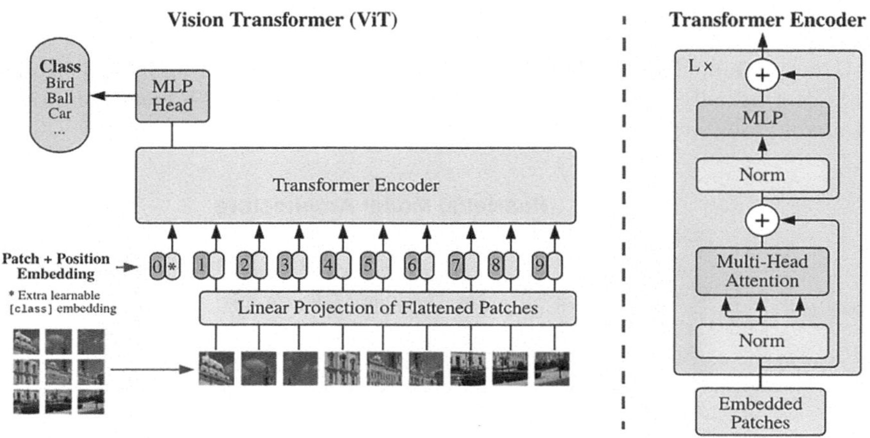

Fig. 5. Vision Transformer (ViT) architecture for image classification

ViT's strength rests in its capacity to represent long-range dependencies and global context using self-attention, making it especially useful for histopathology image analysis, where spatial interactions at several scales are critical. Unlike CNNs, ViT can handle input images of varying sizes and represent interactions across large areas of the image.

Fusion Strategy

We used a hybrid fusion technique that combines ResNet50 and the Vision Transformer decoder. Rather than considering the CNN and Transformer as separate branches, the CNN backbone's feature vector (of shape $[B, C]$, with $C = 2048$) is rearranged to imitate a series of patch embeddings. This is accomplished by expanding the vector along a new dimension and repeating it N times ($N = 16$), producing a tensor of shape $[B, N, C]$.

The Transformer encoder takes this synthetic sequence as input via the `inputs_embeds` interface. The Transformer analyses these embeddings with self-attention, resulting in a more complete picture that contains both local and contextual information. The output sequence is then aggregated over the patch dimension with mean pooling to create a unified global feature vector of shape $[B, D]$, which is then passed to a dense classification head for final prediction.

Dense Classification Head
The output of the ViT decoder is passed through a dense classification head designed to ensure effective discrimination between ovarian cancer subtypes. This head comprises a layer normalisation step to stabilise feature distributions, followed by a fully connected layer with 512 units and ReLU activation to introduce nonlinearity. A dropout layer (rate 0.3) is applied for regularization, and a final linear layer maps the features to six output classes, including a generic Other category. This class includes rare or mixed histological subtypes that are not included in the five main classes. Grouping them into a single class lowers label noise and keeps the model focused on the most clinically important categories.

3.5 Training

Training was conducted with the `AdamW` optimizer, with a learning rate of 0.0001 and a weight decay of 0.01. The learning rate was adjusted over time using a cosine annealing scheduler. Gradient accumulation was limited to two steps, and mixed-precision training (16-bit) was used to increase computational efficiency. Only 50% of available training batches were used every epoch to balance training cost and convergence speed. Table 1 provides a summary of the training setup.

Figure 6 depicts the progression of training and validation loss and accuracy.

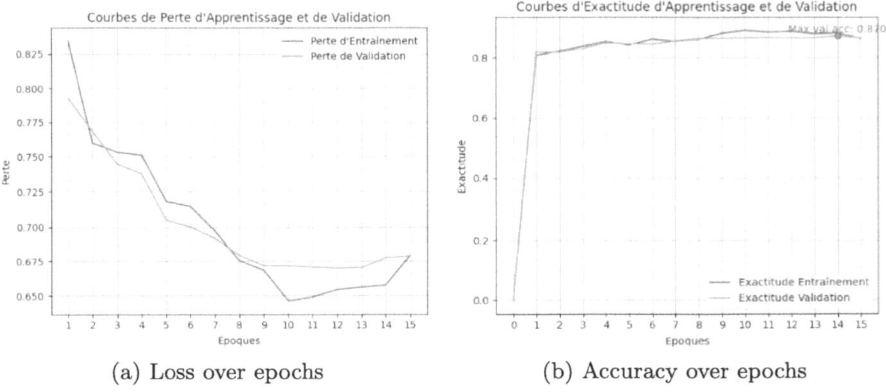

(a) Loss over epochs

(b) Accuracy over epochs

Fig. 6. Training and validation performance across epochs.

3.6 Model Evaluation and Explainability

The hybrid model performed well in multiclass ovarian cancer classification, with an overall accuracy of 87.03% on the test set. A macro F1-score of 0.7356 and a weighted F1-score of 0.8643 were obtained during class-wise performance analysis, demonstrating high performance across the majority of subtypes. Class

Table 1. Model configuration

Parameter	Value
CNN Backbone	Pretrained ResNet50 (last `fc` layer removed)
Transformer Backbone	Pretrained Vision Transformer (ViT Base)
Feature Representation	Reshape ResNet output from $[B, C]$ to $[B, 16, C]$ to simulate patch embeddings
Classification Head	LayerNorm → Linear(512) → ReLU → Dropout(0.3) → Linear(6 classes including *Other*)
Loss Function	`CrossEntropyLoss` with `label_smoothing = 0.1`
Class Weights	Optional (enabled if provided)
Optimizer	`AdamW` (or `Adan` if available)
Learning Rate	0.0001
Weight Decay	0.01
Scheduler	`CosineAnnealingLR` with `T_max = num_epochs`
Monitoring Strategy	`val/acc` with `EarlyStopping` (patience = 15) and `ModelCheckpoint` (best model only)
Precision Mode	16-bit mixed precision (`''16-mixed''`)
Gradient Accumulation	2 steps
Training Batch Limitation	50% of batches per epoch (`limit_train_batches = 0.5`)
Maximum Epochs	50
Hardware	GPU (`accelerator = ''gpu''`)
Logging	`CSVLogger` at `logs/cancer_subtype/`

5 (MC) achieved the highest precision (0.8930) and recall (0.9421). In contrast, classes 0 (CC) and 1 (EC) had lower recall values (0.4852 and 0.4745, respectively). Because the dataset was balanced after augmentation, this result emphasizes the subtypes' inherent morphological similarity and diagnostic difficulty, rather than data scarcity. Addressing such situations may necessitate more advanced feature fusion algorithms or histopathology-specific augmentations (e.g., stain normalization) in future research.

To further understand the model's decision-making procedure, we used Grad-CAM to display class-specific activation maps. Figure 7 shows how the model identified different regions of the tissue based on subtype. The activations in classes 0 (CC) and 1 (EC) were strongly localized around glandular and cytoplasmic structures, which are diagnostically informative but typically invisible, explaining the reduced recall reported in these categories. In contrast, activations in class 3 (LGSC) were more distributed but still corresponded to local regions important for subtype differentiation. Class 5 (MC) had larger and more diffuse heatmaps, corresponding with its varied histological patterns, but maintained a constant concentration on mucin-rich areas. The results presented show that the hybrid CNN-Transformer not only achieves excellent classification accuracy but also pays attention to morphologically significant features, which improves the model's predictions for clinical pathology applications.

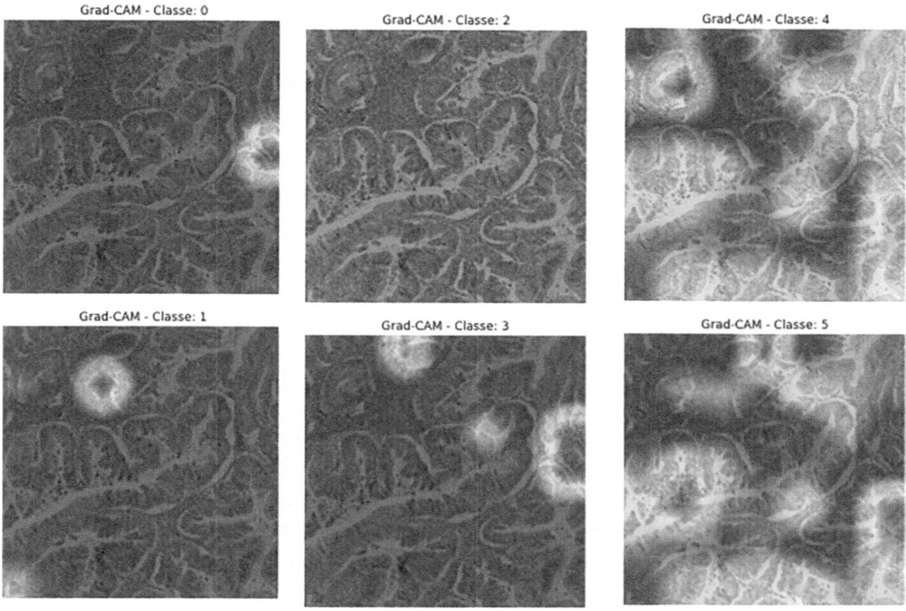

Fig. 7. Grad-CAM visualizations for each class (0–5). Highlighted regions (red/yellow) indicate areas with high contribution to the model's predictions. (Color figure online)

Beyond internal evaluation metrics, systematic benchmarking with recent state-of-the-art methods highlights the competitiveness of our approach. As shown in Table 2, the proposed hybrid CNN–Transformer achieved an accuracy of 86.45% and an AUC of 0.873, outperforming EfficientNet-B0 with a KNN classifier [37] (AUC $\approx$ 0.858) and MIL-based CNNs with outlier detection [39] (balanced accuracy $\approx$ 66%). Compared to Boschman [38], who relied on CNNs with color normalization and augmentation, our architecture delivered more consistent classification across subtypes. Reporting macro and weighted F1-scores further strengthens the comparison, ensuring fairer alignment across methods. While lightweight CNNs and MIL-based strategies remain promising for deployment, our hybrid model more effectively captures both local textures and global tissue context, thereby improving diagnostic accuracy.

Table 2. Comparison of the proposed model with state-of-the-art methods on ovarian cancer classification

Authors/Year	Main Method	Key Results (AUC or Agreement)
Santi Kumari Behera et al., 2024 [37]	EfficientNet-B0 + fine-tuned KNN	AUC ≈ 0.858
Boschman, 2022 [38]	CNN + color normalization + augmentation	Agreement: 80.97% Some AUCs < 0.80
Asadi-Aghbolaghi et al., 2024 [39]	MIL-based CNN + outlier detection	Balanced Accuracy: ≈ 66%
Proposed Solution	**Hybrid (ResNet50 + Transformer)**	**Accuracy = 86.45% AUC = 0.873 Macro F1 = 0.7356 Weighted F1 = 0.8643**

3.7 Discussion

This research demonstrates the efficacy of a hybrid CNN-Transformer architecture for ovarian cancer classification. By integrating ResNet50's local feature extraction power with Vision Transformers' global context modeling, the model outperformed numerous cutting-edge CNN-based approaches. The use of Grad-CAM visuals enhanced the contribution by giving interpretable heatmaps that emphasize diagnostically significant tissue regions, a critical step toward establishing Despite these strengths, certain restrictions must be addressed. First, the study was limited to the UBC-OCEAN dataset, which limits claims of generalizability. Validation on larger and more diversified multi-institutional datasets will be critical to ensuring the approach's resilience. Second, while data augmentation increased resilience and class balance, the changes applied were almost general. Domain-specific augmentation approaches, such as stain normalization, color deconvolution, or morphology-preserving perturbations, may improve performance by better capturing histological diversity. Third, from an ethical and clinical perspective, more testing with pathologists is required to assess the reliability of predictions in real-world diagnostic procedures. Potential biases, such as inter-institutional variations or disparities in patient demographics, must also be carefully considered before clinical implementation.

4 Conclusion

This research proposed a hybrid deep learning strategy combining convolutional neural networks and Vision Transformers to classify ovarian cancer subtypes from histopathology images. The proposed architecture accurately captured the morphological complexity of epithelial ovarian cancers by integrating local feature extraction with global context modeling. Evaluation on the UBC-OCEAN dataset indicated competitive performance, obtaining 87.03% accuracy and reval with thestate-of-the-art approaches. Beyond prediction accuracy, Grad-CAM and attention map studies improved interpretability by consistently highlighting diagnostically significant tissue regions, increasing the model's reliability for future clinical usage. The study shows that hybrid CNN-Transformer frameworks may increase computational pathology by enhancing diagnostic accuracy and model transparency. Nonetheless, clinical integration necessitates additional validation across bigger and more diverse datasets, evaluation of robustness to staining variability and acquisition noise, and collaboration with experienced pathologists to assess usability in real-world diagnostic procedures. Future research should look into lightweight CNN-Transformer variants and optimization methodologies to lower computing costs, making deployment possible in resource-constrained clinical environments. Such initiatives will help to close the gap between experimental research and actual application, opening the path for reliable AI-assisted ovarian cancer diagnosis.

References

1. Lheureux, S., Marsela, B., Oza, A.M.: Epithelial ovarian cancer: evolution of management in the era of precision medicine. CA Cancer J. Clin. **69**(4), 280–304 (2019)
2. Wang, R., et al.: Evaluation of a convolutional neural network for ovarian tumor differentiation based on magnetic resonance imaging. Eur. Radiol. **31**, 4960–4971 (2021)
3. Shoarishoar, S.S., et al.: Comparison of pregnancy outcomes in amniocentesis recipients with normal and abnormal maternal serum analytes. Cell. Mol. Biol. **70**(11), 109–114 (2024)
4. Panigrahi, S., Tripti, S.: Machine learning techniques used for the histopathological image analysis of oral cancer-a review. Open Bioinform. J. **13**(1) (2020)
5. BenTaieb, A., et al.: A structured latent model for ovarian carcinoma subtyping from histopathology slides. Med. Image Anal. **39**, 194–205 (2017)
6. Asadi-Aghbolaghi, M., et al.: Machine learning-driven histotype diagnosis of ovarian carcinoma: insights from the OCEAN AI challenge. medRxiv, pp. 2024-04 (2024)
7. Alahmadi, A.: Towards ovarian cancer diagnostics: a vision transformer-based computer-aided diagnosis framework with enhanced interpretability. Results Eng. **23**, 102651 (2024)
8. Xu, H., et al.: Vision transformers for computational histopathology. IEEE Rev. Biomed. Eng. **17**, 63–79 (2023)
9. Bashashati, A.: UBC ovarian cancer subtype classification and outlier detection (UBCOCEAN) (2023)
10. Aboussaleh, I., et al.: 3DUV-NetR+: a 3D hybrid semantic architecture using transformers for brain tumor segmentation with MultiModal MR images. Results Eng. **21**, 101892 (2024)
11. Breen, J., et al.: Predicting ovarian cancer treatment response in histopathology using hierarchical vision transformers and multiple instance learning. arXiv preprint arXiv:2310.12866 (2023)
12. Mackenzie, R., et al.: Morphologic and molecular characteristics of mixed epithelial ovarian cancers. Am. J. Surg. Pathol. **39**(11), 1548–1557 (2015)
13. Arora, S., Patel, A., Gupta, R.: An explainable machine learning framework for the accurate diagnosis of ovarian cancer. arXiv preprint (2023)

14. Manchanda, R., Gentry-Maharaj, A.: The role of biomarkers in the detection of ovarian cancer. Int. J. Gynecol. Obstetr. **131**(2), 91–97 (2015)
15. Ziyambe, B., Yahya, A., Mushiri, T., et al.: A deep learning framework for the prediction and diagnosis of ovarian cancer. Diagnostics **13**(10), 1703 (2023)
16. Ghoniem, R.M., Algarni, A.D., Refky, B., et al.: Multi-modal evolutionary deep learning model for ovarian cancer diagnosis. Symmetry **13**, 643 (2021)
17. Wang, G., Sun, Y., Jiang, S., et al.: Machine learning-based rapid diagnosis of human borderline ovarian cancer on second-harmonic generation images. Biomed. Opt. Express **12**, 5658–5669 (2021)
18. Chen, H., Yang, B.W., Qian, L., et al.: Deep learning prediction of ovarian malignancy at US compared with O-RADS and expert assessment. Radiology **304**, 106–113 (2022)
19. Gao, Y., Zeng, S., Xu, X., et al.: Deep learning-enabled pelvic ultrasound images for accurate diagnosis of ovarian cancer in China. Lancet Digit. Health **4**, e179–e187 (2022)
20. Wang, X., Li, H., Zheng, P.: Automatic detection and segmentation of ovarian cancer using a multitask model in pelvic CT images. Oxidative medicine and cellular longevity (2022)
21. Boyanapalli, A., Shanthini, A.: Ovarian cancer detection in CT images using ensembled deep optimized learning classifier. Concurr. Comput. **35**, e7716 (2023)
22. Yao, H., Zhang, X.: Prediction model of residual neural network for pathological lymph node metastasis of ovarian cancer. Biomed. Res. Int. **2022**, 9646846 (2022)
23. Maria, H.H., Jossy, A.M., Malarvizhi, S.: A hybrid deep learning approach for detection and segmentation of ovarian tumors. Neural Comput. Appl. **35**, 15805–15819 (2023)
24. Liu, P., Liang, X., Liao, S., et al.: Pattern classification for ovarian tumors by integration of radiomics and deep learning features. Curr. Med. Imaging Rev. **18**, 1486–1502 (2022)
25. Liu, L., Wan, H., Liu, L., et al.: Deep learning provides a new MRI-based prognostic biomarker for recurrence prediction in high-grade serous ovarian cancer. Diagnostics **13**, 748 (2023)
26. CL-Detection 2023 Challenge: Automatic lesion detection and classification from medical images. https://github.com/cwwang1979/MICCAI_ATEC23challenge
27. He, K., Zhang, X., Ren, S., et al.: Deep residual learning for image recognition. In : Proceedings of the IEEE Conference on Computer Vision and Pattern Recognition, pp. 770–778 (2016)
28. Dosovitskiy, A., Beyer, L., Kolesnikov, A., et al.: An image is worth 16x16 words: transformers for image recognition at scale. arXiv preprint arXiv:2010.11929 (2020)
29. Ronneberger, O., Fischer, P., et Brox, T.: U-Net: convolutional networks for biomedical image segmentation. In : Medical Image Computing and Computer-Assisted Intervention–MICCAI 2015: 18th International Conference, Munich, Germany, October 5-9, 2015, proceedings, part III 18. Springer international publishing, pp. 234–241 (2015). https://doi.org/10.1007/978-3-319-24574-4_28
30. Huang, G., Liu, Z., Van Der Maaten, L., et al.: Densely connected convolutional networks. In: Proceedings of the IEEE Conference on Computer Vision and Pattern Recognition, pp. 4700-4708 (2017)

31. Zhou, Z., Rahman Siddiquee, M.M., Tajbakhsh, N., et al.: UNet++: a nested U-Net architecture for medical image segmentation. In : Deep learning in medical image analysis and multimodal learning for clinical decision support: 4th international workshop, DLMIA 2018, and 8th international workshop, ML-CDS 2018, held in conjunction with MICCAI 2018, Granada, Spain, September 20, 2018, proceedings 4, pp. 3–11. Springer International Publishing (2018). https://doi.org/10.1007/978-3-030-00889-5_1
32. Liu, L., Wan, H., Liu, L., et al.: Deep learning provides a new magnetic resonance imaging-based prognostic biomarker for recurrence prediction in high-grade serous ovarian cancer. Diagnostics **13**(4), 748 (2023)
33. Maria, H.H., Jossy, A.M., Malarvizhi, S.: A hybrid deep learning approach for detection and segmentation of ovarian tumors. Neural Comput. Applic. **35**, 15805–15819 (2023)
34. Oktay, O., Schlemper, J., Folgoc, L.L., et al.: Attention U-Net: learning where to look for the pancreas. arXiv preprint arXiv:1804.03999 (2018)
35. Wang, X., Li, H., Zheng, P.: [Retracted] Automatic detection and segmentation of ovarian cancer using a multitask model in pelvic CT images. Oxidative Med. Cell. Longevity **2022**(1), 6009107 (2022)
36. Dosovitskiy, A., Beyer, L., Kolesnikov, A., et al.: An image is worth 1616 words: transformers for image recognition at scale. In: ICLR (2021)
37. Behera, S.K., Das, A., Sethy, P.K.: Deep fine-KNN classification of ovarian cancer subtypes using EfficientNet-B0 extracted features: a comprehensive analysis. J. Cancer Res. Clin. Oncol. **150**(7), 361 (2024). https://doi.org/10.1007/s00432-024-05879-z
38. Boschman, J.: Generalizable deep learning models for epithelial ovarian carcinoma classification. University of British Columbia, Diss (2022)
39. Asadi-Aghbolaghi, M., et al.: Insights from the OCEAN AI challenge, machine learning-driven histotype diagnosis of ovarian carcinoma (2024)

Comparative Analysis of SAC and PPO for Energy Management in Battery Electric Vehicles

Abdelaziz Sahbani[1,2](✉)

[1] Laboratory of Automatic (LARA), National Engineering School of Tunis (ENIT), University of Tunis El Manar, BP37, 1002 Tunis, Tunisia
[2] Faculty of Science of Bizerte, University of Carthage, Bizerte, Tunisia
abdelaziz.sahbani@fsb.ucar.tn

Abstract. Effective energy management in battery electric vehicle (BEV) is required to extend battery life, maximize driving range, and minimize energy consumption. Reinforcement learning (RL) is a model-free and very effective approach to learn adaptive control policies in a variable and uncertain driving environment. This work compares Soft Actor-Critic (SAC) and Proximal Policy Optimization (PPO) for energy management in BEVs. A simulated electric powertrain and battery model is developed and tested in a custom simulation environment and various driving cycles to evaluate the robustness of each algorithm. These RL agents are trained to keep SOC at optimal level and minimize energy consumption. The analysis shows that PPO adopts a more conservative and consistent SOC control strategy, resulting in lower energy consumption. However, it generally falls behind SAC in terms of overall reward performance.

Keywords: Battery Electric Vehicle · Energy Management System · Reinforcement Learning

1 Introduction

As reported by the World Bank, more than four billion people live in cities today and this number is projected to double by 2050. As urban areas continue to expand, urban residents have become increasingly dependent on urban transport systems [1]. As a result, it has become increasingly important to move toward electrified transport as a means of reducing environmental impact and increasing energy efficiency. In turn, many countries have launched incentives and policies to promote the uptake and manufacturing of electric vehicles. Together with efficient lithium-ion batteries, the use of electric motors greatly reduces fossil fuel consumption and greenhouse gas emissions because this energy storage technology allows energy to be used more efficiently and, more importantly, cleanly throughout vehicle operation [2].

T. Ensari et al. (Eds.): ISPR 2025, CCIS 2859, pp. 124–136, 2026.
https://doi.org/10.1007/978-3-032-21585-7_9

Projections indicate that this global shift will continue to intensify. By mid-century, nearly 950 million zero-emission vehicles are expected to be in operation worldwide, including 875 million electric passenger cars, 70 million electric commercial vehicles, and 5 million powered by hydrogen fuel cells. In leading markets such as China, Europe and the United States, more than 55% of all vehicles, and almost half of commercial fleets, are projected to be electric, especially battery electric vehicles (BEV), by 2050, marking a transformative change in the global transportation landscape [3] (see Fig. 1).

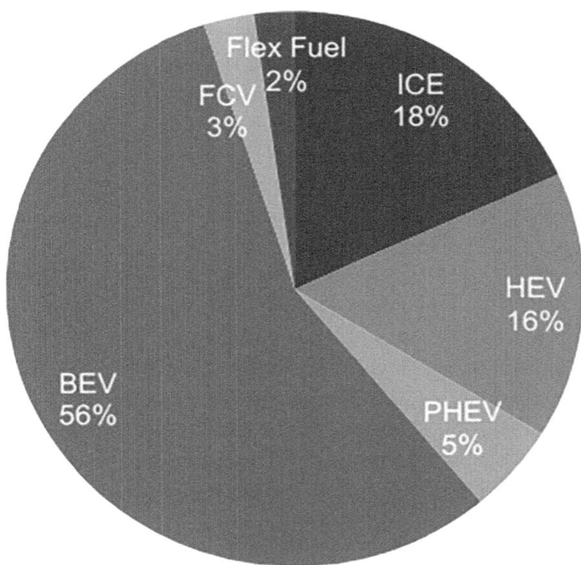

Fig. 1. Projected Global Vehicle Sales by 2050, according to [3].

As the adoption of battery electric vehicles continues to increase, advanced energy management systems (EMS) are becoming increasingly essential. One of the primary tasks of the EMS is to know and control the accurate State of Charge (SOC) of the battery, which tells how much energy is available to use [5,5,6]. SOC is an important parameter not only for energy consumption and driving range, but also for reducing battery aging and improving driving experience. In lithium-ion battery electric vehicles, traditional EMS strategies typically rely on rule-based logic, model predictive control (MPC), and fuzzy logic control to manage power flow and maintain SOC within safe limits [7]. These techniques are often combined with battery aging models, thermal models, and efficiency maps to optimize the system's operation under different driving conditions. However, despite their effectiveness, these approaches can be limited by model accuracy and computational complexity, especially when dealing with dynamic and uncertain environments.

Reinforcement learning (RL) is a paradigm of machine learning in which an agent learns to make decisions by interacting with an environment. It is particularly attractive when it is difficult or impractical to build explicit models of the environment [8]. Reinforcement Learning (RL) methods have significantly advanced over time, evolving from traditional algorithms to modern deep-based techniques and actor-critic-based. These approaches are commonly categorized into two main classes: conventional RL algorithms and advanced RL techniques, each offering unique strengths and trade-offs when applied to energy management systems (EMS) in battery electric vehicles.

Traditional RL algorithms include Temporal Difference (TD) learning [6,8,9], Q-learning [10,11], and SARSA [12,13]. TD learning, which forms the foundation for many RL approaches, blends ideas from dynamic programming and Monte Carlo methods. It is model-free and straightforward, but suffers from biased value estimates during learning. Q-learning is widely used due to its effective exploration capability, yet it is constrained by its reliance on discrete action spaces and is prone to overestimation of action values. SARSA, while employing a more conservative exploration strategy, often leads to longer training times and inherits many of Q-learning's limitations.

To overcome these challenges, modern RL algorithms, such as Deep Q-Networks (DQN) [14,15], Deep Deterministic Policy Gradient (DDPG) [16,17], Actor-Critic frameworks [18], and enhancements like Double Q-learning [19] and Double DQN [20], have been developed. These methods introduce neural function approximators and policy/value separation, enabling greater generalization, stability, and control in high-dimensional or continuous action spaces. For example, while Q-learning-based EMSs may offer faster convergence in simpler environments, DQN and DDPG are more suitable for complex, high-dimensional systems.

Each of these advanced techniques involves trade-offs. DQN provides good generalization but can struggle with control precision and value overestimation. DDPG and Actor-Critic architectures support continuous control and often converge faster, but involve higher training complexity. Double Q learning and Double DQN mitigate overestimation errors and improve stability, though at the cost of increased computational complexity—making them more appropriate for specific precision critical applications.

Further advancements include algorithms such as Proximal Policy Optimization (PPO) [21,22] and Soft Actor-Critic (SAC) [23,24]. Notably, PPO is widely appreciated for its simplicity and stable updates, as it achieves performance comparable to state-of-the-art methods with minimal implementation complexity. SAC, in contrast, incorporates entropy regularization, promoting better exploration and faster convergence, especially effective in stochastic, high-dimensional control problems.

These attributes make both algorithms particularly well-suited for continuous-control tasks within Energy Management Systems (EMS), such as SOC optimization. PPO's clipped policy updates ensure safe and steady learning in embedded or hardware-restricted environments, while SAC's sample efficiency and robustness offer benefits in unpredictable conditions like real driving cycles.

However, comparative studies between SAC and PPO focused on SOC regulation under realistic drive profiles remain limited. Our study fills this gap by providing a direct comparison of SAC and PPO within a unified SOC control framework using real-world drive data, supported by detailed performance metrics.

In this work, we investigate the performance of PPO and SAC for the SOC control problem in battery electric vehicles. Our objective is to evaluate which RL method achieves better control performance, energy efficiency, and SOC stability. Unlike existing works focusing only on policy accuracy or SOC tracking, we perform a multi-criteria evaluation, including energy consumed, regenerative braking usage, SOC deviations, and convergence speed.

This paper is structured as follows: Sect. 2 presents the powertrain and battery models. Section 3 details the RL algorithms and training parameters. Section 4 presents and analyzes the experimental results. Section 5 concludes the paper with future perspectives.

2 Powertrain and Battery System Modeling

2.1 EV Powertrain Traction Modeling

In electric vehicles, the powertrain is the main component that controls energy consumption and dynamic performance under different driving conditions. To properly estimate the required traction, a longitudinal vehicle dynamics model that takes into account the primary resistive forces opposing motion is considered. Among these are the inertial force of acceleration, aerodynamic drag, rolling resistance, and gravitational force on inclines.

The total tractive force required on the wheels, denoted by F_{traction}, is given by:

$$F_{\text{traction}} = m \cdot a + \frac{1}{2} \cdot \rho \cdot C_d \cdot A_f \cdot v^2 + m \cdot g \cdot C_{rr} + m \cdot g \cdot \sin(\theta) \tag{1}$$

where:

- m is the total mass of the vehicle [kg],
- a is the longitudinal acceleration [m/s^2],
- ρ is the air density [kg/m^3],
- C_d is the aerodynamic drag coefficient,
- A_f is the frontal surface area of the vehicle [m^2],
- v is the vehicle speed [m/s],
- g is the gravitational constant ≈ 9.81 [m/s^2],
- C_{rr} is the rolling resistance coefficient,
- θ is the road slope angle [rad].

The mechanical power demand at the wheels can then be expressed as:

$$P_m = \frac{F_{\text{traction}} \cdot v}{\eta_m} \tag{2}$$

where η_m represents the overall efficiency of the motor-inverter drivetrain. Negative values of P_m arise naturally during braking or downhill driving, indicating the potential for energy recovery.

The propulsion system employs a Permanent Magnet Synchronous Motor (PMSM), widely recognized for its high efficiency, power density, and suitability for traction applications. The motor is interfaced with the battery through a bidirectional voltage source inverter, which allows power to flow in both directions depending on the driving state.

During acceleration or uphill motion, the system operates in motoring mode, drawing electrical energy from the battery and converting it into mechanical torque to drive the vehicle. In contrast, when the vehicle decelerates, either by braking or when moving downhill, the mechanical energy is partially recovered. In this regenerative mode, the PMSM acts as a generator, converting kinetic energy back into electrical form, which is redirected to recharge the battery.

This bidirectional flow of energy is required to enhance the overall energy efficiency and vehicle range. The inverter is responsible for motoring and regenerating transitions, synchronization, and safe operation in every condition of driving.

2.2 Battery Model

For the purposes of this research, a lithium-ion battery model is used to accurately reflect the energy storage system of the EV. The pack has a nominal capacity of 60 kWh and a voltage range from 320 V to 420 V. To capture the dynamics of the battery, a second order ECM is chosen for battery modeling, in which there are an open-circuit voltage (OCV) source, resistance, and two RC networks connected in parallel. This model strikes a good compromise of accuracy vs computational effort while allowing for a representation of transient voltage responses and capacity fade effects.

The terminal voltage of the battery is computed by summing the voltage contributions across the internal elements:

$$V_{bat} = V_{OCV}(SOC) - I_{bat} \cdot R_0 - V_{RC1} - V_{RC2} \tag{3}$$

where $V_{OCV}(SOC)$ is the open-circuit voltage as a function of SOC, R_0 is the ohmic resistance, I_{bat} is the instantaneous current, and V_{RC1}, V_{RC2} are the voltage drops across the two RC branches that model dynamic effects.

The State of Charge (SOC) is updated using Coulomb counting, integrating the battery current over time and adjusting for charge and discharge efficiencies:

$$SOC_{k+1} = SOC_k - \frac{I_k \cdot \Delta t}{Q_{\text{nominal}}} \tag{4}$$

where I_k is the battery current at step k (positive for discharge), Δt is the time increment, and Q_{nominal} is the nominal capacity of the battery.

3 Reinforcement Learning Framework and Reward Design

The energy management problem is framed here as a continuous control problem and addressed using two advanced reinforcement learning (RL) algorithms, namely Soft Actor-Critic (SAC) and Proximal Policy Optimization (PPO). The simulated environment is based on a realistic EV model that is exposed to phenomena during several drive cycles. The agent interacts with the environment by selecting control actions based on torque demand.

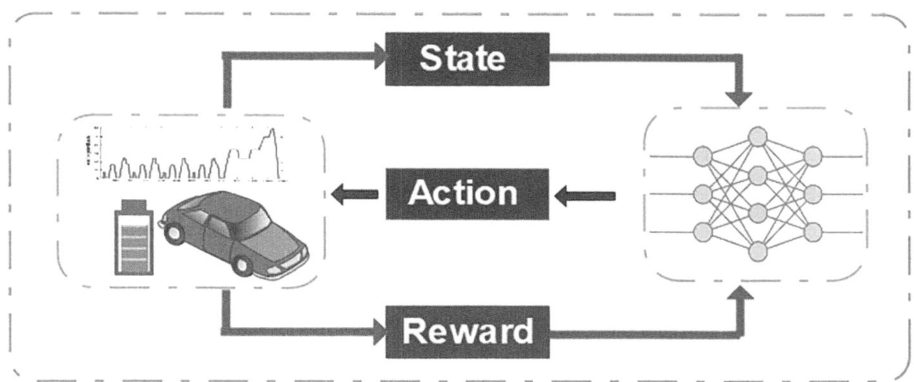

Fig. 2. Block diagram of the proposed agents for the EMS of the BEVs.

The block diagram of the proposed agents for the energy management of electric vehicles is shown in Fig. 2. The state space is a 4-dimensional vector : state $= [v_t, v_{\text{ref},t}, SOC_t, \text{time\_remaining}_t]$

where:

- $v_t \in [0, 40]$: current vehicle speed at time t, in m/s.
- $v_{\text{ref},t} \in [0, 40]$: reference speed at time t, in m/s, from the driving cycle.
- $SOC_t \in [0, 1]$: battery state of charge at time t, normalized between 0 (empty) and 1 (full).
- $\text{time\_remaining}_t \in [0, 1]$: normalized remaining time in the episode.

the current vehicle speed, the current levels of SOC, acceleration, and gradient. The action is acceleration, defined as a continuous scalar in the range $[-3, 3]$ m/s^2.

The objective is to learn a control policy which minimizes energy as possible while keeping the battery at a safe level of charge, so that the driving range can be extended, and battery health can be preserved. A shaped reward function is constructed to train the learning of the agent as follows:

$$r_t = -\alpha \cdot |v_t - v_{\text{ref},t}| - \beta \cdot |SOC_t - SOC_{\text{target}}| + \gamma \cdot E_{\text{regen},t} \tag{5}$$

Here, v_t is the actual speed at time t, $v_{\text{ref},t}$ is the reference speed of the driving cycle, SOC_t is the state of charge of the battery, and SOC_{target} is the desired SOC level (usually mid-range for safety and efficiency). The term $E_{\text{regen},t}$ denotes the energy recovered from regenerative braking. The weights α, β, and γ balance the trade-off between velocity tracking, SOC preservation, and energy recovery. Penalty terms can be added to discourage unsafe SOC levels or excessive torque demand.

This reward signal serves as the foundation for training RL agents to find good control strategies in non-stationary, continuous environments. Building on this, we explore how SAC and PPO agents leverage this formulation to learn robust policies.

Soft Actor-Critic (SAC) is an off-policy, entropy-regularized RL algorithm that learns both a stochastic policy and a Q-function through maximum entropy optimization. Its key innovation lies in incorporating an entropy term into the objective, encouraging exploration by preventing premature policy convergence. The SAC objective maximizes both expected return and policy entropy:

$$J(\pi) = \mathbb{E}_{(s,a)\sim D}\left[Q(s,a) - \alpha \log \pi(a|s)\right] \tag{6}$$

where α is the temperature coefficient that balances reward and entropy, $Q(s,a)$ is the estimated return of taking action a in state s, and $\pi(a|s)$ is the policy distribution. This structure makes SAC particularly effective in handling continuous, high-dimensional action spaces like those in EV control.

In contrast, Proximal Policy Optimization (PPO) is an on-policy method based on trust-region optimization. The newest vesrion of PPO, called Ratio Clipping Proximal Policy Optimization (RC-PPO), updates the policy using a clipped surrogate objective, which prevents large, destabilizing policy updates and ensures steady improvement [19]. The PPO objective is given by:

$$J^{\text{PPO}}(\theta) = \mathbb{E}_t\left[\min\left(r_t(\theta)\hat{A}_t, \text{clip}(r_t(\theta), 1-\epsilon, 1+\epsilon)\hat{A}_t\right)\right] \tag{7}$$

where $r_t(\theta) = \frac{\pi_\theta(a_t|s_t)}{\pi_{\theta_{\text{old}}}(a_t|s_t)}$ is the probability ratio between the new and old policies, $\hat{A}_t$ is the advantage estimate, and ϵ is the clipping threshold. This conservative update strategy improves robustness and sample efficiency in environments with variable dynamics such as EVs under real-world driving conditions.

4 Experimental Results

Both SAC and PPO agents are implemented using Stable-Baselines3 with the MlpPolicy architecture, which employs multilayer perceptrons for function approximation. The agents are trained using eight parallel environments for sample efficiency. Key hyper-parameters such as learning rate, training duration, and optimizer settings are consistent across both agents to ensure a fair comparison. These hyper-parameters are summarized in Table 1.

Table 1. Simulation Parameters for Vehicle Dynamics and RL Agents

Parameter	Description	Value
Vehicle and Battery Parameters		
battery_kWh	Battery capacity	60.0 kWh
motor_efficiency	Motor energy efficiency	0.9
regen_efficiency	Regenerative braking efficiency	0.6
vehicle_mass	Vehicle mass	1600 kg
drag_coefficient	Aerodynamic drag coefficient C_d	0.28
frontal_area	Frontal area A	2.2 m^2
rolling_resistance	Rolling resistance coefficient C_{rr}	0.015
air_density	Air density ρ	1.225 kg/m^3
gravity	Gravitational acceleration g	9.81 m/s^2
v_max	Maximum vehicle speed	40.0 m/s
RL Agent Hyper-parameters (SAC and PPO)		
policy	Neural network policy architecture	MlpPolicy
learning_rate	Learning rate for optimizer	3×10^{-4}
timesteps	Total training steps per cycle	300,000
clip_range	PPO clipping range (epsilon)	0.2
init_soc	Initial state of charge (SOC)	0.8

Moreover, the agents are evaluated on a set of standardized and synthetic driving cycles that reflect diverse real-world conditions, including urban (UDDS: Urban Dynamometer Driving Schedule), suburban (NEDC: New European Driving Cycle and WLTP: Worldwide harmonized Light vehicles Test Procedure), and aggressive driving profiles. The aggressive driving profile simulates rapid acceleration, frequent braking, and high-speed fluctuations, representing real-world scenarios such as stop-and-go traffic or impatient driver behavior.

The trained SAC and PPO agents are evaluated on each driving cycle across five independent runs. Metrics include total reward, energy consumed, distance traveled, final SOC, and energy efficiency measured as kilometers per kWh.

4.1 Learning Curves

Figures 3 and 4 illustrate the training performance of the SAC and PPO algorithms across various driving cycles by showing their smoothed reward curves and loss trajectories. The SAC agent demonstrates a more consistent and faster convergence in most driving scenarios, achieving higher mean rewards with reduced variability. This behavior suggests stronger sample efficiency and improved robustness in continuous control settings.

On the other hand, PPO initially shows greater variability and a slower rise in performance, particularly during the early training phase. Although PPO

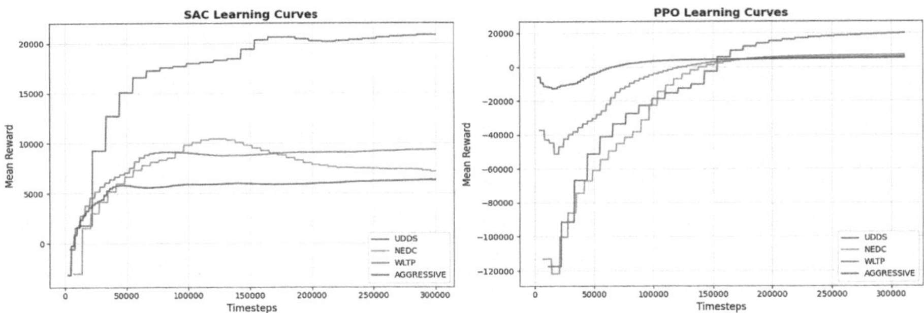

Fig. 3. Mean rewards for SAC and PPO agents per driving cycle.

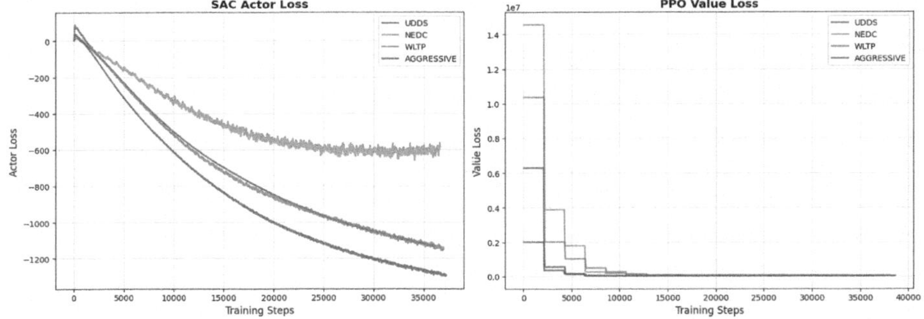

Fig. 4. Losses for SAC and PPO agents per driving cycle.

eventually stabilizes, its learning curve remains less smooth and generally lags behind SAC in terms of both convergence speed and reward magnitude.

4.2 SOC Evolution

Figure 5 illustrates the evolution of the State of Charge over time for both SAC and PPO agents across the four distinct driving cycles: UDDS, NEDC, WLTP, and aggressive. Each subplot presents a comparison between the two reinforcement learning strategies, with SAC shown in orange and PPO in blue.

In addition, Fig. 6 shows the average final SOC for both PPO and SAC agents in the four driving cycles. PPO performs a slightly higher final SOC than SAC in all scenarios overall. The largest difference is in the NEDC cycle, where the SAC shows a lower end of charge state, which corresponds to more aggressive depleting of energy.

4.3 Energy Consumption over Trajectories

Figure 7 compares the average travel distance and energy consumption for both agents. SAC consistently covers longer distances than PPO in the aggressive,

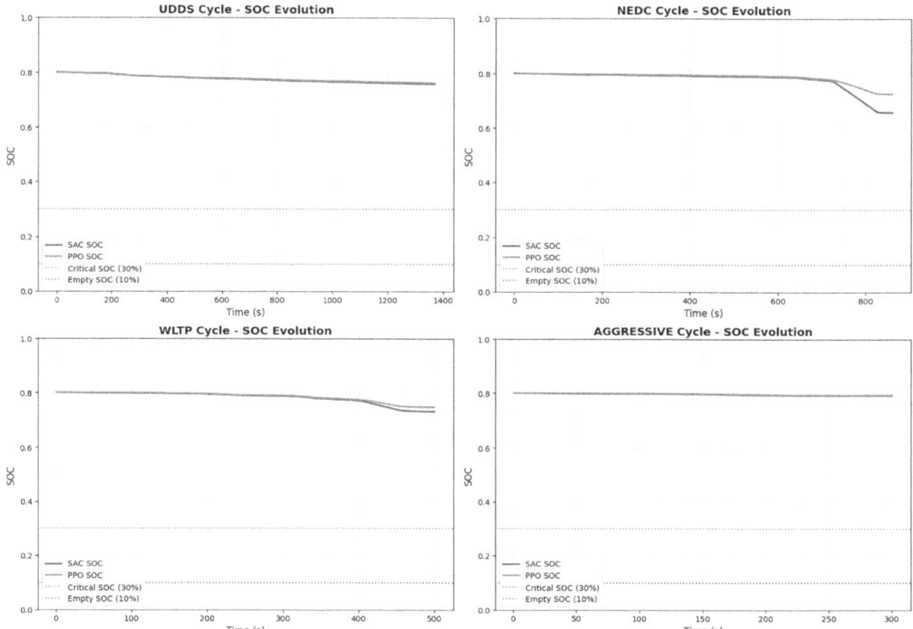

Fig. 5. SOC evolution over time for both agents.

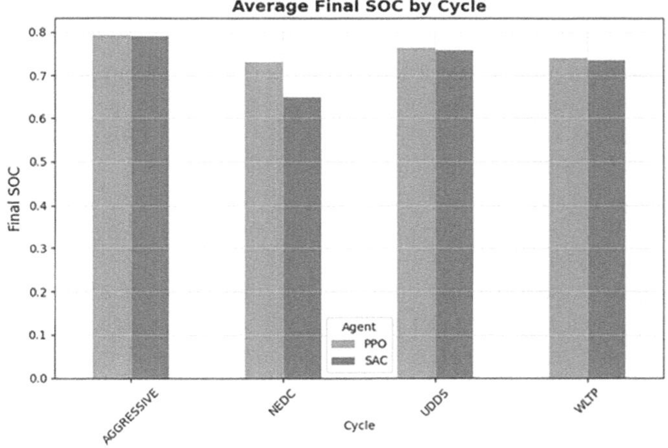

Fig. 6. average final SOC for both agents.

NEDC, and UDDS cycles, making it a suitable choice for applications where extended range is important. However, this advantage comes with a drawback: SAC also consumes more energy and shows greater variability in its consumption, as indicated by the large standard deviation.

In contrast, PPO demonstrates better energy efficiency, with a lower and more consistent energy use, even if it occasionally covers slightly shorter distances. This reflects a clear trade-off between the two agents: SAC favors longer range, while PPO prioritizes energy savings. The choice between them should depend on the specific goals of the application.

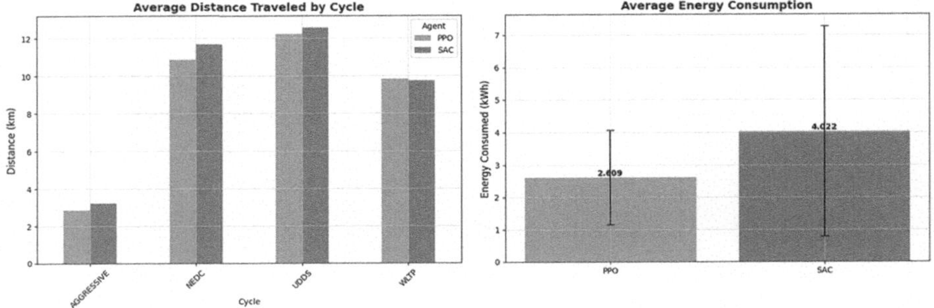

Fig. 7. average traveled distance and average energy consumption for both agents.

5 Conclusion

This study evaluated the performance of Soft Actor-Critic (SAC) and Proximal Policy Optimization (PPO) algorithms for energy management in electric vehicles, using battery State of Charge (SOC) as a key metric across various driving cycles. The findings reveal that SAC generally enables longer travel distances, particularly in the aggressive, NEDC, and UDDS cycles. It also converges more quickly and demonstrates greater stability in dynamic environments, which highlights its suitability for applications requiring extensive range and adaptability to complex or noisy scenarios.

However, these benefits come at a notable cost. SAC consistently consumed more energy and exhibited greater variability in energy usage compared to PPO. PPO, by contrast, maintained a lower and more stable energy consumption across all cycles, making it the most energy-efficient.

Both agents successfully maintained the battery SOC within acceptable bounds under different driving conditions and provided effective energy recuperation, particularly during aggressive driving. The choice between SAC and PPO should therefore be guided by specific application priorities, whether the goal is to extend driving range or to minimize energy usage.

Future work will aim to enhance the realism of the simulation by incorporating additional factors such as battery thermal behavior, long-term degradation, and internal energy losses during charging and discharging. These improvements will further refine the models and bring them closer to real-world applicability.

References

1. Overview (2025). https://www.worldbank.org/en/topic/urbandevelopment/overview. Accessed 7 July 2025
2. Lopez, I., Ibarra, E., Matallana, A., Andreu, J., Kortabarri, I.: Next generation electric drives for HEV/EV propulsion systems: technology, trends and challenges. Renew. Sustain. Energy Rev. **114**, 1–23 (2019)
3. Sutton, R.S., Barto, A.G.: Reinforcement Learning: An Introduction, 2nd edn. MIT Press, Cambridge (2018)
4. Baumler, A., Benterki, A., Meng, J., Azib, T., Boukhnifer, M.: Energy management strategies based on soft actor critic reinforcement learning with a proper reward function design based on battery state of charge constraints. J. Energy Storage **90**, Part A (2024). https://doi.org/10.1016/j.est.2024.111797
5. Peng, J., et al.: Multiple electric components health-aware eco-driving strategy for fuel cell hybrid electric vehicle based on soft actor-critic algorithm. IEEE Trans. Transport. Electrificat. **10**(3), 6242–6257 (2024). https://doi.org/10.1109/TTE.2023.3339490
6. Sutton, R.S., Barto, A.G.: Reinforcement Learning: An Introduction, 2nd edn. MIT Press, Cambridge (2018)
7. Kermansaravi, A., Refaat, S., Trabelsi, M., Vahedi, H.: AI-based energy management strategies for electric vehicles: challenges and future directions. Energy Rep. **13** (2025). https://doi.org/10.1016/j.egyr.2025.04.053
8. Wu, P., Partridge, J., Anderlini, E., Liu, Y., Bucknall, R.: Near-optimal energy management for plug-in hybrid fuel cell and battery propulsion using deep reinforcement learning. Int. J. Hydrogen Energy **46**(80), 40022–40040 (2021). https://doi.org/10.1016/j.ijhydene.2021.09.196
9. Haarnoja, T., Zhou, A., Abbeel, P., Levine, S.: Soft actor-critic: off-policy maximum entropy deep reinforcement learning with a stochastic actor. In: Proceedings of the 35th International Conference on Machine Learning, vol. 80, pp. 1861–1870 (2018)
10. Guo, L., Li, Z., Outbib, R.: An unbiased fuzzy double q-learning based energy management for fuel cell hybrid electric vehicles. In: IEEE 32nd International Symposium on Industrial Electronics (ISIE), Finland, 19–21 June 2023 (2023). https://doi.org/10.1109/ISIE51358.2023.10227988
11. Haarnoja, T., Zhou, A., Abbeel, P., Levine, S.: Soft actor-critic: off-policy maximum entropy deep reinforcement learning with a stochastic actor. In: Proceedings of the 35th International Conference on Machine Learning, vol. 80, pp. 1861–1870 (2018)
12. Hu, J., et al.: Performance analysis of AI-based energy management in electric vehicles: a case study on classic reinforcement learning. Energy Conv. Manag. **300** (2024). https://doi.org/10.1016/j.enconman.2023.117964
13. Haarnoja, T., Zhou, A., Abbeel, P., Levine, S.: Soft actor-critic: off-policy maximum entropy deep reinforcement learning with a stochastic actor. In: Proceedings of the 35th International Conference on Machine Learning, vol. 80, pp. 1861–1870 (2018)
14. Li, M., Wang, Y., Yu, P., Sun, Z., Chen, Z.: Online adaptive energy management strategy for fuel cell hybrid vehicles based on improved cluster and regression learner. Energy Conv. Manag. **292** (2023). https://doi.org/10.1016/j.enconman.2023.117388

15. Li, K., Zhou, J., Jia, C., Yi, F., Zhang, C.: Energy sources durability energy management for fuel cell hybrid electric bus based on deep reinforcement learning considering future terrain information. Int. J. Hydrogen Energy **52**(D), 821–833 (2024). https://doi.org/10.1016/j.ijhydene.2023.05.311
16. Chen, W., Yin, G., Fan, Y., Zhuang, W., Zhang, H., Peng, J.: Ecological driving strategy for fuel cell hybrid electric vehicle based on continuous deep reinforcement learning. In: 2022 6th CAA International Conference on Vehicular Control and Intelligence (CVCI), China, 28–30 October 2022. https://doi.org/10.1109/CVCI56766.2022.9964786
17. Xiao, Y., Fu, S., Choi, J., Zheng, C.: A collaborative energy management strategy based on multi-agent reinforcement learning for fuel cell hybrid electric vehicles. In: Proceedings of IEEE 98th Vehicular Technology Conference (VTC2023-Fall), Hong Kong (2023). https://doi.org/10.1109/VTC2023-Fall60731.2023.10333636
18. Kim, D., Hong, S., Cui, S., Joe, I.: Deep reinforcement learning-based real-time joint optimal power split for battery–ultracapacitor–fuel cell hybrid electric vehicles. Electronics **11** (2022). https://doi.org/10.3390/electronics11121850
19. Acquarone, M., Maino, C., Misul, D., Spessa, E., Mastropietro, A., Sorrentino, L., Busto, E.: Influence of the reward function on the selection of reinforcement learning agents for hybrid electric vehicles real-time control. Energies **16** (2023). https://doi.org/10.3390/en16062749
20. Wu, P., Partridge, J., Anderlini, E., Liu, Y., Bucknall, R.: Near-optimal energy management for plug-in hybrid fuel cell and battery propulsion using deep reinforcement learning. Int. J. Hydrogen Energy **46**(80), 40022–40040 (2021). https://doi.org/10.1016/j.ijhydene.2021.09.196
21. Schulman, J., Wolski, F., Dhariwal, P., Radford, A., Klimov, O.: Proximal Policy Optimization Algorithms. arXiv:1707.06347 (2017). https://doi.org/10.48550/arXiv.1707.06347
22. Baumler, A., Azib, T., Meng, J., Benterki, A., Boukhnifer, M.: Reward function design for reinforcement learning based energy management strategy for a fuel cell/supercapacitor hybrid electric vehicle. In: 2024 International Conference on Control, Automation and Diagnosis (ICCAD), Paris, 15–17 May 2024 (2024). https://doi.org/10.1109/ICCAD60883.2024.10554034
23. Haarnoja, T., Zhou, A., Abbeel, P., Levine, S.: Soft actor-critic: off-policy maximum entropy deep reinforcement learning with a stochastic actor. In: Proceedings of the 35th International Conference on Machine Learning, vol. 80, pp. 1861–1870 (2018)
24. Baumler, A., Azib, T., Meng, J., Benterki, A., Boukhnifer, M.: Reward function design for reinforcement learning based energy management strategy for a fuel cell/supercapacitor hybrid electric vehicle. In: 2024 International Conference on Control, Automation and Diagnosis (ICCAD), Paris, 15–17 May 2024 (2024). https://doi.org/10.1109/ICCAD60883.2024.10554034

On-Device Lightweight CNNs for Industrial Thermal Fault Diagnosis and Monitoring

A. Kherbachi(✉), M. Aouache, and Z. Guettatfi

Telecom Division, Center for Development of Advanced Technologies (CDTA), Baba Hassen, 16303 Algiers, Algeria
{akherbachi,maouache,zguettatfi}@cdta.dz

Abstract. Predictive maintenance plays a critical role in reducing unplanned downtime and extending the operational life of industrial equipment. Thermal imaging, with its ability to detect early-stage faults via abnormal heat signatures, offers a powerful non-contact diagnostic tool. However, existing solutions often rely on cloud-based processing or manual inspection, which limits their responsiveness and scalability. This paper presents an embedded AI system for real-time thermal image analysis aimed at fault diagnosis and monitoring, deployed on a low-power ESP32-S3 microcontroller. Two lightweight convolutional neural network (CNN) architectures, Sequential Shallow Model (SSM) and Deep Structured Model (DSM) were developed and trained on a labeled dataset of motor fault thermal images. The models were converted to TensorFlow Lite and deployed in both float32 and int8 formats for on-device inference. A built-in web interface supports direct image upload, preprocessing, and fault classification visualization, all without reliance on external infrastructure. Quantized models achieved up to an 8 $\times$ speedup in inference (as low as 23.8 ms), with memory usage reduced to under 20 KB, while maintaining classification accuracy (F1-score up to (98.6%)) on par with their full-precision counterparts. The proposed system demonstrates the feasibility of fast, compact, and autonomous AI-driven thermal monitoring at the edge, offering a scalable solution for industrial predictive maintenance aligned with Industry 4.0 goals.

Keywords: Predictive maintenance · Embedded AI · Thermal imaging · Edge computing · CNNs · ESP32-S3 · Fault Detection · Industrial Monitoring

1 Introduction

In recent years, the integration of embedded systems with artificial intelligence (AI) has facilitated the development of autonomous, low-power, and real-time monitoring solutions for industrial applications. Among these advancements, predictive maintenance has emerged as a critical strategy, offering the potential to

T. Ensari et al. (Eds.): ISPR 2025, CCIS 2859, pp. 137–152, 2026.
https://doi.org/10.1007/978-3-032-21585-7_10

significantly reduce unplanned downtime and associated operational costs by enabling the early detection of system anomalies prior to catastrophic failure. Within this context, thermal imaging has proven to be an indispensable tool. As a non-contact and highly sensitive diagnostic technique, it enables the identification of localized temperature deviations that often serve as precursors to mechanical or electrical faults, thereby supporting more informed and timely maintenance decisions [2,14,18].

Despite their effectiveness, most existing AI-based thermal analysis systems depend heavily on cloud infrastructure or external processing units to execute computationally intensive tasks such as image classification and anomaly detection [6,16]. While these architectures offer significant processing power, they are inherently constrained by issues such as increased latency, reliance on stable network bandwidth, and potential risks to data privacy. In response to these limitations, recent research has increasingly turned toward the deployment of AI models directly on edge devices particularly resource-constrained platforms such as microcontrollers through lightweight frameworks like TensorFlow Lite for Microcontrollers (TFLM) [7,8]. This paradigm shift has led to the emergence of two dominant approaches: partial integration, wherein selected inference tasks are offloaded to the edge, and full integration, which entails executing the complete AI pipeline locally on the embedded device.

In the partial integration approach, microcontrollers primarily handle sensor interfacing, data acquisition, and basic preprocessing, while the computationally demanding inference tasks are delegated to external platforms such as cloud servers or nearby edge nodes [1,20]. In contrast, full integration entails embedding the entire AI inference pipeline including signal preprocessing, model execution, and result interpretation directly within the microcontroller [3,4,19]. This fully embedded approach offers several key advantages: enhanced system autonomy, reduced energy consumption, improved data privacy, and increased resilience to network interruptions, making it particularly suitable for mission critical or resource constrained industrial applications.

A growing body of research has addressed the challenges associated with deploying lightweight convolutional neural networks (CNNs) on resource-constrained embedded platforms. Notably, efficient architectures such as MobileNets [12], LightConvNet [15], and MicroNets [3] have been specifically designed to reduce model size and computational complexity, making them suitable for edge deployment. In parallel, model optimization techniques particularly quantization have been extensively explored to lower memory footprint and inference latency while maintaining acceptable levels of accuracy [11,13]. These advancements have enabled the successful application of edge AI across various industrial scenarios, including real-time object detection, fault classification, and embedded thermographic analysis [5,9,10], thereby demonstrating the practical viability of on-device intelligence in demanding environments.

In this study, we present an embedded AI system tailored for real-time thermal image classification, fully implemented on the ESP32-S3 microcontroller. The system utilizes optimized convolutional neural network (CNN) models, trained on publicly available thermal image datasets and converted into quantized TensorFlow Lite (TFLite) format, to facilitate efficient on-device inference. Unlike hybrid architectures that depend on external computing resources, the proposed solution performs the entire AI pipeline including image acquisition, pre-processing, inference, and result communication directly on the microcontroller. By eliminating reliance on cloud connectivity, this fully embedded approach ensures low-latency performance, reduced energy consumption, and enhanced deployment flexibility in resource-constrained environments. These characteristics make the system particularly suitable for predictive maintenance and other industrial monitoring applications requiring real-time, autonomous operation.

The remainder of the paper is organized as follows: Sect. 2 describes the overall system architecture, including hardware specifications of the ESP32-S3 microcontroller and the embedded software framework. Section 3 presents the thermal image dataset, preprocessing pipeline, and the design of two lightweight CNN architectures (SSM and DSM). Section 5 details the model training methodology, evaluation metrics, and performance analysis across different input resolutions. Section 6 covers the TensorFlow Lite conversion process, quantization techniques, and post-conversion performance assessment. Section 7 presents the on-device deployment results, including inference latency, memory utilization, and real-time performance evaluation on the ESP32-S3 platform. Finally, Sect. 8 concludes the paper with a summary of key findings and outlines future research directions for embedded AI in industrial fault diagnosis.

2 System Architecture

This work proposes an embedded artificial intelligence (AI) system architecture designed towards and perform real-time thermal image classification for predictive maintenance in industrial environments. The system is based on the ESP32-S3 WROOM-1 N16R8 module, which integrates a dual-core Xtensa LX7 processor, 16MB of Flash memory, and 8MB of PSRAM. The ESP32-S3 is equipped with Single Instruction Multiple Data (SIMD) vector instructions and dedicated hardware accelerators for neural network operations, making it particularly well-suited for low-power, real-time inference on the edge. These capabilities allow the system to execute the complete AI pipeline locally, thereby reducing latency, minimizing power consumption, and eliminating dependency on external computing resources.

At this stage, the proposed system is architected to function independently of a physical thermal imaging sensor. Instead, thermal images are provisioned through a web-based interface, allowing users to upload image data directly to the device. This approach facilitates flexible experimentation, supports rapid prototyping, and enables seamless integration of diverse thermal datasets. The ESP32-S3 microcontroller leverages its onboard peripheral subsystems, including general purpose input/output (GPIO) interfaces, hardware timers, and an integrated Wi-Fi module to manage communication protocols and task scheduling.

Convolutional neural network (CNN) models are trained offline using conventional machine learning frameworks, subsequently quantized into TensorFlow Lite (TFLite) representations using either 32-bit floating-point (float32) or 8-bit integer (INT8) precision formats.

The quantized models are subsequently integrated into the firmware as static **C** arrays using the **xxd** tool, which converts binary files into a text-based format suitable for direct inclusion in the source code. This approach eliminates the need for external storage and enables efficient, low-latency inference directly on the microcontroller.

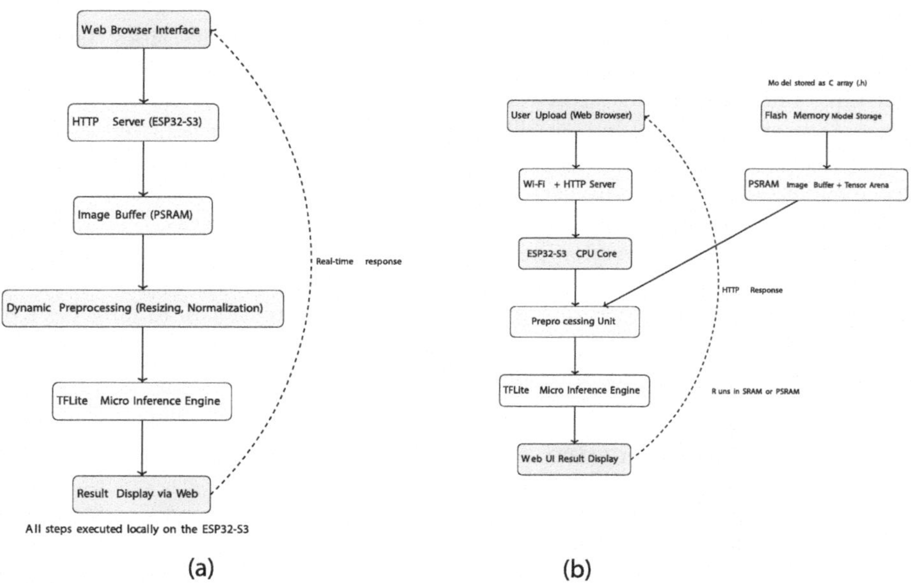

Fig. 1. Combined view: (a) Software processing pipeline and (b) Hardware/software architecture for thermal image classification on ESP32-S3.

Figure 1 illustrates the integrated hardware and software architecture of the proposed system. The diagram captures the complete data flow, beginning with image acquisition via a web-based interface, followed by temporary buffering in PSRAM, pre-processing operations, including resizing and normalization on-device inference, and final result visualization through a browser-accessible interface. The system is optimized for low-latency processing, leveraging compact convolutional neural network (CNN) models tailored to the memory and computational constraints of the ESP32-S3 microcontroller. The embedded software is implemented using the ESP-IDF development framework atop the FreeRTOS real-time operating system, which enables deterministic task scheduling and efficient memory management. The software stack consists of the following key components: 1) lightweight HTTP server that facilitates user interaction and image upload; 2) A preprocessing module that performs grayscale conversion, spatial downsampling, and normalization; TensorFlow Lite Micro (TFLM) inference engine responsible for executing the CNN model on-device; and A web-based visualization interface rendered using embedded HTML, CSS, and JavaScript assets stored in program memory (PROGMEM).

Thermal images uploaded via the web interface are initially buffered in the external PSRAM of the ESP32-S3 microcontroller. Each image subsequently undergoes normalization based on the standard score transformation, defined as:

$$x_{\text{norm}} = \frac{x - \mu}{\sigma} \tag{1}$$

where x denotes the raw pixel intensity, μ is the dataset mean, and σ is the standard deviation. This normalization ensures numerical stability and consistency across varying thermal image distributions. In scenarios where post-training quantization is applied, the normalized values are further mapped to discrete integer levels using uniform affine quantization:

$$x_{\text{quant}} = \text{round}\left(\frac{x_{\text{norm}}}{\text{scale}}\right) + \text{zero\_point} \tag{2}$$

where, `scale` and *zero_point* are quantization parameters derived during the model calibration phase, enabling efficient integer-only inference while preserving the representational fidelity of the original floating-point model. Inference is carried out by the TensorFlow Lite Micro (TFLM) interpreter, which operates within a statically allocated memory region known as the tensor arena. The memory requirement for this buffer can be approximated as:

$$\text{TENSOR\_ARENA\_SIZE} = (\text{I/O Size} \times \text{sizeof(dtype)}) + \text{Overhead} \tag{3}$$

This fixed-size allocation guarantees deterministic memory usage during inference, a critical requirement for real-time embedded systems operating under stringent resource constraints.

A fixed memory allocation of 10,KB is reserved to accommodate intermediate tensor buffers and operator-specific metadata required during inference. Depending on the model's memory footprint and numerical precision (e.g., `float32` versus `INT8`), the system dynamically selects either internal SRAM or external PSRAM for tensor arena allocation to optimize memory utilization and computational efficiency. Classification outputs are mapped to fault categories, such as overheating, imbalance) and displayed via the integrated web interface. The modular architecture supports runtime model switching across input resolutions and quantization formats, enabling flexible in-field benchmarking of accuracy, memory usage, and inference latency.

3 DL Model Design and Training

The careful selection of design of neural network architectures are critical for the deployment of AI-based thermal imaging (TI) systems in resource-constrained environments, especially in safety-critical applications such as induction motor fault detection. This study focuses on the development and evaluation of lightweight CNN models optimized for real-world deployment constraints, including inference latency, limited memory availability, energy efficiency, and the accuracy requirements necessary for reliable predictive maintenance.

To explore the trade-offs between model complexity, input resolution, and embedded performance, two distinct CNN architectures were developed. Each architecture was trained and evaluated using three input image resolutions, denoted as configurations (A), (B), and (C). These configurations correspond to different spatial dimensions of the thermal input images, enabling a systematic analysis of the impact of input size on model accuracy, inference latency, and memory consumption.

The architectural structures and corresponding input configurations are illustrated in Fig. 2. A detailed description and design rationale for each architecture are presented in the following sections.

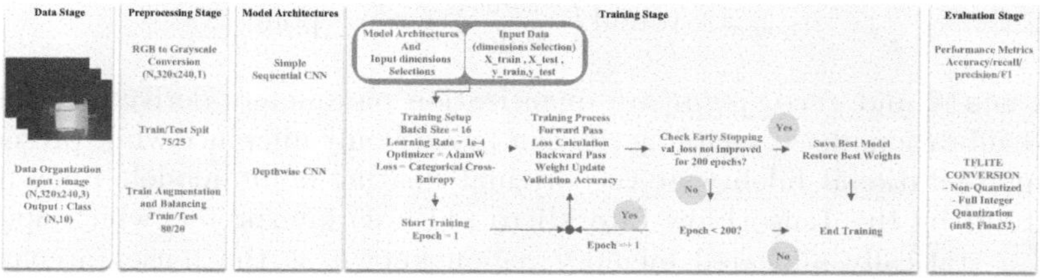

Fig. 2. System-level overview and fault classification pipeline.

3.1 Dataset Preparation and Processing Pipeline

The thermal image dataset utilized in this study consists of infrared (IR) images of electric motors captured under various fault conditions [17]. The images are stored in BMP format and systematically organized into class-specific folders within the IR-Motor-bmp directory. To prepare the dataset for CNN-based training, a structured preprocessing pipeline was applied, comprising grayscale conversion, normalization, data augmentation, and class-balanced dataset partitioning. The dataset was originally acquired using a Dali-Tech T4/T8 infrared thermal camera at an ambient temperature of approximately 23°C. Image acquisition was conducted by the Electrical Machines Laboratory at Babol Noshirvani University of Technology. Specifically, the dataset encompasses thermal images corresponding to the following fault scenarios: 1) Eight distinct stator short-circuit conditions; 2) Rotor mechanical jamming (stuck rotor), and Cooling fan malfunction. In the file naming convention, `num%-stator` denotes the percentage of short-circuit severity per stator phase, whereas `num-phase` indicates the number of phases affected by the fault (Fig. 3).

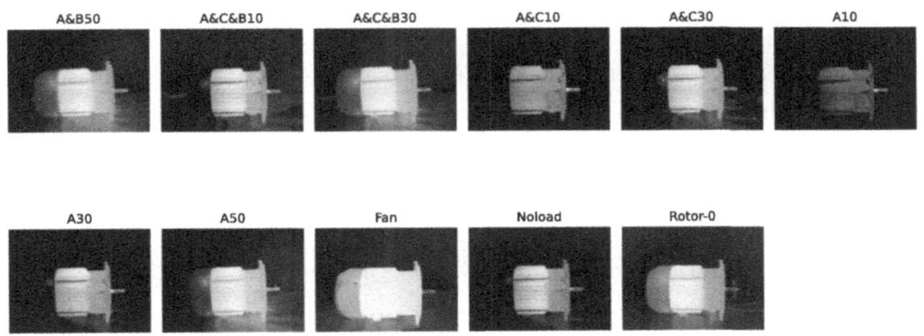

Fig. 3. Representative thermal images for each fault condition in the induction motor dataset.

Table 1 summarizes the fault categories along with the corresponding number of thermal images per class. The preprocessing phase begins by parsing the dataset directory structure to enumerate all available fault classes. Image dimensions are extracted from a representative sample and used to uniformly resize all images, ensuring consistency across the dataset. Since thermal data is primarily conveyed through intensity variations, RGB images are converted to grayscale using a weighted transformation:

$$G_{\text{thermal}}(x, y) = 0.5 \cdot R(x, y) + 0.3 \cdot G(x, y) + 0.2 \cdot B(x, y) \tag{4}$$

This formulation emphasizes the red channel, which often carries the most salient thermal features in infrared imagery.

Table 1. Image distribution for motor fault conditions. `num%-stator`: short-circuit severity; `num-phase`: affected phases.

Healthy	Cooling	Rotor	50%-stator		30%-stator			10%-stator		
			2-Phase	1-Phase	3-Phase	2-Phase	1-Phase	3-Phase	2-Phase	1-Phase
25	28	30	38	35	42	38	37	31	31	34
Total of 369 images										

To establish a consistent and balanced dataset for model training, the preprocessing pipeline was organized as follows: The original dataset was first partitioned into training and testing subsets using a 75:25 ratio. Stratified sampling was employed to maintain comparable class distributions across both subsets:

$$\begin{aligned} X_{\text{train}}, X_{\text{test}}, y_{\text{train}}, y_{\text{test}} = \text{split}(&X, y, \text{test\_size} = 0.25, \\ &\text{random\_state} = 42, \text{stratify} = y) \end{aligned} \tag{5}$$

To address class imbalance and expand the training set, data augmentation was selectively applied only to the training subset. This procedure ensured that each fault category in the training data contained the required number of samples while maintaining proportionality with the testing subset. The augmentation process was carried out by applying sequential geometric transformations on the normalized grayscale images, including random rotations within the range [–20°, 20°], width and height shifts of up to ±10%, as well as horizontal and vertical flipping. Formally:

$$I_{aug} = T_{rot}(T_{shift}(T_{flip}(I_{normalized}))) \tag{6}$$

The augmentation process was repeated until the training subset reached the target ratio of 80% / 20% relative to the testing subset, thereby yielding a balanced distribution across all fault categories.

After augmentation, the dataset comprised evenly distributed samples among all fault categories, with proportions adjusted to reflect a 80% training and 20% testing split. All images were standardized by converting to single-channel grayscale, normalized to the pixel intensity range [0, 255], and resized to uniform dimensions ($W_{final} \times H_{final}$).

4 Model Architecture and Training Strategy

This study presents two convolutional neural network (CNN) architectures specifically optimized for deployment on the ESP32-S3 microcontroller unit (μ C). Both models are designed to facilitate real-time classification of thermal images while addressing the constraints typical of embedded platforms, including limited memory availability, low-power operation, and the need for fast inference.

The first model, referred to as the **SSM_Model** (Sequential Shallow Model), adopts a compact and sequential design optimized for resource-efficient execution. It consists of a series of convolutional layers with progressively increasing filter sizes, each followed by max-pooling to reduce spatial dimensions. The architecture concludes with a Global Average Pooling layer and a dense output layer with softmax activation to perform multi-class classification. This configuration balances simplicity and performance, making it suitable for low-latency deployment in edge environments.

- **Input:** Grayscale thermal images of size $[H \times W \times 1]$
- **Convolutional Blocks:**
  - 5×5 Conv2D with 2 filters, followed by 2×2 MaxPooling
  - 5×5 Conv2D with 16 filters, followed by 2×2 MaxPooling
  - 3×3 Conv2D with 32 filters, followed by 2×2 MaxPooling
  - 3×3 Conv2D with 64 filters, followed by 2×2 MaxPooling
- **Pooling Layer:** GlobalAveragePooling2D
- **Output Layer:** Fully connected (Dense) layer with softmax activation

The second architecture, designated as the **DSM_Model** (Deep Structured Model), extends the classification head of the baseline SSM_Model to enhance its capacity for generalization. While retaining the same convolutional backbone, the DSM_Model introduces a fully connected head composed of additional regularization and nonlinearity layers. Specifically, it includes a flattening operation, dropout layers to mitigate overfitting, and an intermediate dense layer with ReLU activation prior to the softmax output layer. This enhanced head structure increases the model's expressive power while maintaining a compact footprint suitable for embedded deployment.

- **Dense Head:**
  - Flatten Layer
  - Dropout Layer (rate = 0.1)
  - Dense Layer (64 neurons, ReLU activation)
  - Dropout Layer (rate = 0.1)
  - Final Dense Layer with Softmax Activation

To evaluate deployment feasibility, both models are benchmarked using three different input dimensions (labeled A, B, and C), with comparisons made across key performance metrics: memory consumption, inference latency, and classification accuracy under real-time constraints.

The training configuration, including optimizer selection, loss function, label encoding scheme, and evaluation metrics is described in detail in Sect. 5. These settings are uniformly applied across all models to ensure consistency and fairness in performance comparison and deployment readiness.

5 Model Training and Evaluation

After defining and compiling the model architectures, training is carried out using mini-batch gradient descent, supported by tailored input preprocessing, label encoding, and real-time performance monitoring to ensure optimal convergence and generalization.

5.1 Training Procedure and Input Scaling

To evaluate the impact of input resolution on model performance, thermal images are rescaled using gain factors $G \in \{0.5, 0.375, 0.25\}$. For an original image of dimensions $H \times W$, the scaled resolution is computed as:

$$H_G = H \cdot G, \quad W_G = W \cdot G \tag{7}$$

All grayscale images are normalized to the range $[0, 1]$, and class labels are converted to one-hot vectors via:

$$y_i = \texttt{to\_categorical}(c_i, C) \tag{8}$$

where $c_i \in \{0, 1, \ldots, C-1\}$ denotes the class index and $C = 10$ is the total number of classes.

Gradient Descent and Optimization

Training is conducted with a batch size of $B = 32$ over 800 epochs. Each iteration consists of the following steps:

- **Forward Pass:** Compute predictions $\hat{y} = f(x; \theta)$
- **Loss Computation:** Employing label smoothing, the loss is defined as:

$$\mathcal{L} = -\sum_{i=1}^{C} [(1 - \epsilon) \cdot y_i + \epsilon / C] \cdot \log(p_i) \tag{9}$$

- **Gradient Calculation:**

$$\partial \mathcal{L} / \partial \theta \tag{10}$$

- **Parameter Update (AdamW):**

$$\theta_{t+1} = \theta_t - \eta \cdot \left(\hat{m}_t / \sqrt{\hat{v}_t} + \epsilon + \lambda \cdot \theta_t \right) \tag{11}$$

To ensure stable convergence, training employs a learning rate of $\eta = 10^{-3}$, weight decay $\lambda = 10^{-4}$, and gradient clipping with a maximum norm of 1.0.

5.2 Validation and Checkpointing

At the end of each training epoch, validation loss and accuracy are evaluated. To mitigate overfitting and promote model generalizability, the weights corresponding to the highest validation accuracy are preserved using model checkpointing. This mechanism also enhances reproducibility across experimental runs.

5.3 Learning Curve Analysis

Figure 4 depicts the joint evolution of accuracy and loss for both training and validation across the six evaluated models (SSM-A, SSM-B, SSM-C, DSM-A, DSM-B, and DSM-C). In general, DSM-based architectures exhibited faster convergence with smoother training trajectories, whereas SSM variants revealed more gradual improvements. Among the SSM family, SSM-B maintained a favorable compromise, sustaining relatively high validation accuracy with controlled loss, thereby reducing overfitting risks. Conversely, DSM-C achieved elevated training accuracy but experienced notable instability in validation loss, highlighting the susceptibility of deeper models to overfitting under aggressive downscaling (Gain C). Moreover, the DSM architectures demonstrated heightened sensitivity to input resolution reduction, emphasizing the trade-off between model depth and generalization capacity in resource-constrained embedded contexts.

5.4 Final Evaluation Results

Table 2 summarizes the performance metrics of SSM and DSM models evaluated under three input resolutions (A, B, and C). Across all metrics, both architectures demonstrate high performance, with consistently strong precision, recall, and F1-scores. Notably, the SSM variants maintain a stable balance between training and validation, achieving validation accuracies above 97% across all resolutions. In particular, SSM-B and SSM-C reach near-perfect F1-scores (0.986), coupled with validation accuracies of 98.65%, indicating robust generalization even when input dimensions are reduced. SSM-A also performs competitively, with slightly lower validation accuracy (97.30%) but still confirming the effectiveness of lightweight configurations.

On the other hand, DSM models attain perfect training performance (100% accuracy across all resolutions), yet exhibit marginally lower validation accuracies compared to SSM, dropping to 97.30% in the case of DSM-C. This suggests that the deeper DSM architecture may be more prone to subtle overfitting effects under aggressive downscaling. Overall, these findings highlight SSM-B and SSM-C as the most reliable trade-offs between accuracy and efficiency, reinforcing their suitability for real-time deployment on resource-constrained platforms such as the ESP32-S3.

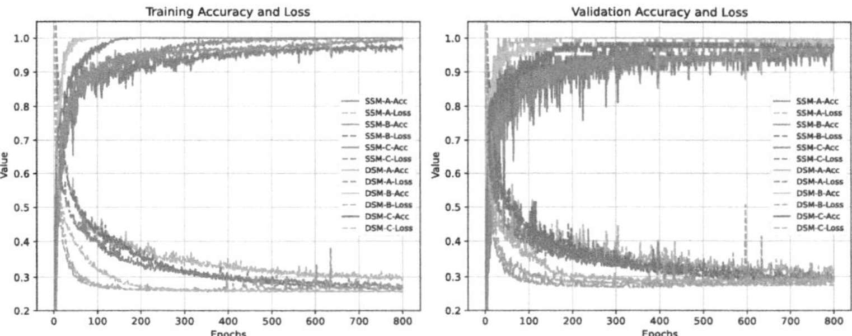

Fig. 4. Training and validation learning curves for all CNN model variants evaluated in this study.

Table 2. Performance Metrics of SSM and DSM Models at Different Input Resolutions

Model	Res	F1	Recall	Prec.	Acc.	Val Acc.
SSM	A	0.943050	0.945946	0.947072	96.694213	97.297299
SSM	B	0.986359	0.986486	0.987526	100.000000	98.648649
SSM	C	0.986548	0.986486	0.987838	99.449039	98.648649
DSM	A	1.000000	1.000000	1.000000	100.000000	98.648649
DSM	B	0.986548	0.986486	0.987838	100.000000	98.648649
DSM	C	0.986548	0.986486	0.987838	100.000000	97.297299

6 TFLite Conversion and Quantized Deployment

To support deployment on resource-constrained microcontrollers such as the ESP32-S3, the trained CNN models were converted into TensorFlow Lite (TFLite) format. Two conversion strategies were employed: full-precision (`float32`) and full-integer quantization (`int8`). This conversion significantly reduces model memory footprint and computational overhead, while preserving acceptable inference accuracy, thereby facilitating real-time embedded fault classification.

6.1 Conversion Process

The TensorFlow Lite (TFLite) conversion process incorporates key optimization strategies such as graph pruning, operator fusion, and quantization to enable efficient deployment on resource-constrained hardware. Two conversion pathways are employed: full-precision (float32) and full-integer quantization (int8). In the float32 variant, models preserve their original numerical precision throughout the computational graph, ensuring negligible accuracy loss but resulting in relatively higher memory and computational requirements.

For the quantized variant, post-training quantization is applied using representative datasets to derive optimal scale (s) and zero-point (z) parameters for each layer. This enables all model weights and activations to be transformed from 32-bit floating-point to 8-bit integers according to:

$$q = \text{round}\left(\frac{r}{s}\right) + z, \qquad \text{and its inverse:} \quad r = s \cdot (q - z) \tag{12}$$

6.2 Post-conversion Evaluation

To evaluate the effects of TFLite conversion, both float32 and int8 quantized models were assessed using the TFLite Python interpreter in a desktop environment. Key performance indicators including accuracy, precision, recall, and F1-score, were compared against their original TensorFlow counterparts using the relative performance ratio:

$$\text{XRatio} = \text{X}_{\text{TFLite}}/\text{X}_{\text{Original}} \tag{13}$$

Table 3 presents a comparative analysis of model performance before and after quantization. Overall, the SSM models exhibit strong robustness, maintaining accuracy ratios very close to 1.0 while achieving substantial memory savings. Notably, SSM_A_int8 preserves nearly the same accuracy as its floating-point counterpart (accuracy ratio 0.975) while reducing the memory footprint by more than 50% (from 0.047 MB to 0.019 MB). Similarly, SSM_B_int8 and SSM_C_int8 sustain high validation accuracy and F1-scores with negligible degradation, confirming the suitability of the SSM family for deployment under strict memory constraints. In contrast, DSM models show higher sensitivity to quantization, especially DSM_C_int8, which experienced a drop in F1-score

Table 3. Evaluation of Quantized and Non-Quantized CNN Models: Performance and Memory Efficiency

Model	Model Name	Gain	Model Type	TFLite Accuracy	Accuracy Ratio	Precision Ratio	Recall Ratio	F1-Score Ratio	Model Size (MB)
SSM_A_f32	SSM	A	f32	0.939697	0.997680	1.000000	1.000000	1.000000	0.047379
SSM_A_int8	SSM	A	int8	0.914601	0.975840	0.992951	0.966863	0.963378	0.019707
SSM_B_f32	SSM	B	f32	0.977245	0.997211	1.000000	1.000000	1.000000	0.047436
SSM_B_int8	SSM	B	int8	0.967245	0.997232	1.000000	1.000000	1.000000	0.019737
SSM_C_f32	SSM	C	f32	0.975207	0.975589	0.990485	0.988566	0.988347	0.047443
SSM_C_int8	SSM	C	int8	0.961736	0.997230	1.000000	1.000000	1.000000	0.019745
DSM_A_f32	DSM	A	f32	0.989486	1.000000	1.000000	1.000000	1.000000	0.256824
DSM_A_int8	DSM	A	int8	0.968137	0.997854	1.000000	1.000000	1.000000	0.072075
DSM_B_f32	DSM	B	f32	0.986486	1.000000	1.000000	1.000000	1.000000	0.122562
DSM_B_int8	DSM	B	int8	0.962572	0.992951	1.000000	1.000000	1.000000	0.038528
DSM_C_f32	DSM	C	f32	0.983168	0.995582	0.979555	0.966226	0.966535	0.074223
DSM_C_int8	DSM	C	int8	0.969394	0.989658	0.972171	0.952263	0.952551	0.026443

ratio to 0.952 and accuracy ratio below 0.99. Although DSM_A and DSM_B preserved relatively stable performance, their memory footprints remained significantly larger than those of SSM counterparts, limiting their efficiency in embedded scenarios. Collectively, these findings highlight that SSM quantized models, particularly SSM_A_int8 and SSM_B_int8, offer the most favorable balance between accuracy retention and memory reduction, reinforcing their advantage for real-time deployment on microcontrollers such as the ESP32-S3.

6.3 Header File Generation for Deployment

To enable seamless integration of the trained models into the ESP32-S3 firmware, each *tflite* file was converted into a C-compatible header file using the **xxd** utility. This conversion process transforms the binary model into a byte array format that can be embedded directly within the firmware source code.

The generated header files contain three essential components: a byte array representation of the TFLite model, definitions specifying the model's size and memory footprint, and declarations for accessing or initializing the embedded model. These elements collectively allow the model to be statically linked during compilation.

This method eliminates the need for external storage or dynamic model loading at runtime, making it particularly suited for real-time, offline AI inference on resource-constrained embedded systems. As a result, the approach supports fully self-contained, low-latency deployments for edge intelligence applications.

7 On-Device Evaluation and Timing Analysis

To validate real-world applicability, the optimized TFLite models were deployed on the ESP32-S3 microcontroller for on-device inference. This evaluation focused on both functional performance and system-level efficiency, including inference latency, memory utilization, and classification accuracy under operational conditions. Figure 5 illustrates the embedded inference pipeline, detailing key stages from input signal acquisition to final fault classification.

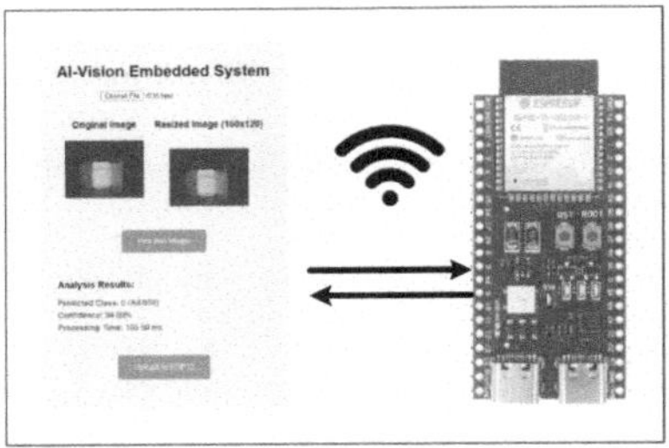

Fig. 5. System-level workflow of real-time inference on the ESP32-S3 MCU.

Table 4. On-device inference results for TFLite models on ESP32-S3 MCU.

Model	Name	Input	Quant.	Size (KB)	Time (ms)
SSM_A_f32	SSM	A	f32	047.0	875.00
SSM_A_int8	SSM	A	int8	020.0	105.00
SSM_B_f32	SSM	B	f32	047.0	451.45
SSM_B_int8	SSM	B	int8	020.0	055.40
SSM_C_f32	SSM	C	f32	047.0	178.43
SSM_C_int8	SSM	C	int8	020.0	024.19
DSM_A_f32	DSM	A	f32	237.0	887.53
DSM_A_int8	DSM	A	int8	068.7	104.40
DSM_B_f32	DSM	B	f32	116.0	452.81
DSM_B_int8	DSM	B	int8	037.5	055.64
DSM_C_f32	DSM	C	f32	073.0	167.10
DSM_C_int8	DSM	C	int8	026.3	023.80

Inference latency was measured using the ESP32's internal microsecond timer, encapsulating the invoke() function of the TFLite Micro interpreter. The results, summarized in Table 4, include input resolution (A, B, or C), quantization format (float32 or int8), model size (in kilobytes), and average inference time (in milliseconds).

The results consistently highlight the efficiency of quantized int8 models, which not only reduced memory size by more than 50% but also delivered a drastic reduction in inference latency. For instance, SSM-B's execution time decreased from 451.45 ms (f32) to 55.40 ms (int8), while DSM-A improved from 887.53 ms to 104.40 ms. The smallest configuration, SSM-C-int8, required only 24.19 ms with a memory footprint of just 20 KB, making it highly suitable for real-time applications.

Input resolution also influenced performance: lower resolutions (Type C) consistently yielded faster inference. DSM-C-int8, for example, reached 23.80 ms compared to 104.40 ms for DSM-A-int8. Model complexity further impacted results, as DSM variants, due to their deeper architectures, required larger memory footprints and longer runtimes even after quantization (e.g., DSM-A-int8 at 68.7 KB versus SSM-A-int8 at 20 KB).

Overall, float32 models imposed significant computational and memory burdens, with limited accuracy gains, while quantized int8 models offered the best trade-off between efficiency and performance. These findings confirm that SSM-C-int8 and SSM-B-int8 provide the most favorable configurations for real-time deployment on resource-constrained platforms such as the ESP32-S3.

8 Conclusion

This study presented an embedded AI framework for real-time fault classification in induction motors using thermal imagery. Using lightweight CNNs, the system performs complete on-device processing on the ESP32-S3 WROOM-1 microcontroller, supporting float32 and int8 TFLite formats, and offering a web interface for image upload and result visualization without cloud reliance. Two CNN architectures, SSM and DSM, were trained on a public thermal dataset covering various fault conditions. Results showed that SSM achieved consistently higher validation accuracy and stronger generalization than DSM, with SSM-B and SSM-C providing the best balance between efficiency and accuracy. Quantization to int8 reduced model size by over 60% and accelerated inference latency up to tenfold, with SSM-C-int8 reaching 24.19 ms and 20 KB memory use, confirming its suitability for embedded real-time deployment. Conversely, DSM models were more sensitive to quantization and reduced input resolution, with DSM-C exhibiting unstable validation behavior. These findings confirm the feasibility of embedded AI for real-time thermal fault diagnosis on low-power microcontrollers and highlight the need to balance input resolution with architectural complexity to ensure efficiency without compromising accuracy. The framework establishes a scalable and energy-efficient solution for predictive maintenance and monitoring. Nonetheless, limitations persist: reliance on controlled datasets may not fully capture real-world variability, and quantized models may degrade under very low-resolution inputs or rare faults. Future work will extend datasets, explore advanced augmentation, integrate temporal-spatial cues, and exploit NPU-enabled hardware to support deeper and more complex edge models.

References

1. Ahmed, T.: Design patterns for edge AI: from partial to full inference. IEEE Embed. Syst. Lett. (2022)
2. Bagavathiappan, S., Lahiri, B.B., Saravanan, T., Philip, J., Jayakumar, T.: Infrared thermography for condition monitoring–a review. Infrared Phys. Technol. **60**, 35–55 (2013)
3. Banbury, C.e.a.: Efficient convolutional neural networks for microcontroller-based systems. NeurIPS TinyML Workshop (2021)
4. Biglari, N., et al.: Embedded deep learning for real-time fault detection in industrial machines. Sensors **23**(4), 2131 (2023)
5. Binaee, A., Chen, L.: Ai at the edge: power-efficient strategies for real-time inference in thermal monitoring systems. IEEE Trans. Industr. Inf. **19**(5), 5143–5151 (2023)
6. Chen, J., Ran, X.: Edge AI: machine learning at the edge of the network. IEEE Internet Things J. **7**(8), 7457–7471 (2020)
7. David, R., et al.: Tensorflow lite micro: Embedded machine learning on tinyml systems. arXiv preprint arXiv:2010.08678, (2020). https://arxiv.org/abs/2010.08678
8. David, R.: Tensorflow lite for microcontrollers. arXiv preprint arXiv:2010.08678 (2020)
9. Esmaeilzadeh, R.e.a.: Anomaly detection for predictive maintenance using embedded neural networks. Sensors **21**(12) (2021)
10. Garcia, R., et al.: An embedded solution for pv fault diagnosis using thermographic images. IEEE Trans. Industr. Inf. **18**(9), 6042–6053 (2022)
11. Gholami, A., Kim, S., Dong, Z., Yao, Z., Mahoney, M.W., Keutzer, K.: A survey of quantization methods for efficient neural network inference. In: Low-Power Computer Vision, pp. 291–326 (2022)
12. Howard, A.G., et al.: Mobilenets: efficient convolutional neural networks for mobile vision applications. In: arXiv preprint arXiv:1704.04861 (2017)

13. Jacob, B., et al.: Quantization and training of neural networks for efficient integer-arithmetic-only inference. In: Proceedings of the IEEE Conference on Computer Vision and Pattern Recognition (CVPR), pp. 2704–2713 (2018)
14. Kaplanis, S., Panayiotou, C.: Thermal imaging in predictive maintenance: a review. Measurement **135**, 250–261 (2019)
15. Li, M., et al.: Lightconvnet: Lightweight and efficient cnn architecture for edge AI. In: IEEE ICASSP (2022)
16. Mukherjee, R., Singh, A.: Embedded ai for real-time industrial fault detection: a case study on thermal imagery. J. Real-Time Image Proc. **20**, 103–117 (2023)
17. Najafi, M., Baleghi, Y., Mirimani, S.M.: Thermal image of equipment (induction motor) + 40 ground truths added (2023). https://doi.org/10.17632/m4sbt8hbvk.3, https://data.mendeley.com/datasets/m4sbt8hbvk/3
18. Sharma, A., Chhabra, M.: Thermal image processing and analysis: a review. Infrared Phys. Technol. **104**, 103137 (2020)
19. Warden, P., Situnayake, D.: Tinyml: Machine Learning on Ultra-low-Power Hardware. Inc, O'Reilly Media (2020)
20. Zhang, Y., Wang, Y., Liu, Y.: Real-time object detection on edge devices: a review. IEEE Access **9**, 119538–119556 (2021)

Multi-views Knee MRI Classification Using Vision Transformers for Automated Meniscal Tear Detection

Amel Sbita[1](✉), Ekram Chameseddine[2], Lotfi Tlig[2], Afef Kacem Echi[1], and Mounir Sayadi[2]

[1] ENSIT, Labo LaTICE, University of Tunis, Av Taha Hussein, 1008 Tunis, Tunisia
amel.sbita@isimg.tn, afef.kacem@ensit.u-tunis.tn
[2] ENSIT, Labo SIME, University of Tunis, Av Taha Hussein, 1008 Tunis, Tunisia
ekram.chamseddine@enetcom.u-sfax.tn, lotfi.tlig@isimg.tn, mounir.sayadi@ensit.u-tunis.tn

Abstract. The knee is one of the most complex and heavily utilized joints in the human body. Within the knee, the menisci are crescent-shaped fibrocartilaginous structures that act as shock absorbers, stabilize the joint, and distribute load during movement. Meniscal tears are common injuries, particularly among athletes and older individuals, and their accurate diagnosis is essential for preventing long-term joint damage and early-onset osteoarthritis. Magnetic resonance imaging is the preferred non-invasive technique for evaluating meniscal integrity, but interpreting these images remains challenging due to subtle pathological variations and inter-reader variability. In recent years, Deep Learning has shown promise in automating image interpretation, with convolutional neural networks being the most widely used. In this paper, we developed a Vision Transformer model for the classification of meniscal tears using 3D scans. To preserve the spatial integrity of the knee anatomy, the model incorporates the sagittal, coronal, and axial planes from the publicly available MRNet data set. The transformer-based model is trained to differentiate between healthy and torn menisci. Our approach achieved a classification accuracy of 92.89% on the test set, surpassing several state-of-the-art methods reported in the literature. These findings demonstrate that our proposed model can successfully capture long-range spatial dependencies across multiple imaging planes, leading to more accurate and reliable diagnostic performance in meniscal tear detection.

Keywords: Meniscal Tears · Vision Transformer · Multi-head-attention · Image Classification

1 Introduction

The knee is a complex joint essential for various everyday movements such as walking, running, and jumping [1]. Central to its function are the menisci, two crescent-shaped fibrocartilaginous tissues (medial and lateral) that help absorb

T. Ensari et al. (Eds.): ISPR 2025, CCIS 2859, pp. 153–165, 2026.
https://doi.org/10.1007/978-3-032-21585-7_11

shocks, stabilize the joint, and evenly distribute loads across the knee surfaces [2]. Meniscal tears rank among the most common knee injuries, with an incidence estimated at about 60 to 70 cases per 100,000 people each year. These injuries frequently affect athletes and older adults, resulting from either trauma or degenerative processes [1,2]. Without proper diagnosis and treatment, such tears may lead to chronic pain, limited mobility, and premature osteoarthritis.

Prompt and accurate detection of meniscal injuries is vital for effective treatment and to prevent progression to long-term joint damage [1]. However, clinical examinations alone are often insufficient due to symptom overlap with other knee conditions [3]. Consequently, there is a growing reliance on advanced imaging techniques and computer-aided diagnosis (CAD) systems to assist clinicians in identifying difficult or subtle meniscal lesions [4,5].

Magnetic Resonance Imaging (MRI) remains the preferred non-invasive method for evaluating meniscal integrity, offering high-resolution images across multiple planes [3,6]. The meniscus is typically assessed in sagittal, coronal, and axial views; the sagittal plane is particularly useful for examining the anterior and posterior horns, the coronal plane for the meniscal body and root attachments, while the axial view, though less commonly used, can be beneficial in complex assessments [3,7]. Accurate interpretation often requires understanding the three-dimensional anatomy as visualized across these planes [4].

Despite MRI's effectiveness, interpretation can be difficult due to subtle tear appearances, anatomical variations, and variability among radiologists [4,8]. This has driven efforts to develop robust CAD tools to support radiologists, minimize diagnostic errors, and improve efficiency [4].

Recent breakthroughs in Artificial Intelligence (AI), particularly in Machine Learning (ML) and Deep Learning (DL), have greatly impacted medical imaging analysis [9]. Earlier CAD methods depended on handcrafted features and conventional ML models, which were limited in their ability to represent complex spatial relationships in MRI data [10]. The introduction of convolutional neural networks (CNNs) enabled automatic hierarchical feature extraction from images, achieving expert-level results in detecting meniscal tears [4,11]. However, CNNs mainly focus on local spatial information and may have difficulty capturing the global 3D context across multiple MRI planes [12].

Inspired by advances in natural language processing, Vision Transformers (ViTs) have emerged as a powerful alternative for image analysis [12]. Their self-attention mechanisms allow them to model long-range dependencies and capture contextual information throughout entire images or volumes, making them particularly suited to tasks like meniscal tear detection that benefit from integrating sagittal, coronal, and axial plane data [12,13].

In this paper, we investigate the use of attention-driven Vision Transformers for meniscal tear classification, demonstrating their potential advantages compared to previous methods and their promise in enhancing knee MRI interpretation and clinical decision-making. The key contributions of this study are the following.

- Preserve the spatial and anatomical integrity of the knee joint by leveraging the self-attention mechanism of Vision Transformers, which effectively captures long-range dependencies across sagittal, coronal, and axial MRI views
- An efficient training strategy that fine-tunes the classification head. This significantly reduces training time and computational overhead while maintaining strong predictive performance.
- Comprehensive experiments on the MRNet public dataset and achieve a test accuracy of 92.89%, outperforming several existing state-of-the-art methods. Our findings confirm the efficacy of transformer-based models in enhancing diagnostic performance for knee MRI analysis.

The remainder of this paper is organized as follows. Section 2 presents the related works. Section 3 outlines the proposed approach. Section 4 displays the experimental setup. Section 5 discusses the results and their implications. Finally, Sect. 6 concludes the study and gives some prospects.

2 Related Works

In recent years, several deep learning-based methods have been proposed for automatic detection and classification of meniscal tears in knee MRI, leveraging the MRNet dataset and its variants. Table 1 summarizes the key characteristics, merits, and limitations of these approaches.

The work by Bien et al. [4] introduced the MRNet framework, based on a custom 2D CNN applied to the MRNet dataset. This study was significant as it established the first large-scale benchmark for knee MRI interpretation using deep learning, showing promising AUC results (84.6% internal, 82.3% external) and sensitivity close to radiologist performance. However, the model suffered from lower specificity and did not integrate attention mechanisms, which are now known to enhance feature focus and interoperability.

Chen et al. [11] improved upon this approach by using an ensemble of 2D CNNs, achieving a higher AUC of 90.0% for meniscus tear detection. Ensembling boosted performance and reduced overfitting, demonstrating the benefit of model diversity. Nonetheless, the increased complexity and computational demands of deploying multiple models make this solution less suitable for real-time clinical settings.

Rizk et al. [8] emphasized the importance of external validation by testing their CNN on both MRNet and external datasets. Their model achieved an AUC of 83.0% on external data, which increased to 89.0% after fine-tuning, underlining the necessity of domain adaptation. Still, the method faced challenges due to dataset imbalance and the need for retraining when applied to different clinical settings. In a more recent work, Ying et al. [14] explored knowledge distillation between CNN models to enhance efficiency and performance. Their method yielded satisfactory AUC values (83.4% for medial, 74.6% for lateral meniscus tears) with improved F1-scores, especially under limited data regimes. However, their approach was constrained by the size of the distillation dataset.

Finally, Gungor et al. [15] proposed a hybrid framework combining YOLOv8 for localization and EfficientNetV2 for classification, reporting very high AUC scores (97.0% sagittal, 98.0% coronal). This model demonstrated excellent accuracy and fast inference, paving the way for real-time and structured reporting. Despite these achievements, the study was conducted on a relatively small dataset, raising questions about its generalizability to broader clinical populations.

In summary, the literature reflects a clear evolution from conventional CNN-based models toward more efficient, externally validated, and attention-augmented systems. However, key challenges persist, particularly in terms of domain generalization, dataset imbalance, and the limited integration of multi-view representations and denoising strategies aspects which are critical for robust diagnosis in clinical settings. These observations motivate the need for more comprehensive and generalizable models capable of handling noisy MRI data and leveraging complementary anatomical views.

Table 1. Summary of the related works.

Ref.	Model	Dataset	Results (%)	Merits	Limitations
[4] 2019	Custom 2D CNN (MRNet)	MRNet (Stanford)	AUC: 84.6 (internal), 82.3 (external); Sensitivity: 71; Specificity: 82	First large-scale, open benchmark; performance comparable to radiologists	Lower specificity than radiologists; lacks attention mechanisms
[11] 2020	Ensemble of 2D CNNs	MRNet	AUC: 90.0 (meniscus)	Improved performance via ensembling; robust to overfitting	Increased computational cost; complex deployment
[8] 2021	Deep CNN with external validation	MRNet + external	AUC: 83.0 (external), improved to 89.0 with fine-tuning	External validation; performance close to radiologists	Requires fine-tuning for domain shifts; Imbalanced data
[14] 2023	Knowledge Distillation (CNNs)	MRNet (subset)	AUC: 83.4 (medial), 74.6 (lateral); Accuracy: 76.4/73.4	Knowledge distillation improves efficiency and F1-score	Small distillation dataset; no transformers used
[15] 2024	YOLOv8 + EfficientNetV2	MRNet	AUC: 97.0 (sagittal), 98.0 (coronal)	High accuracy; fast; supports structured reporting	Limited dataset size; generalizability unproven

3 Proposed Approach

3.1 Overview

The proposed method employs a Vision Transformer architecture for the classification of knee MRI scans to detect meniscal tears. The dataset comprises multi-planar MRI images (sagittal, coronal, and axial views), which undergo a preprocessing stage to standardize and prepare the data. These processed images are then input into the ViT model, where they are divided into patches and linearly projected before being passed through a transformer encoder. The encoded features are subsequently processed by a multi-layer perceptron (MLP) head, consisting of batch normalization, linear transformations, ReLU activations, and dropout layers. The output of the MLP head is fed into a classification block to categorize the input as either healthy "Normal" or Meniscal Tears "Meniscus". During the testing phase, unseen MRI scans follow the same pipeline to predict the corresponding class. Figure 1 illustrates the overall workflow of the proposed approach, which is detailed as follows:

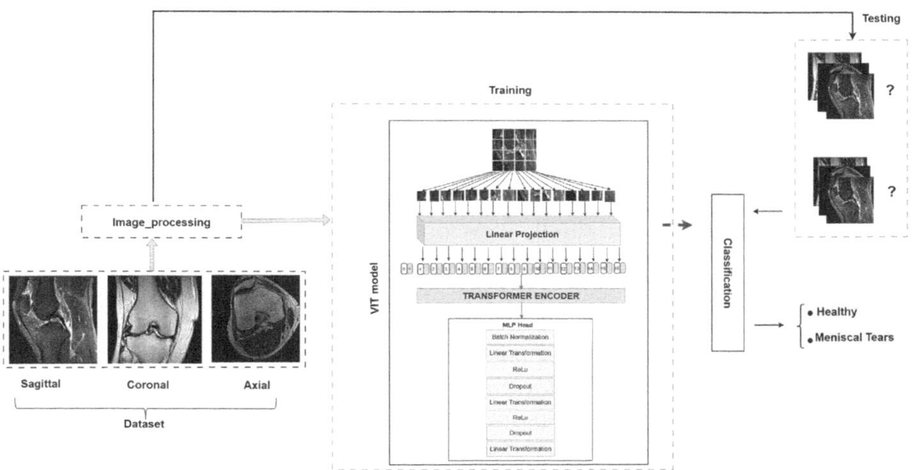

Fig. 1. Overview of the proposed approach.

3.2 Data Description

This study utilizes the MRNet dataset, which comprises 1,370 knee MRI examinations performed at Stanford University Medical Center [16]. The imaging protocol includes multiple standard non-contrast knee MRI sequences acquired on GE scanners with either 1.5-T or 3.0-T magnetic fields. The sequences cover three orthogonal planes: coronal T1-weighted, coronal T2-weighted with fat saturation, sagittal proton density–weighted, sagittal T2-weighted with fat saturation, and axial proton density–weighted with fat saturation. The most frequent clinical indications for the exams were acute or chronic knee pain, injury, trauma, or preoperative evaluation.

Our study focuses on the detection of meniscal tears, which are present in 508 exams (37.1% of the dataset). Labels were manually extracted from clinical radiology reports, ensuring accurate ground truth. All available imaging planes are used to maximize diagnostic information while maintaining a comprehensive approach. Figure 2 shows a sample MRI from MRNet dataset.

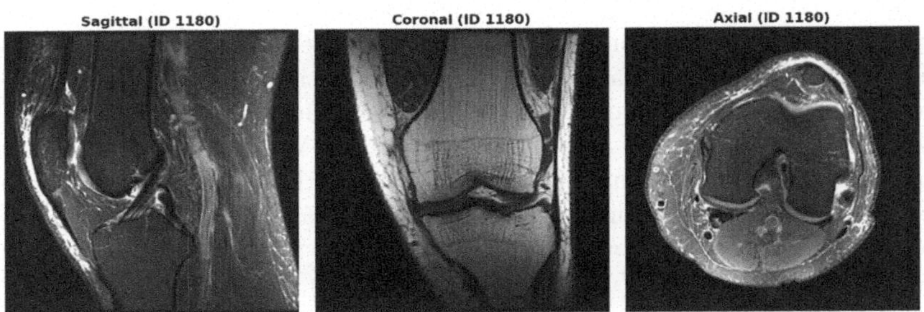

Fig. 2. A sample of MRI from MRNet dataset in different planes.

3.3 Vision Transformer

The Vision Transformer (ViT) is a deep learning (DL) architecture that applies the Transformer model, originally developed for natural language processing, to image data [12]. Unlike conventional convolutional neural networks (CNNs), which process images through local convolutional filters, ViT splits an image into fixed-size patches and treats each patch as a token in a sequence, enabling it to model both local and global patterns.

In this work, we adapt ViT for grayscale knee MRI images by dividing each single-channel input image into non-overlapping patches of size $P \times P$. Each patch is flattened and projected into a vector through a trainable linear layer, forming a sequence of patch embeddings. A learnable class token is prepended to this sequence, and positional embeddings are added to preserve spatial information, as described in Eq. 1:

$$\mathbf{z}_0 = [\mathbf{I}_{\text{class}}; x_1^p\mathbf{E}; \ldots; x_N^p\mathbf{E}] + \mathbf{E}_{\text{pos}} \tag{1}$$

where $\mathbf{I}_{\text{class}}$ is the learnable class token, x_i^p denotes the i-th image patch, $\mathbf{E}$ is the patch embedding projection matrix, and $\mathbf{E}_{\text{pos}}$ represents the positional embeddings.

The resulting sequence $\mathbf{z}_0$ is then passed through a stack of Transformer encoder layers. Each layer consists of a multi-head self-attention (MSA) mechanism followed by a feed-forward multilayer perceptron (MLP), both wrapped with residual connections and layer normalization. This operation is described in Eqs. 2 and 3:

$$\mathbf{z}'_l = \mathrm{MSA}(\mathrm{LN}(\mathbf{z}_{l-1})) + \mathbf{z}_{l-1} \quad (2)$$

$$\mathbf{z}_l = \mathrm{MLP}(\mathrm{LN}(\mathbf{z}'_l)) + \mathbf{z}'_l \quad (3)$$

where LN denotes layer normalization, and MSA and MLP refer to the self-attention and feed-forward components, respectively. The final output corresponding to the class token serves as a global representation of the image and is passed to a classification head for binary prediction (normal vs. meniscus tear). The overall architecture is illustrated in Fig. 3.

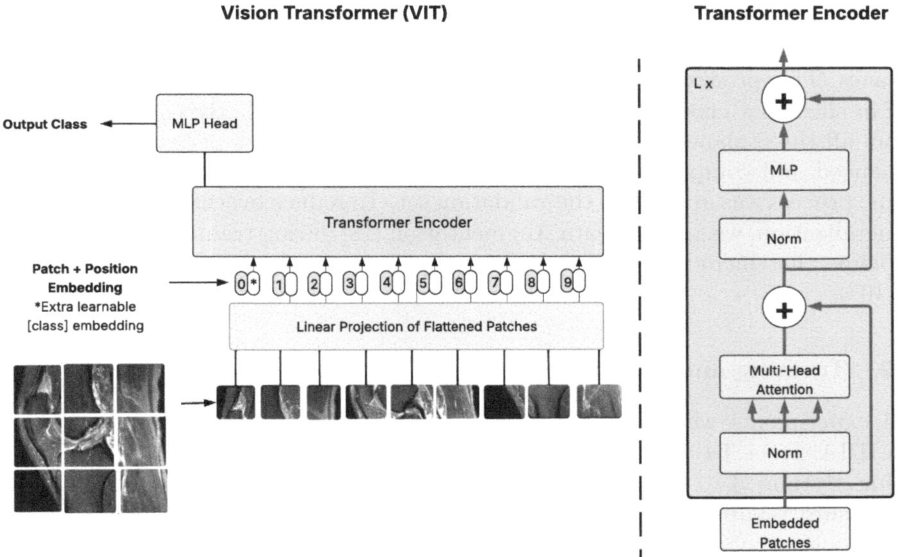

Fig. 3. Vision transformer model adopted for classification of meniscal tears from MRI. MLP: multilayer perceptron. * is the extra learnable patch embedding to be used by the final classification head, The vision transformer encoder with multi-head self-attention

This customized ViT architecture is well-suited for the meniscus classification task. Its self-attention mechanism enables the model to capture complex spatial dependencies across different regions of the grayscale MRI scans, improving its diagnostic performance.

4 Experimental Setup

4.1 Data Preprocessing and Augmentation

Training a ViT from scratch typically requires a large amount of data. To overcome this limitation, we followed a strategy designed to maximize the available information from the MRNet dataset. We extracted 2D slices from the 3D MRI volumes across all three anatomical planes: sagittal, coronal, and axial. Each MRI scan was associated with a binary label indicating the presence or absence of a meniscus tear, as provided in the dataset's accompanying CSV file.

The original training set contained 733 normal and 397 meniscus tear cases, while the validation set included 68 normal and 52 meniscus tear cases, indicating a notable class imbalance. To address this, we balanced the dataset by randomly selecting 397 normal cases for training and 52 for validation, matching the number of meniscus tear cases in each set and ensuring equal representation during model training and evaluation [17].

Since MRI volumes vary in the number of slices, we selected the minimum common depth across all volumes and extracted 17 slices centered around the middle of each scan to ensure consistency and preserve anatomical relevance. These slices were normalized to the range [0, 255] and saved as grayscale JPEG images. This process was applied separately to each imaging plane, resulting in 6,749 slices per class per plane (397 volumes $\times$ 17 slices). By combining slices from all three planes, we obtained a total of 20,247 images per class, forming a balanced and comprehensive dataset for training the classification model. The same process was applied to the validation set. To reduce overfitting and improve generalization, we applied Data Augmentation [18] during training using a simple pipeline that included random horizontal flipping and small random rotations ($\pm 10^{\circ}$).

4.2 Training and Evaluation Details

All experiments were performed on a system with 128 GiB of RAM and an NVIDIA Tesla T4 GPU with 16 GiB of memory. The models were developed using Python 3.10.16 and TensorFlow 2.13.1, with support for CUDA 12.2. To ensure reliable training and avoid overfitting, we used model checkpointing to save the version with the lowest validation loss. This approach ensured the best-performing model was retained. Although training was conducted over 500 epochs, the best performance was achieved at epoch 378, corresponding to a validation loss of 0.3225. Table 2 outlines the key hyperparameters used during implementation. The source code is publicly available for research use.[1]

After training, we employed several key evaluation metrics to assess the performance of our model in the task of meniscal tears classification. These metrics offer valuable insights into the model's ability to make accurate and reliable predictions. Specifically, we used Accuracy, AUC, Precision, Recall and F1-sco to comprehensively evaluate the classification outcomes.

[1] https://www.kaggle.com/code/numanshabbir/brain-tumor-using-vision-transformer.

Table 2. Proposed Vision Transformer Hyperparameters for Meniscus Tear Classification

Hyperparameter	Value
Loss Function	Binary Crossentropy
Optimizer	Adam
Learning Rate	0.0001
Weight Decay	0.0001
Batch Size	128
Number of Classes	2 (meniscus, normal)
Image Size	256 × 256 (grayscale)
Patch Size	16 × 16
Number of Patches	256
Projection Dimension	64
Number of Attention Heads	4
Transformer Layers	15
Transformer Units	[128, 64]
Hidden Size	64
MLP Head Units	1024
Number of Epochs	500

5 Results and Discussion

This section presents the experimental results obtained using our proposed Vision Transformer-based architecture for meniscal tear classification with multi-view MRI inputs. By jointly processing sagittal, coronal, and axial slices, the model leverages complementary anatomical information across planes.

The proposed multi-view ViT achieved an average accuracy of 92.8% (±1.3) and an AUC of 92.1% (±1.1, % CI: 91.2–94.1) across 5-fold cross-validation. Table 3 details the performance metrics, including precision, recall, and F1-score, confirming that integrating multi-planar information enhances diagnostic performance. All experiments were initially conducted on a Tesla T4 GPU. To assess robustness, we repeated training on an NVIDIA RTX 3090, obtaining consistent results within ±1%, which confirms hardware independence.

Table 3. Performance of the proposed Multi-view ViT model.

Model	Accuracy (%)	AUC (%)	Precision (%)	Recall (%)	F1-score (%)
Multi-view ViT	92.8 ± 1.3	92.1 ± 1.1	91.6 ± 1.4	91.2 ± 1.2	91.2 ± 1.2

The confusion matrix in Fig. 4 illustrates classification outcomes on the held-out test set (2,025 slices, representing 10% of the dataset).: the model correctly identified 1,879 cases, with 73 false positives and 73 false negatives. Errors were thus balanced between false alarms and missed detections. Specifically, we distinguished between false positives (healthy knees misclassified as torn) and false negatives (torn menisci misclassified as healthy). Table 4 summarizes the frequency of these errors along with potential contributing factors. Since the training was performed on a balanced dataset, we also evaluated the model under the real prevalence of meniscal tears in the MRNet dataset (approximately 37% positive cases). In this scenario, the model achieved an accuracy of 86% with a sensitivity of 83% and specificity of 83%. These results confirm that, although the overall accuracy decreases compared to the balanced setting, the proposed approach remains robust under realistic clinical prevalence conditions. Figure 5 depicts the training and validation curves. Training accuracy reached 99%, while validation stabilized around 92%, with losses converging smoothly. Qualitative results are shown in Fig. 6, including correctly and incorrectly classified examples. Most errors correspond to subtle or ambiguous tear patterns. Table 5 compares our method with state-of-the-art approaches. Our model achieved an AUC of 92.1%, outperforming most CNN-based models. Although Gungor et al. reported a higher AUC (97.5%) using a hybrid YOLOv8 and EfficientNetV2 architecture, their model was restricted to two planes and did not integrate axial slices, which may limit anatomical coverage and generalizability compared to our tri-planar approach. In contrast, our method leverages sagittal, coronal, and axial views simultaneously, providing a more comprehensive diagnostic tool.

Table 4. Quantitative analysis of misclassifications

Error Type	Count	Possible Cause
False Positives	73	Motion artifacts, noisy slices, ambiguous hyperintensity
False Negatives	73	Small or subtle tears, poor contrast, atypical anatomy

Table 5. Comparative analysis of state-of-the-art methods for meniscus tear detection.

Model	Planes Used	AUC (%)	Accuracy (%)
Bien et al. [4]	Sagittal only	83.4	82.5
Chen et al. [11]	Sagittal + Coronal	90.0	–.–
Rizk et al. [8]	Sagittal only	86.0	86.0
Ying et al. [14]	Medial or Lateral only	79.0	76.4
Gungor et al. [15]	Sagittal + Coronal	97.5	93.2
Proposed Model	**Sagittal + Coronal + Axial**	**92.1**	**92.8**

Overall, the results confirm that multi-view ViTs capture richer anatomical dependencies compared to single-plane or dual-plane CNNs. Nevertheless,

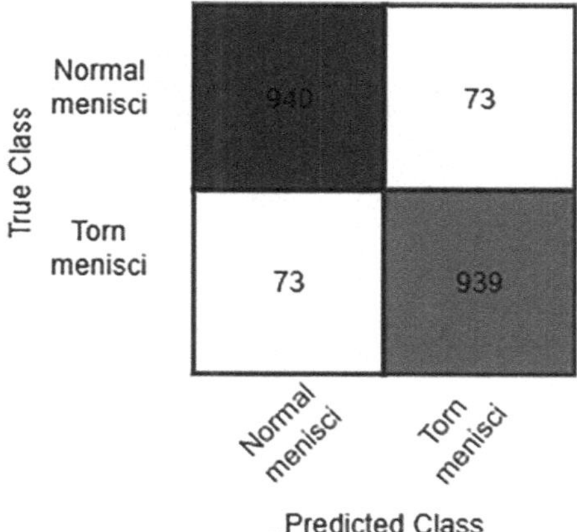

Fig. 4. Confusion matrices for meniscal tear classification.

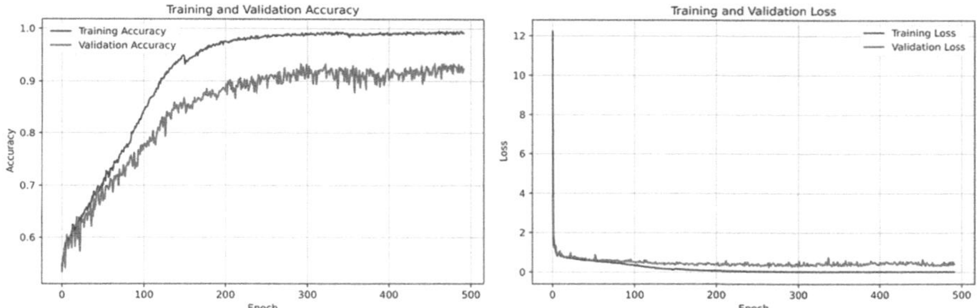

Fig. 5. Training and validation accuracy and loss curves.

this study has several limitations. First, the MRNet dataset lacks demographic diversity, which may reduce generalizability. Second, external validation on independent cohorts is still required to confirm robustness. Finally, while class balancing improved training stability, it does not fully reflect real-world prevalence distributions, which should be addressed in future evaluations.

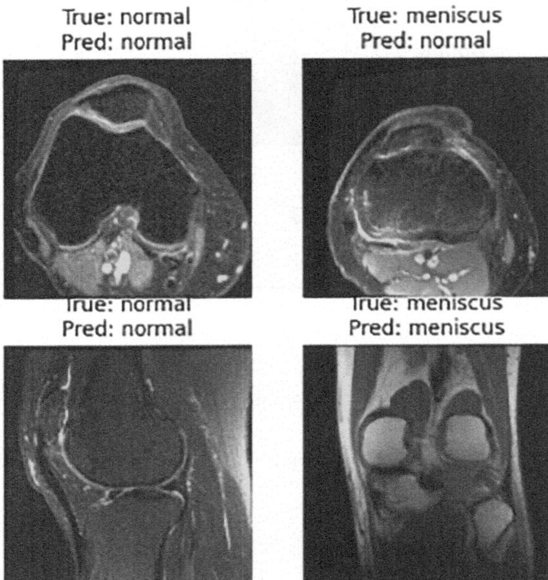

Fig. 6. Qualitative examples of model predictions.

6 Conclusion and Future Work

The proposed multi-view Vision Transformer achieved competitive performance for meniscal tear classification, surpassing most CNN-based baselines and confirming its potential as a reliable diagnostic aid. Future work will extend this approach to multi-label knee pathologies, integrate attention-based explainability for clinical interpretability, and validate performance on larger, more diverse external cohorts to ensure generalizability.

Acknowledgements. Experiments presented in this paper were carried out using the Grid'5000 testbed, supported by a scientific interest group hosted by Inria and including CNRS, RENATER and several universities as well as other (see https://www.grid5000.fr).

Data Availability. The data utilized in this study are publicly available. The dataset can be accessed at: https://stanfordmlgroup.github.io/competitions/mrnet/. The dataset includes knee MRI exams performed at Stanford University Medical Center and is provided under specific terms of use outlined on the website. The authors affirm compliance with all relevant ethical regulations for data usage.

References

1. Makris, E.A., Hadidi, P., Athanasiou, K.A.: The knee meniscus: structure–function, pathophysiology, current repair techniques, and prospects for regeneration. Biomaterials (2011)
2. Englund, M., Roos, E.M., Lohmander, L.S.: Impact of type of meniscal tear on radiographic and symptomatic knee osteoarthritis: a sixteen-year followup of meniscectomy with matched controls. Arthritis Rheum. (2003)
3. Stoller, D.W.: Magnetic Resonance Imaging in Orthopaedics and Sports Medicine. 4th edn (2017). DOI not applicable
4. Bien, N., Rajpurkar, P., Ball, R.L., et al.: Deep-learning-assisted diagnosis for knee magnetic resonance imaging: development and retrospective validation of mrnet. PLoS Med. **15**(11), e1002699 (2018)
5. Chamseddine, E., Tlig, L., Sayadi, M., Bouchouicha, M.: Segnet architecture for dermoscopic image segmentation (2022)
6. Chamseddine, E., Tlig, L., Chaari, L., Sayadi, M.: A gabor-enhanced deep learning approach with dual-attention for 3D MRI brain tumor segmentation. Comput. Biol. Med. **197**, 111047 (2025)
7. Crema, M.D., Roemer, F.W., Marra, M.D., et al.: Articular cartilage in the knee: current MR imaging techniques and applications in clinical practice and research. Radiographics (2011)
8. Rizk, M., Patel, S., Weber, A.E., et al.: Deep learning for detection of meniscal tears on knee MRI: external validation and comparison to radiologists. Radiol.: Artif. Intell. **3**(2), e200134 (2021)
9. Lundervold, A.S., Lundervold, A.: An overview of deep learning in medical imaging focusing on MRI. Zeitschrift für Medizinische Physik (2019)
10. Fu, J.-C., Cheng, H.-L., Chang, C.-K., et al.: Computer-aided diagnosis for knee meniscus tears in magnetic resonance imaging. J. Ind. Prod. Eng. **30**(2), 67–77 (2013)
11. Chen, Y., Zhang, X., Cheng, T., et al.: Deep learning approach for meniscus tear detection using MRI. J. Digit. Imaging **33**(4), 748–753 (2020)
12. Dosovitskiy, A., Beyer, L., Kolesnikov, A., et al.: An image is worth 16x16 words: transformers for image recognition at scale. arXiv preprint arXiv:2010.11929 (2020)
13. Ko, H., Chung, H., Kang, W.S., et al.: Vision transformer for classification of chest radiographs for lung disease detection. Comput. Biol. Med. (2025)
14. Ying, M., Lin, Y., Wang, Q., et al.: A deep learning knowledge distillation framework using knee MRI and arthroscopy data for meniscus tear detection. Radiol.: Artif. Intelli. **5**(1), e220221 (2023)
15. Güngör, E., Aydin, A., Kose, C., et al.: Achieving high accuracy in meniscus tear detection using advanced deep learning models. Procedia Comput. Sci. **231**, 1110–1116 (2024)
16. Stanford University Center for Artificial Intelligence in Medicine & Imaging. Mrnet: Knee MRIs (2018). https://doi.org/10.71718/rcbp-8c35. Accessed 3 May 2025
17. Chamseddine, E., Mansouri, N., Soui, M., Abed, M.: Handling class imbalance in COVID-19 chest x-ray images classification: using smote and weighted loss. Appl. Soft Comput. **129**, 109588 (2022)
18. Chamseddine, E., Tlig, L., Sayadi, M.: Enhanced deep learning framework for automatic MRI brain tumor segmentation with data augmentation. In: 7th IEEE International Conference on Advanced Technologies, Signal and Image Processing, ATSIP 2024, pp. 409–413. Institute of Electrical and Electronics Engineers Inc (IEEE) (2024)

A Case Study on Applying StyleGAN2-ADA for Fashion Image Synthesis with Limited Data

Ali H. Shareef[1,2,3](✉), Hajer Ghodhbani[1,5], Tarek M. Hamdani[1,4], Habib Chabchoub[6], and Adel M. Alimi[1,5]

[1] Research Groups in Intelligent Machines (REGIM Lab), University of Sfax, National Engineering School of Sfax (ENIS), BP 1173, 3038 Sfax, Tunisia
[2] National School of Electronics and Telecommunications of Sfax (ENET'Com), University of Sfax, 10587 Sfax, Tunisia
[3] University of Thi-Qar, 64001 Nasiriyah, Thi-Qar, Iraq
ali.handhel@utq.edu.iq
[4] Higher Institute of Computer Science Mahdia, University of Monastir, 5111 Mahdia, Tunisia
[5] Department of Electrical and Electronic Engineering Science, University of Johannesburg, Auckland Park, 2006, Johannesburg, South Africa
[6] College of Business, Al Ain University of Science and Technology, Abu Dhabi, United Arab Emirates

Abstract. This paper provides a dedicated case study of the stylegan2-ada implementation on fashion images generation with limited data. We tested the capability of the model using the ZEN dataset to produce high-quality and diverse fashion images with a comparatively small training corpus. Compared to the baseline with StyleGAN2, it is shown that the model is unstable in the absence of ADA and fails to produce high-quality images (FID> 80), whereas the StyleGAN2-ADA baseline converges steadily with significantly better results (FID = 10.8). We also point out instances of failure that are common like poor fabric textures and color bleeding and we also recognize the weaknesses of the model together with its strength. Presenting this work as a case study, we will offer useful information as to the usefulness of StyleGAN2-ADA in fashion synthesis tasks that are conducted using limited data, and critically assess this to inform future research and advancements.

Keywords: StyleGAN2-ADA · Fashion Image Generation · Image Synthesis

1 Introduction

With the rise of generative adversarial networks (GANs), new principles have arisen in image synthesis and computer vision [1]. To quickly develop new fashion ideas, designers require high quality, mixed, and accurate images of their

T. Ensari et al. (Eds.): ISPR 2025, CCIS 2859, pp. 166–177, 2026.
https://doi.org/10.1007/978-3-032-21585-7_12

products [2]. Most conventional methods, including IR, have limitations when it comes to matching clothes to images available in extensive collections [3]. Adjusting the design of these pieces in unique ways or creating brand new looks is often a challenge. In Masukawa et al. [2], it is highlighted that the process of designing fashionable garments is time consuming and requires a lot of skill effort.

Over the past few years, conditional GANs have captured interest because they make it easier to create images based on certain attributes [4]. It is by using the StyleGAN2-ADA model that you find that using adaptive methods for data augmentation allows it to function effectively with fewer data [5]. The study is aimed at assessing how good the clothing images are that are created by the StyleGAN2-ADA system which highlights clear details within the visuals.

The motivation underpinning this research is rooted in the fashion industry's burgeoning need for efficient computational methods that alleviate the labor-intensive facets of design creation [6]. Designers stand to benefit substantially from virtual design assistants capable of generating realistic and attribute-consistent fashion images, thereby accelerating the ideation process and mitigating the need for manual editing. Generated images are quantitatively evaluated using the Fréchet Inception Distance (FID) [7].

2 Background and Related Works

The computational synthesis of fashion images using GANs has become a focal point of research interest across various applications, including virtual try-on systems [8], fashion item recommendations [9], and design generation [10]. Traditional approaches have primarily relied on information retrieval methods, wherein images are identified from extensive databases using feature matching techniques [3]. However, these techniques are not flexible enough to enhance or alter every detail on a piece of clothing in an image.

Recently, GAN models have become more popular among researchers since they can create new images and offer means to manage different elements of the results by using latent space [1]. Using AdaIN, StyleGAN became a significant model that made it possible to generate highly realistic images [11]. The updates in StyleGAN2 [12] and StyleGAN2-ADA [5] have made it possible to both improve image quality and train using fewer data thanks to their augmentation methods. Besides, researchers have found that GANSpace can point out understandable ways to control GANs in generating images [13].

Besides, a number of works use feature extraction modules to make details clearer than simple GAN approaches allow [14–16]. A swapping autoencoder module has proven useful in generating detailed patterns for garments by separating the textural and the structural aspects in an image [17]. The Tailorgan technique by Chen et al. [18] relies on GANs to adjust the width of a shirt's sleeves and collar.

3 Methodology

In this section, the methods and approaches used for making and reviewing the fashion images through StyleGAN2-ADA are defined. It was selected for this task because it can produce high-quality and highly varied images. The first step was to assemble and preprocess a set of fashion images so that information remains consistent and reliable. The data set was next employed to train StyleGAN2-ADA, while the hyperparameters were carefully adjusted to improve its performance. During training, some methods, including growing the GAN and adapting the discriminator, were used to strengthen the model's ability to create realistic and appropriate fashion images according to the style. After the model was trained, to judged how high-quality and diversified they were. The way is to use a metric like the Frechet Inception Distance (FID), Learned Perceptual Image Patch Similarity (LPIPS), and Inception Score (IS). Experiments showed that StyleGAN2-ADA is successful in generating fashion images similar to the real world while also exhibiting new ideas in fashion design. See Fig. 1.

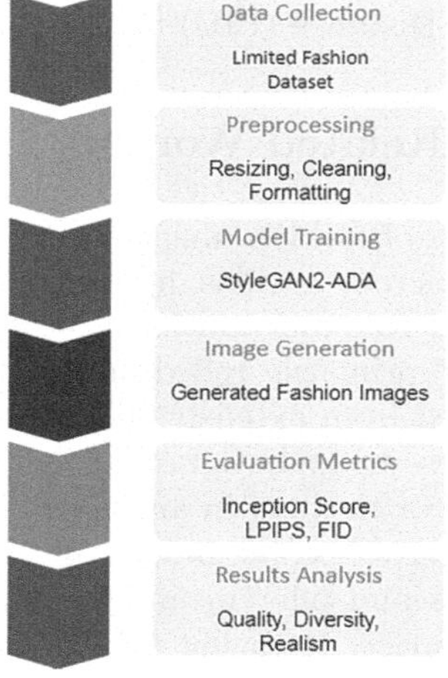

Fig. 1. The phases of the proposed methodology

3.1 StyleGAN2-ADA for Fashion Image Generation

StyleGAN2-ADA forms the main component in the workflow for producing images. It gives the generator the ability to model different pieces of clothing depending on various aspects like their style, shape, length, and color. Given a latent code called $\mathbf{z}$, the mapping network projects it to an intermediate code called $\mathbf{w}$ and uses the embedding vectors of the attributes to do so. The output is used to generate the final fashion images.

Training is performed on a dataset of 9,000 clothing images (256 × 256 resolution) provided from ZEN Company, a ready-to-wear Tunisia brand [19], where each image is provided with multi attribute labels. During its training, StyleGAN2-ADA uses the Adam optimizer and is updated with a learning rate of 0.002. StyleGAN2-ADA makes maximum use of adaptive data augmentation to protect against overfitting in situations when the data is limited.

4 Experimental Setup and Evaluation Metrics

This section details the experimental parameters, dataset characteristics, and evaluation metrics used to assess the performance of the generation system.

4.1 Dataset

Our dataset [19] consists of 9,000 high-quality fashion images with a resolution of 256 × 256, collected from ZEN Company[1], a Tunisian ready-to-wear brand. The dataset composition is as follows:

- **Garment categories:** dresses (28%), pants (22%), shirts (19%), skirts (17%), coats (14%)
- **Color distribution:** solid colors (65%), patterns (35% with floral 15%, geometric 12%, abstract 8%)
- **Model demographics:** female models (85%), male models (15%)
- **Pose variation:** front view (60%), side view (25%), back view (15%)
- **Backgrounds:** studio (70%), outdoor (20%), indoor (10%)
- **Age distribution of models:** 20–30 years (45%), 30–40 years (35%), 40–50 years (20%)

This distribution reflects the actual product mix of the brand and provides sufficient diversity for training a robust generative model. To ensure dataset quality, we implemented a three-stage filtering process:

1. Automated removal of low-quality images using a quality assessment CNN.
2. Manual verification by fashion domain experts to ensure proper garment representation.
3. Balancing of categories to prevent model bias toward dominant classes.

The dataset was split into training (8,100 images), validation (450 images), and test (450 images) sets. This split allows for proper model selection and unbiased evaluation of generated images against a held-out test set. The validation and test sets maintain the same category distribution as the training set to ensure representative evaluation. Each image includes detailed annotations such as garment type, color, pattern type, fabric material, and key design elements, which were used to verify the semantic correctness of generated images during evaluation.

[1] https://zen.com.tn/fr.

Fig. 2. An array of the sample images of the dataset of 5 x 5 elements. The choice of images is characterized by the variety of visuality in stylistics, color, pose, and background as this is necessary to train and examine generative models.

This visual evaluation proves the quality of the dataset and its heterogeneity, which is essential with regard to training generative models effectively. The data variance, in particular, is brought to the fore in the organized presentation of samples in Fig. 2, seeing as the samples can be viewed directly through an appreciation of the coverage of the dataset in an instant.

4.2 Generation Mechanism

To generate fashion images without relying on explicit attribute conditioning, we employ version of StyleGAN2-ADA. In this setup, the generator synthesizes images based solely on a latent input vector, without receiving any semantic labels or attribute guidance. A latent vector $z \sim \mathcal{N}(0, I)$ is sampled from a standard Gaussian distribution and passed through a mapping network (MLP), which transforms it into an intermediate style vector.

This style vector captures the underlying structure and diversity of the fashion domain and is fed into the generator to produce high-resolution images. By omitting attribute-based conditioning, the model relies entirely on the latent space to represent and explore variations in clothing type, color, texture, and pose.

Figure 3 illustrates the overall pipeline of the image generation process. Despite the lack of explicit control signals, the model is capable of generating diverse and realistic fashion images by leveraging the rich representation learned from the training data.

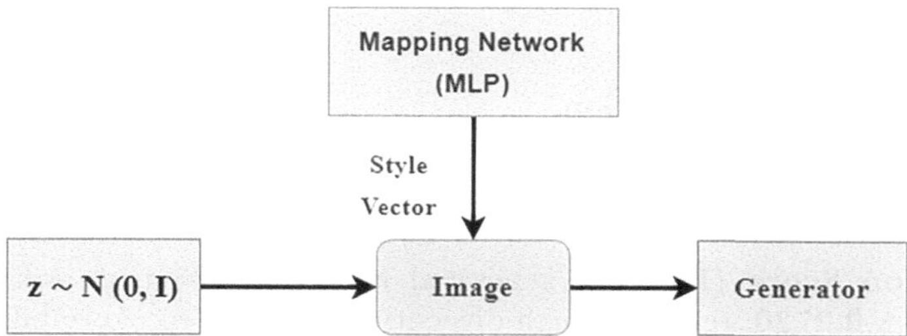

Fig. 3. Fashion image generation pipeline using StyleGAN2-ADA. The generator produces realistic fashion images solely from latent vectors z, without using any semantic attributes or additional conditioning information.

4.3 Training and Implementation Details

The proposed model was trained using the official StyleGAN2-ADA PyTorch implementation provided by NVlabs [5]. The training process was conducted on a single NVIDIA Tesla T4 GPU using the standard Google Colab environment, under Ubuntu 20.04 with PyTorch 1.13 and CUDA 11.8.

We trained the model for 25000 kimg using a dataset of 9,000 fashion images resized to 256 × 256 resolution. Adaptive Discriminator Augmentation (ADA) was enabled throughout the training. The key training parameters are summarized in Table 1.

Table 1. Training hyperparameters for StyleGAN2-ADA model

Parameter	Value
Input Resolution	256 × 256
Number of Training Images	9,000
Training Steps	25000 kimg
Batch Size	32
Learning Rate	0.002
Optimizer	Adam (β_1=0.0, β_2=0.99)
Augmentation	ADA (Adaptive Discriminator Augmentation)
Style Mixing Probability	0.9
Latent Space Dimension	512
Discriminator Iterations	1:1 with Generator
Training Environment	Google Colab (Tesla T4 GPU)

5 Results and Discussion

This section discusses the experimental results in terms of quantitative metrics (FID) and qualitative assessments. This discussion elaborates on the performance of the integrated system, addresses its advantages over traditional methods, and identifies existing challenges.

5.1 Quantitative Evaluation

To provide a rigorous evaluation, we employed three widely accepted image quality metrics.

- **Inception Score (IS)**: The generated images achieved an average IS of **3.4295 ± 0.2720**, reflecting high diversity and object recognizability.
- **Fréchet Inception Distance (FID)**: The model reached a FID score of **10.8**, indicating a close resemblance between the generated and real image distributions.
- **Learned Perceptual Image Patch Similarity (LPIPS)**: The LPIPS score was **0.4383**, suggesting strong perceptual similarity and realism.

5.2 Qualitative Analysis and Visual Assessment

In addition to quantitative evaluation, qualitative assessment plays a crucial role in understanding the aesthetic and perceptual nuances of the generated images. Visual inspection confirms that the generated clothing images consistently reflect the combination of color. Moreover, the generated images with the identical combination of the differ from each other, which indicates that the model achieves diverse generation. See Fig. 4.

5.3 Analysis of Generated Fashion Images

It regularly produced images that satisfied the input given, covering the type, shape, length and color of clothing. Tests have revealed that the system can produce images that are both realistic and varied. In this step, the resulting images look real by showing clothing shapes and patterns. regularly can use the rules it is given, but it also adds touches that make everything look realistic For instance, unfolds clothing in a lifelike way and deals with light and shadows. Because of this detail, every image generated looks different and genuine, so StyleGAN2-ADA is valuable for those in fashion who prefer hi-tech to traditional methods.

Fig. 4. Sample images generated by the trained StyleGAN2-ADA model on the ZEN fashion dataset. The model successfully captures diverse clothing styles, poses, and textures, highlighting its ability to synthesize high-quality and visually plausible fashion images. These samples were generated using a truncation value of $\psi = 0.7$.

5.4 Baseline Comparison with StyleGAN2

In order to strictly assess the importance of adaptive discriminator augmentation (ADA) we created a control experiment with the standard StyleGAN2 which worked under the same preprocessing and training conditions, only omitting ADA on the same dataset, ZEN. The original model was shown to converge not steadily, but instead converged quickly and produced poor quality and unrealistic pictures. The initial StyleGAN2 had a baseline FID score of over 80, which shows a serious degradation of generative performance. By comparison, StyleGAN2-ADA model successfully converged and generated high quality results, with a score of 10.8 on the FID. This empirical evidence of performance gap is the solid support of ADA playing a critical role at stabilizing training in limited data conditions and has a good evidence of the effectiveness of ADA in tasks related to fashion image synthesis.

5.5 Comparison with Traditional Retrieval Methods and GAN Architectures

Traditionally, information retrieval (IR) systems only find photos that are already stored in the database. Although these approaches are efficient for finding similar things, they cannot generate new styles or add a lot of detail to images of clothing. Rather, the GAN system allows for many types of editing and generation that changes with each input set. Not only does this help with creative design, but it also makes the manual workload simpler.

5.6 Evaluation

These results were compared with those reported in several recent studies on fashion synthesis using GANs. As shown in Table 2, our model outperformed baseline models such as BC-GAN, DFDGAN, SCGAN in terms of FID and LPIPS.

Table 2. Quantitative comparison with previous methods.

Method	Dataset Size	**FID ↓**	LPIPS ↓
BC-GAN [20]	31,631	52.370	0.504
DFDGAN [21]	21889	80.9	0.642
SCGAN [22]	800,000	19.8	N/A
Ours	**9000**	**10.8**	**0.4383**

These findings demonstrate the capability of the StyleGAN2-ADA model, not only in generating visually convincing samples, but also in capturing perceptual and statistical properties comparable to real fashion images. The improved LPIPS and FID scores indicate enhanced realism and structural consistency.

5.7 Failure Analysis

While StyleGAN2-ADA demonstrates strong performance in generating realistic fashion images, certain limitations are observed in specific scenarios. In particular, the model sometimes struggles to generate coherent textures for complex garments, and color bleeding occurs on items with intricate patterns.

To provide a more complete and credible evaluation, we present visual examples of these failure cases in Fig. 5. These examples highlight instances where the generated outputs deviate from realistic appearance, revealing challenges associated with limited data diversity and rare attribute combinations in the dataset.

Analyzing these failure cases offers valuable insights for future improvements, suggesting directions such as augmenting the training dataset, enhancing latent space regularization, or incorporating specialized modules to handle complex patterns and color distributions.

5.8 Strengths and Limitations

The experimental results affirm several strengths of the proposed system. Key strengths include:

Fig. 5. Representative failure cases of StyleGAN2-ADA. The model occasionally produces incoherent textures and color bleeding in complex garments.

- **High realism of images:** The images produced are real and persuasive visually, enabling them to be used in fashion-oriented scenarios.
- **Versatile editing:** With the system, it is possible to manipulate flexibly, such as transformations of explore routes or combinations of colors by journeying through the latent space.
- **Reduced manual workload:** Reduces the need for manual design or sketching, saving time and effort for fashion designers.
- **Novel design inspiration:** The model generates innovative and out-of-the-box products that could also be an inspiration for new fashion ideas.

Despite these strengths, the system also has some limitations:

- **Complexity of training:** Model training requires high-end technical knowledge.
- **Dataset dependency:** The performance of the model is highly dependent on the quality and diversity of the training dataset.
- **Pattern accuracy:** It can be poor at producing fine-grained patterns or consistent where it has to impersonate complicated patterns and texture.
- **Computational resources:** Training and inference processes require significant computationally intensive tasks such as full-featured GPUs and long runtimes.

6 Conclusion and Future Directions

In this paper, we investigated the application of StyleGAN2-ADA to produce fashion images in the limited data scenario. The experiment proves that ADA is an essential part of the process of stabilizing the training and providing the creation of good-quality images under the conditions when only a comparatively small amount of data is at hand. Though some minor drawbacks were noticed e.g. some texture inconsistencies and slight differences in the color rendering, these drawbacks are still marginal in comparison with the overall performance improvements. They emphasize the natural complexity of reproducing fine-grained details of fashion instead of being one of the underlying flaws of the process. It can be anticipated that future research can work on adding conditional labels, investigating hybrid architectures, or more sophisticated refinement strategies to make the design more realistic and applicable to real-world fashion design scenarios.

Acknowledgements. The research leading to these results has received funding from the Ministry of Higher Education and Scientific Research of Tunisia under the grant agreement number LR11ES48.

Declaration

Financial Interests. The authors declare they have no financial interests.

Conflict of Interest. The authors declare that they have no conflict of interest.

References

1. Goodfellow, I., et al.: Generative adversarial nets. Adv. Neural Inform. Process. Syst. **27** (2014)
2. Tepe, J.: Approaching generative adversarial network systems for design in fashion (2023)
3. Ning, C., Di, Y., Menglu, L.: Survey on clothing image retrieval with cross-domain. Complex Intell. Syst. **8**(6), 5531–5544 (2022)
4. Mirza, M., Osindero, S.: Conditional generative adversarial nets. arXiv preprint arXiv:1411.1784 (2014)
5. Karras, T., Aittala, M., Hellsten, J., Laine, S., Lehtinen, J., Aila, T.: Training generative adversarial networks with limited data. Adv. Neural. Inf. Process. Syst. **33**, 12104–12114 (2020)
6. Lee, Y.K.: How complex systems get engaged in fashion design creation: using artificial intelligence. Thinking Skills Creativity **46**, 101137 (2022)
7. Heusel, M., Ramsauer, H., Unterthiner, T., Nessler, B., Hochreiter, S.: Gans trained by a two time-scale update rule converge to a local nash equilibrium. Adv. Neural. Inf. Process. Syst. **30** (2017)
8. Pandey, N., Savakis, A.: Poly-gan: multi-conditioned gan for fashion synthesis. Neurocomputing **414**, 356–364 (2020)

9. Kumar, S., Gupta, M.D.: c^+ gan: complementary fashion item recommendation. arXiv preprint arXiv:1906.05596 (2019)
10. Jin, Y., Yoon, J., Andrew Self, J., Lee, K.: Ai as a catalyst for creativity: exploring the use of generative approach in fashion design for improving their inspiration (2024)
11. Karras, T., Laine, S., Aila, T.: A style-based generator architecture for generative adversarial networks. In: Proceedings of the IEEE/CVF Conference on Computer Vision and Pattern Recognition, pp. 4401–4410 (2019)
12. Karras, T., Laine, S., Aittala, M., Hellsten, J., Lehtinen, J., Aila, T.: Analyzing and improving the image quality of stylegan. In: Proceedings of the IEEE/CVF Conference on Computer Vision and Pattern Recognition, pp. 8110–8119 (2020)
13. Härkönen, E., Hertzmann, A., Lehtinen, J., Paris, S.: Ganspace: discovering interpretable gan controls. Adv. Neural. Inf. Process. Syst. **33**, 9841–9850 (2020)
14. Odena, A., Olah, C., Shlens, J.: Conditional image synthesis with auxiliary classifier gans. In: International Conference on Machine Learning, pp. 2642–2651. PMLR (2017)
15. Zhang, H., Goodfellow, I., Metaxas, D., Odena, A.: Self-attention generative adversarial networks. In: International Conference on Machine Learning, pp. 7354–7363. PMLR (2019)
16. Hinz, T., Fisher, M., Wang, O., Wermter, S.: Improved techniques for training single-image gans. In: Proceedings of the IEEE/CVF Winter Conference on Applications of Computer Vision, pp. 1300–1309 (2021)
17. Park, T., Zhu, J.-Y., Wang, O., Lu, J., Shechtman, E., Efros, A., Zhang, R.: Swapping autoencoder for deep image manipulation. Adv. Neural. Inf. Process. Syst. **33**, 7198–7211 (2020)
18. Chen, L., et al.: Tailorgan: making user-defined fashion designs. In: Proceedings of the IEEE/CVF Winter Conference on Applications of Computer Vision, pp. 3241–3250 (2020)
19. zen, I.: zen (2024). https://zen.com.tn/fr/overleaf Accessed: (2023)
20. Zhou, D., Zhang, H., Ma, J., Shi, J.: Bc-gan: a generative adversarial network for synthesizing a batch of collocated clothing. IEEE Trans. Circuits Syst. Video Technol. **34**(5), 3245–3259 (2023)
21. Jung, J., Kim, H., Park, J.: Deep fashion designer: generative adversarial networks for fashion item generation based on many-to-one image translation. Electronics **14**(2), 220 (2025)
22. Jiang, S., Liu, H., Wu, Y., Fu, Y.: Spatially constrained gan for face and fashion synthesis. In: 2021 16th IEEE International Conference on Automatic Face and Gesture Recognition (FG 2021), pp. 01–08 (2021). IEEE

A U-Net GNN Hybrid Approach for MRI-Based Brain Tumor Segmentation

Salaheddine Addoune[1](✉), Mohammed Islam Ouahbi[1], Khadra Bouanane[1], and Mohammed Lamine Kherfi[1,2]

[1] Laboratory of Artificial Intelligenceand Information Technologies LINATI, Computer Science and IT Department, Kasdi Merbah University, Ouargla, Algeria
{addoune.salaheddine,ouahbi.mohamedislam,bouanane.khadra}@univ.ouargla.dz
[2] Department of Computer Science, SQU, Muscat, Oman
m.kherfi@squ.edu.om

Abstract. Accurate brain tumor segmentation from MRI is essential for diagnosis and treatment planning. While most exsting graph neural network (GNN)-based approaches use GNNs as the primary segmentation module applied to the full image, we propose a novel hybrid strategy where GNNs act as a targeted refinement of U-Net predictions. By focusing only on regions of uncertainty, the framework combines U-Net's local spatial precision with GNN-based long-range context, reducing computational overhead. Experiments on the BraTS 2020 dataset show Dice score improvements of 2.63% for whole tumor, 2.18% for tumor core, and 3.71% for enhancing tumor compared to a baseline U-Net. This refinement strategy highlights the potential of graph-enhanced CNNs as an efficient and clinically meaningful alternative for advancing automated brain tumor analysis.

Keywords: Brain tumor segmentation · MRI · U-Net · Graph neural networks · Medical image analysis · Deep learning

1 Introduction

Accurate brain tumor segmentation in MRI is vital for clinical tasks such as diagnosis and treatment planning [13]. Deep learning models, particularly U-Net variants, have become the standard due to their ability to capture spatial details. However, their limited receptive fields hinder the modeling of long-range dependencies, often resulting in suboptimal segmentation of complex tumor regions [11].

While Vision Transformers (ViTs) address this limitation through global attention [5], they require large datasets and high computational resources [1], limiting their practicality. Graph Neural Networks (GNNs) offer a more efficient alternative by modeling spatial dependencies over sparse graph structures.

T. Ensari et al. (Eds.): ISPR 2025, CCIS 2859, pp. 178–193, 2026.
https://doi.org/10.1007/978-3-032-21585-7_13

They are particularly suited for medical images, where abnormalities are often irregular and disconnected [11].

In this work, we propose a hybrid segmentation framework combining U-Net with a GraphSAGE-based GNN that selectively refines predictions in low-confidence regions. Unlike prior GNN based approaches that apply graph processing to the entire image, our GNN-UNet selectively refines only low-confidence regions, yielding higher efficiency and targeted accuracy improvements. We introduce three refinement strategies and evaluate them on the BraTS 2020 dataset, showing consistent improvements across all tumor subregions.

The remainder of the paper is organized as follows: Sect. 2 reviews related work. Section 3 details the proposed hybrid framework. Section 4 presents the experimental setup and results while Sect. 5 concludes the paper and outlines the limitations of the current work and directions for future research.

2 Related Work

Graph Neural Networks (GNNs) have recently emerged as a promising approach for brain tumor segmentation, improving feature extraction and multimodal data integration [11]. In [12], a hybrid GNN-CNN model is proposed for brain tumor segmentation. Each 3D scan is over-segmented into supervoxels to build a graph, where a GraphSAGE-based GNN predicts node-level labels using features from four MRI modalities. The GNN outputs are reprojected and fused with voxel features, then refined by a shallow 3D CNN. This two-stage design combines global context with local detail, improving segmentation on the BraTS 2021 dataset, especially for whole and core tumor regions. Gammoudi et al. [4] proposed SCGNN-3DBUNet, a hybrid architecture that combines a 3D U-Net with Graph Neural Networks with a spectral clustering and pooling module. The 3DBUNet extracts volumetric features from MRI scans, which are then used to construct region adjacency graphs processed by a GNN. This combination enables multi-scale feature learning and spatial relationship modeling. Evaluated on the BraTS 2020 dataset, the method achieved a Dice score of up to 0.89 for whole tumor segmentation, outperforming the baseline 3D U-Net and several state-of-the-art approaches. In [9], the authors propose a multi-class brain tumor segmentation method using a Graph Attention Network (GAT). MRI scans are converted into region adjacency graphs processed by a multi-head GAT, capturing both local and global context. The method reduces computational cost compared to voxel-wise CNNs while achieving high accuracy and robustness across tumor subtypes. Mohammadi and Allali [7] proposed a Spectral-Spatial GNN (SSGNN) that combines GraphSAGE for local spatial features with Chebyshev convolutions for global spectral patterns. This fusion captures multi-scale context and improves segmentation, especially in complex tumor subregions, achieving up to 99.2% accuracy and 98.9% Dice score, outperforming CNNs and standalone GNNs. In [10], a hybrid framework combines Variational Spatial Attention with a Graph CNN (VSA-GCNN) to segment brain tumors with varying shapes and sizes. Post-segmentation, AlexNet extracts features, and a Bi-GRU classifies tumor types. The model achieves high accuracy and robustness across BraTS 2019–2021 datasets. Vatanpour et al. [14] proposed a hybrid framework that sequentially combines a simple 3D CNN with a Graph Neural Network (GraphSAGE) for brain tumor segmentation. The CNN extracts local spatial representations from multi-modal MRI volumes, which are then transformed into graph nodes for the GNN to capture long-range dependencies and global structure. Deconvolution layers reconstruct the segmentation masks. Evaluated on the BraTS 2021 dataset, the method achieved higher Dice scores and competitive Hausdorff distances, outperforming several CNN- and GNN-based baselines. Ziaeetabar [15] proposed EfficientGFormer, a graph-augmented transformer for

Table 1. Summary of recent graph-based approaches for brain tumor segmentation.

Work	Main Idea
Saueressig et al. [11]	Exploration of different GNNs for brain tumor segmentation, highlighting their potential for multimodal integration and improved feature extraction.
Saueressig et al. [12]	Hybrid GNN–CNN model: supervoxel graphs processed by GraphSAGE, fused with voxel features, then refined by a shallow 3D CNN. Improves segmentation on BraTS 2021, especially WT and TC.
Patel et al. [9]	Multi-class segmentation using a multi-head Graph Attention Network (GAT). Region adjacency graphs encode context while reducing computational demands relative to voxel-wise CNNs.
Gammoudi et al. [4]	SCGNN-3DBUNet: hybrid model combining 3D U-Net feature extraction with GNN processing of spectral clustering-based graphs.
Mohammadi & Allali [7]	Spectral-Spatial GNN (SSGNN) combining GraphSAGE for spatial locality with Chebyshev filters for spectral context, improving accuracy in complex tumor subregions.
Pratap et al. [10]	VSA-GCNN: integrates Variational Spatial Attention with Graph CNNs; AlexNet and Bi-GRU used post-segmentation for tumor type classification across BraTS datasets.
Vatanpour et al. [14]	Hybrid framework combining a 3D CNN for local feature extraction with a GraphSAGE-based GNN to capture long-range dependencies.
Ziaeetabar [15]	EfficientGFormer: Combines nnFormer-based feature extraction with a dual-edge Graph Attention Network and knowledge distillation for multimodal brain tumor segmentation. Achieved state-of-the-art Dice scores on BraTS 2021.

multimodal brain tumor segmentation. The model combines nnFormer-based feature extraction with a dual-edge Graph Attention Network, incorporating knowledge distillation for efficiency. On BraTS 2021, it achieved state-of-the-art Dice scores with reduced computational cost. While accurate and efficient, the dual-edge graph and pruning mechanisms increase implementation complexity.

Table 1 provides a summary of the key contributions of existing works.

As seen in prior studies, GNNs are typically employed either as the primary segmentation module or as a global refinement stage applied to the entire MRI volume. While these approaches have shown strong performance, they are computationally demanding since every region is processed equally, regardless of prediction confidence. In contrast, our framework is unique in how the GNN is applied: rather than operating globally, it focuses only on regions where the U-Net produces uncertain predictions. By targeting refinement to these challenging areas, we aim to improve segmentation accuracy in critical subregions while reducing unnecessary computation, resulting in a more efficient and scalable solution.

In the following section, we detail the architecture, graph construction strategy, and enhancement methods designed to address this challenge.

3 Methodology

The proposed method, illustrated in Fig. 1, enhances U-Net segmentation by first generating model predictions and constructing a graph from the dataset in parallel. A mask is then derived, targeting poorly segmented regions, which serves as the foundation for the three enhancement strategies described below:

1. **GNN-Based Enhancement:** Applies a GNN to directly refine weak regions in the U-Net output.
2. **1D CNN + GNN:** Encodes U-Net predictions using 1D convolution, which are then combined with graph features and passed to a GNN for refinement.
3. **GNN Encoding of Predictions:** Uses a GNN to jointly encode U-Net prediction labels and graph features, focusing refinement on under-segmented regions.

3.1 Data Preprocessing and U-Net Training

Before training the U-Net model, we selected the T1CE and FLAIR modalities. Each MRI volume was cropped to focus on the brain, yielding tensors of size $100 \times 160 \times 208 times 2$. To make training computationally feasible, the cropped 3D volumes were decomposed into axial 2D slices of size $160 \times 208 \times 2$, each paired with its ground-truth mask. Although this slice-based strategy does not capture full volumetric context, it allows detailed modeling of intra-slice tumor structures and has been widely adopted as a practical compromise. These slices served as inputs for the training of a 2D U-Net described in Sect. 4.

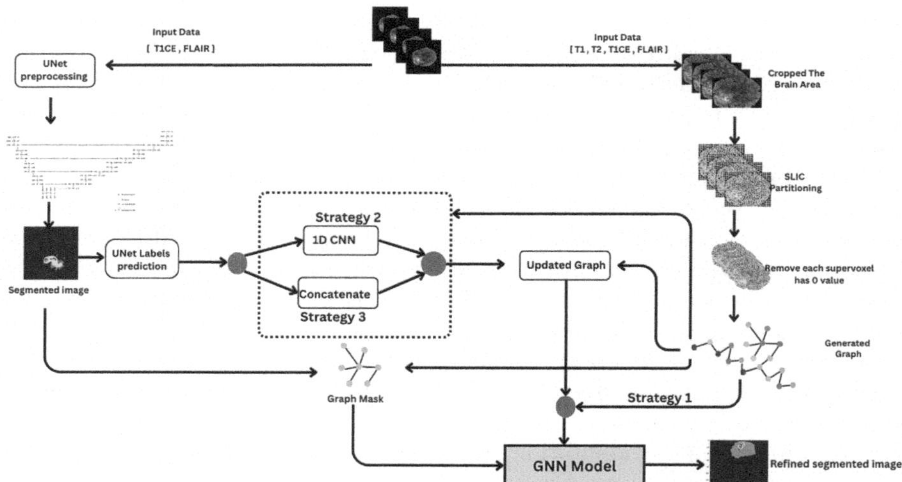

Fig. 1. Overview of the proposed methodology for enhancing U-Net-based image segmentation using Graph Neural Networks (GNNs).

3.2 Graph Generation

In the preprocessing stage, each MRI volume is converted into a graph representation $\mathcal{G} = (\mathcal{V}, \mathcal{E})$, where nodes correspond to homogeneous image regions and edges capture their spatial adjacency. The brain is first isolated by cropping, after which the SLIC algorithm [2] is applied to partition the scan into supervoxels. Supervoxels containing only background values are discarded in order to reduce node count and redundancy.

Each remaining supervoxel forms a node in the graph, and connections are established with neighboring supervoxels that share a boundary. Node attributes are derived from voxel intensity distributions: for a supervoxel i with values X_i, we compute its five intensity quintiles $(Q1, Q2, Q3, Q4, Q5)$. Repeating this process for all four MRI modalities produces a 20-dimensional feature vector per node, aggregated in an $N \times F$ feature matrix.

3.3 Graph Mask for Region-of-Interest Identification

Predictions from the baseline U-Net are used to determine a binary mask that highlights regions with low confidence. Specifically, each supervoxel is examined with respect to the voxel-level probability scores predicted by the CNN. If at least one voxel falls below a confidence threshold T, the supervoxel is flagged as uncertain and included in the mask. This guarantees that the mask covers all potentially misclassified regions.

Formally, let $P(x, y, z)$ denote the U-Net probability for voxel (x, y, z). For a given threshold T, voxels with $P(x, y, z) < T$ are considered unreliable and can be substituted with their corresponding ground-truth labels $G(x, y, z)$ during validation. By varying T over a candidate set $\{T_1, \ldots, T_n\}$, we obtain a series of refined segmentations $R(T_i)$. The optimal threshold T^* is selected as the value that maximizes Dice performance while limiting unnecessary replacements.

After threshold selection, the graph is divided into two categories of nodes: those above the threshold (considered reliable) and those below it (considered uncertain). The uncertain subset defines the region of interest and becomes the focus of GNN refinement.

3.4 GNN Model Architecture

We adopt GraphSAGE as the GNN backbone due to its inductive learning capability, scalability, and flexibility in aggregating neighborhood information [11].

The graph is constructed from supervoxels obtained via SLIC, with each node representing a supervoxel and edges defined by spatial proximity.

The GNN processes these graphs by aggregating information from neighboring nodes through a series of graph convolutional layers. ReLU activations are used after each convolution, followed by a fully connected layer and a softmax output. Only nodes within uncertain (masked) regions contribute to the training loss, while the model has access to the full graph context during inference. This approach enables the model to leverage global context while focusing the learning on regions of interest.

Specific architectural parameters, training settings, and loss weighting are presented in Sect. 4.

3.5 Training Strategies

We detail three training strategies to improve segmentation: (i) GNN-based enhancement, (ii) 1D CNN encoding of U-Net predictions with GNN, and (iii) GNN encoding of prediction labels, as illustrated in Fig. 2.

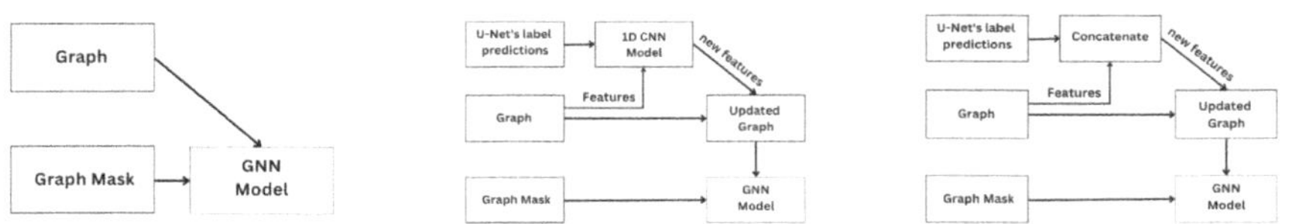

(a) Strategy 1: GNN-based enhancement (b) Strategy 2: 1D-conv encoded predictions (c) Strategy 3: CNN prediction encoding

Fig. 2. Overview of the three training strategies combining GNN and CNN-based information.

Strategy 1: GNN-Based Enhancement. In this first strategy (Fig. 2a), we enhance the segmentation results of the U-Net model using a Graph Neural

Network (GNN) trained on a graph constructed from supervoxel representations of the input image. The purpose of this strategy is to refine uncertain regions in the CNN predictions by propagating contextual information through the graph structure.

During training, each node is assigned a single label L_i, and this label is propagated to all voxels v_j within the corresponding supervoxel s_i. However, a limitation arises when supervoxels span voxels with heterogeneous labels, which can lead to inconsistencies in voxel-level reconstruction. To address this, we propose an updated SLIC-based variant: for any supervoxel s_i containing voxels with multiple predicted labels from the U-Net, the region is split into individual voxels. Each voxel is then treated as its own node in the graph. This refinement removes the constraint of label homogeneity per supervoxel and allows the GNN to predict at a finer spatial resolution.

Thus, we evaluate two variants under this strategy:

- **Original SLIC Graph:** Each node represents a supervoxel; all voxels within a supervoxel share the same predicted label.
- **Updated SLIC Graph:** Supervoxels with label heterogeneity are decomposed into voxel-level nodes to avoid label inconsistencies and improve reconstruction.

This strategy provides a direct and lightweight means of improving the segmentation outputs without introducing additional encoding modules. The graph-level predictions are ultimately projected back to the voxel space to obtain the final segmentation result.

Strategy 2: 1D CNN for Encoding CNN-Predicted Labels. In this strategy (Fig. 2b), we integrate a 1D Convolutional Neural Network (1D CNN) into the segmentation pipeline to encode U-Net predictions alongside graph-based features. This enriched representation is used to enhance the node features in the GNN, particularly in regions where U-Net predictions are uncertain.

For each graph node, we construct a feature vector by concatenating the original node attributes with a label representation derived from U-Net outputs. To ensure compatibility with the 1D CNN input format, the predicted label is repeated to match the dimensionality of the feature vector. The combined vector is passed through the 1D CNN, which outputs a compact and expressive encoding. The output of the penultimate CNN layer is used as the updated node feature vector.

The motivation for using a 1D CNN stems from its ability to efficiently capture local dependencies in structured 1D sequences while maintaining a lightweight architecture. Compared to fully connected layers, 1D convolutions are parameter-efficient and better preserve feature locality, making them suitable for embedding structured semantic and spatial information into graph node representations.

We explore two variants under this strategy:

- **Training on Ground Truth Labels:** The CNN learns to encode predicted inputs using supervision from the dataset labels.

- **Training on CNN-Predicted Labels:** The CNN is trained using the U-Net's own predictions, enabling the model to operate in a weakly supervised regime.

Detailed training settings, loss function choices, and class weighting schemes are provided in Sect. 4.

Strategy 3: GNN Encoding of U-Net Predicted Labels. In this strategy (Fig. 2c), we directly integrate the U-Net predicted label with each node's original graph features by simple concatenation, forming a 21-dimensional feature vector per node (20 from graph features + 1 from label). The GNN then processes the resulting graph without any additional encoding step.

Unlike Strategy 2, which learns a joint feature representation through a 1D CNN, this approach relies on the GNN itself to extract meaningful interactions between raw features and labels. This simplification makes the pipeline more efficient while still leveraging the predictive power of the U-Net.

The key challenge in this strategy arises from the presence of masked regions, areas where U-Net predictions are uncertain or unreliable. To address this, we explore several initialization techniques for the label component of the node features in these uncertain regions:

1. **Keep Original Labels**
   In this baseline approach, we concatenate the original labels to the node features for all nodes, including those in masked regions:
   $$\mathbf{h}_i = [\mathbf{f}_i; y_i]$$
   where $\mathbf{f}_i$ is the feature vector and y_i is the label of node i.
2. **Use a Dummy (Non-existent) Label**
   In this approach, nodes in masked regions are assigned a placeholder label d, while non-masked nodes retain their ground truth labels:
   $$\mathbf{h}_i = \begin{cases} [\mathbf{f}_i; y_i] & \text{if } i \text{ is not masked} \\ [\mathbf{f}_i; d] & \text{if } i \text{ is masked} \end{cases}$$
3. **Mean of Neighboring Labels**
   Here, each masked node is initialized using the average label of its neighbors, assuming local consistency:
   $$\mathbf{h}_i = \begin{cases} [\mathbf{f}_i; y_i] & \text{if } i \text{ is not masked} \\ \left[\mathbf{f}_i; \frac{1}{|\mathcal{N}(i)|} \sum_{j \in \mathcal{N}(i)} y_j\right] & \text{if } i \text{ is masked} \end{cases}$$
   where $\mathcal{N}(i)$ denotes the neighbors of node i.
4. **Weighted Mean Based on Label Distribution**
   In this approach, masked nodes take a weighted average of their neighbors'

labels. The weights reflect the distribution of labels in the training data:

$$\mathbf{h}_i = \begin{cases} [\mathbf{f}_i; y_i] & \text{if } i \text{ is not masked} \\ \left[\mathbf{f}_i; \frac{\sum_{j \in \mathcal{N}(i)} w_j \cdot y_j}{\sum_{j \in \mathcal{N}(i)} w_j}\right] & \text{if } i \text{ is masked} \end{cases}$$

where w_j is the weight associated with label y_j, typically defined based on label frequency in the dataset.

These initialization strategies allow the GNN to reason contextually in uncertain areas, making it possible to recover more accurate segmentation boundaries by propagating confident information from neighboring regions.

The specific design choices, training variants, and experimental configurations associated with each strategy are presented and analyzed in Sect. 4.

4 Experiments and Results

This section presents our experimental setup, baseline performance, and a detailed evaluation of the proposed strategies for enhancing U-Net segmentation using GNNs. We conclude with a comparative analysis across all strategies.

4.1 Experimental Setup

Dataset. Our experiments are based on the BraTS 2020 benchmark dataset [3], which collects pre-operative multi-modal MRI scans of patients diagnosed with glioblastoma (GBM) and lower-grade glioma (LGG) from multiple clinical sites. Each subject is provided with four imaging modalities: native T1-weighted, T2-weighted, contrast-enhanced T1 (T1CE), and FLAIR, offering complementary views of tumor structure and surrounding tissue.

The dataset includes voxel-wise annotations prepared by expert raters, distinguishing normal brain from pathological regions such as peritumoral edema, necrotic or non-enhancing tumor core, and active enhancing tumor. For evaluation, the BraTS challenge defines three aggregated tumor subregions: whole tumor (WT), tumor core (TC), and enhancing tumor (ET) (Fig. 3).

In total, the dataset provides 369 training cases with publicly available ground truth and 125 test cases without annotations. For our study, we relied on the labeled training set, further splitting it into 295 volumes for training and 74 volumes for testing.

Performance Metrics for Evaluation. Segmentation accuracy was evaluated using the Dice similarity coefficient (DSC), a widely used overlap measure in medical image analysis. To provide a more complete assessment, we report both micro- and macro-averaged Dice scores. The micro Dice reflects overall voxel-level agreement across all tumor regions, while the macro Dice averages the score per class, highlighting class-specific performance [6,8].

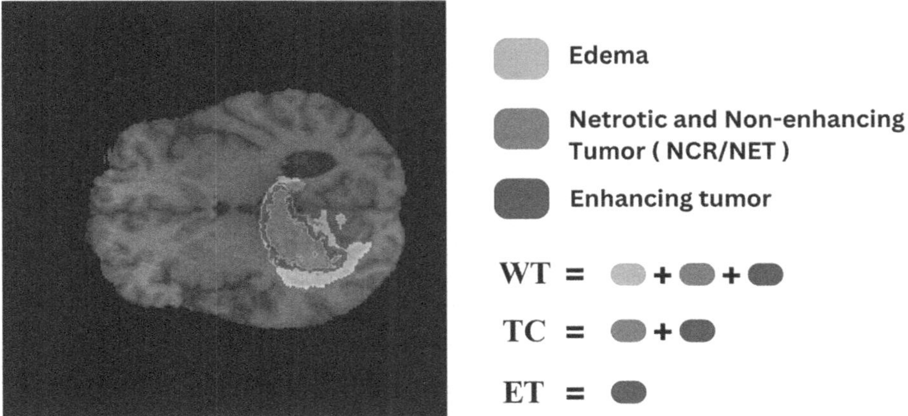

Fig. 3. Segmentation targets used in BraTS 2020: Whole Tumor (WT), Tumor Core (TC), and Enhancing Tumor (ET).

In our experiments, we evaluate performance at two levels: the node level, where predictions are made directly on the graph representation (supervoxels), and the image level, where graph-based predictions are projected back to produce full-resolution segmentation masks for comparison with the ground truth.

4.2 Platform and Environment

All experiments were conducted using the Google Colab Pro and Kaggle platforms, selected for their accessibility and availability of GPU-accelerated computing resources. Model training and inference were carried out on GPU-enabled instances, equipped with either NVIDIA Tesla P100 or T4 GPUs, depending on resource availability at runtime.

U-Net and GNN Implementation Details. The U-Net model consists of four encoder and decoder blocks with increasing filter depths (32, 64, 128, 256), ReLU activations, and 2D convolutions of kernel size 3×3. MaxPooling is used in the encoder for downsampling, and upsampling is performed using transposed convolutions. The model is trained using the Adam optimizer with a learning rate of 0.001 for 77 epochs, using categorical cross-entropy loss.

The GNN is composed of one input layer and four GraphSAGE layers with 128 hidden units, followed by a linear layer to predict class logits. ReLU activations are applied after each graph layer. The model is trained using the AdamW optimizer with a learning rate of 1×10^{-4} and a weighted cross-entropy loss to handle class imbalance. The class weights used are $[0.1900, 0.9970, 0.9100, 0.9930]$ (Table 2).

4.3 U-Net Baseline Performance

Table 2 reports the U-Net performance on the test set using standard segmentation metrics.

Table 2. U-Net model performance on the test set

Metric	Loss	Dice Macro	Dice Micro	Dice WT	Dice TC	Dice ET
Value	0.0326	0.8294	0.7104	0.7320	0.7037	0.6956

4.4 Evaluation of Strategy 1: GNN-Based Enhancement of U-Net Predictions

In this experiment, we assess how a GNN can refine U-Net outputs using graph-based representations built from SLIC supervoxels. We compare two graph types: one based on the original SLIC and another refined by splitting heterogeneous supervoxels using U-Net predictions (see Subsect. 3.5).

Table 3 shows that the original graph performs well at the node level, but suffers at the image level due to supervoxel boundary mismatches. In contrast, the refined graph improves image-level accuracy by better capturing complex regions, despite a slight drop in node-level Dice due to reduced regularization.

Though the U-Net model remains the same, node-level scores differ due to graph topology changes. However, its image-level Dice remains constant, as it's derived from full voxel predictions.

Figure 4 highlights these differences, showing cleaner segmentations from the updated graph, especially near label transitions.

Table 3. Performance Comparison of Strategy 1: GNN with Original vs. Updated SLIC Partitioning

Model	Dice Micro	Dice Macro	Dice WT	Dice TC	Dice ET
Node-Level Evaluation					
U-Net (Original SLIC)	0.9903	0.7793	0.7557	0.7297	0.7386
U-Net (Updated SLIC)	0.9654	0.7153	0.6686	0.6430	0.6617
U-Net+GNN (Original SLIC)	**0.9904**	**0.7922**	**0.7674**	**0.7464**	**0.8187**
U-Net+GNN (Updated SLIC)	0.9746	0.7668	0.7515	0.7179	0.7318
Image-Level Evaluation					
U-Net	0.8294	0.7105	0.7320	0.7037	0.6957
U-Net GNN (Original SLIC)	0.8258	0.6483	0.6592	0.6064	0.6538
U-Net GNN (Updated SLIC)	**0.8297**	**0.7202**	**0.7466**	**0.7207**	**0.7271**

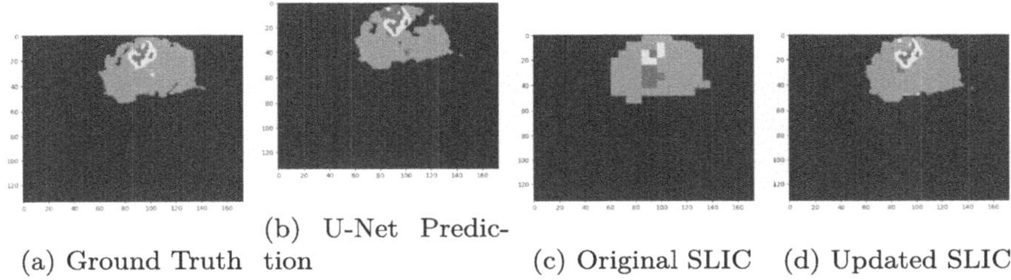

(a) Ground Truth (b) U-Net Prediction (c) Original SLIC (d) Updated SLIC

Fig. 4. Qualitative comparison of U-Net GNN predictions using original vs. updated SLIC graph representations.

4.5 Strategy 2: Using 1D CNN to Encode CNN Model Prediction with Graph Features

This strategy enriches graph node features using a 1D CNN to encode high-confidence U-Net predictions, aiming to guide GNN refinement in weakly predicted regions. We experiment with the two aforementioned supervision signals during the 1D CNN training: Training with Ground Truth Labels and Training with Predicted Labels. The 1D CNN was trained using the Adam optimizer with a learning rate of 0.001. A ReLU activation was applied after each convolutional layer. To address class imbalance, a weighted cross-entropy loss was employed with class weights [0.1900,0.9970,0.9100,0.9930]. During inference, we extracted the output from the penultimate layer of the 1D CNN to serve as the encoded feature vector for each node, yielding 20-dimensional embeddings.

Table 4 summarizes the segmentation results at both node and image levels for three model variants: the baseline U-Net, U-Net GNN with 1D CNN trained on ground truth labels, and U-Net+GNN with 1D CNN trained on predicted labels.

Table 4. Comparison of U-Net and U-Net GNN models using 1D CNN with different training strategies at node and image levels

Model (Node Level)	Dice Micro	Dice Macro	Dice WT	Dice TC	Dice ET
U-Net (Node)	0.9655	0.7153	0.6686	0.6430	0.6617
U-Net GNN + 1D CNN (GT Labels)	0.9679	0.7393	0.7080	0.6763	**0.6714**
U-Net GNN + 1D CNN (Pred. Labels)	**0.9715**	**0.7500**	**0.7228**	**0.6852**	0.6658
Model (Image Level)	**Dice Micro**	**Dice Macro**	**Dice WT**	**Dice TC**	**Dice ET**
U-Net (Image)	0.8294	0.7105	0.7320	0.7037	**0.6957**
U-Net GNN + 1D CNN (GT Labels)	0.8284	0.7061	0.7316	0.7046	0.6940
U-Net GNN + 1D CNN (Pred. Labels)	**0.8298**	**0.7141**	**0.7413**	**0.7070**	0.6844

Using predicted labels with the 1D CNN yielded the best node-level Dice Micro (0.9715) and Macro (0.7500), showing benefits from confidence-aligned features. At the image level, it also led in Dice Micro (0.8298) and Macro (0.7141), with gains in WT and TC, though U-Net remained better for ET. Overall, prediction-based encoding effectively enhances segmentation by aligning features with model confidence.

4.6 Strategy 3: GNN Encoding U-Net Prediction Labels

We evaluate the four GNN label encoding strategies described in Sect. 3.5, comparing their performance at both node and image levels (Table 5).

Table 5. Comparison of GNN Encoding Strategies Using CNN Prediction Labels

Level	Model	Dice Micro	Dice Macro	Dice WT	Dice TC	Dice ET
Node	U-Net	0.9655	0.7153	0.6686	0.6430	0.6617
	GNN Keep Labels	0.9766	**0.7929**	0.7731	0.7427	0.7483
	GNN Non-Existent Label	**0.9780**	0.7844	0.7728	0.7340	0.7466
	GNN Neighbor Mean	0.9770	0.7892	0.7685	0.7454	0.7484
	GNN Weighted Neighbor Mean	0.9776	0.7905	**0.7749**	**0.7489**	**0.7553**
Image	U-Net	0.8294	0.7105	0.7320	0.7037	0.6957
	GNN Keep Labels	0.8305	0.7410	0.7654	0.7443	0.7420
	GNN Non-Existent Label	0.8310	0.7328	0.7640	0.7353	0.7369
	GNN Neighbor Mean	0.8308	0.7405	0.7621	0.7430	0.7451
	GNN Weighted Neighbor Mean	**0.8311**	**0.7419**	**0.7676**	**0.7477**	**0.7525**

Experimental results show that integrating U-Net labels into GNN node features significantly boosts segmentation, especially at the node level. All four initialization methods outperform the U-Net baseline, with the weighted neighbor mean achieving the best overall Dice scores. This highlights the value of combining local context with global class distribution to better guide GNN refinement in uncertain regions.

4.7 Performance Comparison Across Training Strategies

Tables 6 and 7 summarize the best-performing variant from each strategy at node and image levels, respectively. These results confirm that Strategy 3 with weighted neighbor initialization performs best overall.

At the image level (Table 7), Strategy 3 achieves the best results across all metrics and subregions, confirming the benefit of integrating U-Net predictions with weighted neighbor labels. These improvements also have clinical relevance. The ET Dice increases from 0.6957 with U-Net to 0.7525 with Strategy 3, a meaningful gain since enhancing tumor guides surgery and radiotherapy planning; better delineation here can reduce the risk of residual tumor. Similarly, TC (0.7477 vs. 0.7037) and WT (0.7676 vs. 0.7320) improvements support more

Table 6. Node-Level Comparison of Best-Performing Models from Each Strategy

Model	Dice Micro	Dice Macro	Dice WT	Dice TC	Dice ET
U-Net (Node)	0.9655	0.7153	0.6686	0.6430	0.6617
Strategy 1 (Updated SLIC)	0.9746	0.7668	0.7515	0.7179	0.7318
Strategy 2 (Pred. Labels)	0.9715	0.7500	0.7228	0.6852	0.6658
Strategy 3 (Weighted Mean)	**0.9776**	**0.7905**	**0.7749**	**0.7489**	**0.7553**

Table 7. Image-Level Comparison of Best-Performing Models from Each Strategy

Model	Dice Micro	Dice Macro	Dice WT	Dice TC	Dice ET
U-Net (Image)	0.8294	0.7105	0.7320	0.7037	0.6957
Strategy 1 (Updated SLIC)	0.8297	0.7202	0.7466	0.7207	0.7271
Strategy 2 (Pred. Labels)	0.8298	0.7141	0.7413	0.7070	0.6844
Strategy 3 (Weighted Mean)	**0.8311**	**0.7419**	**0.7676**	**0.7477**	**0.7525**

accurate estimation of tumor extent and edema, which are critical for defining resection margins and radiation fields. Strategy 1, using refined SLIC partitioning, also improves over U-Net and approaches Strategy 3 in WT and TC, showing that finer supervoxels are particularly useful at class boundaries where infiltration is ambiguous on MRI. This suggests potential to reduce boundary-related errors that complicate treatment planning. Strategy 2, which encodes U-Net outputs with a 1D CNN, performs comparably overall but falls short in ET (0.6844), a clinically important limitation given the role of enhancing tumor in guiding therapy. At the node level (Table 6), Strategy 3 again leads across all metrics, followed by Strategy 2, while Strategy 1 lags due to coarse supervoxels. Although node-level outputs are not directly used clinically, consistent gains here show that the refinement is already effective in the graph space, providing a methodological basis for future explainability, such as tracing how uncertain regions are corrected through neighborhood interactions.

5 Conclusion

In this work, we proposed a hybrid U-Net GNN framework for brain tumor segmentation, where graph refinement is applied selectively to regions of uncertainty in CNN predictions. Unlike traditional approaches that process the entire image uniformly, our method focuses GNN refinement specifically on regions where the U-Net model exhibits uncertainty. We designed and evaluated three strategies: graph refinement via SLIC partitioning, node feature augmentation using a 1D CNN encoder, and contextual label propagation in masked regions. The results demonstrate consistent improvements over the baseline U-Net, particularly in the most challenging subregions. Although our performance does not yet reach the level of large-scale state-of-the-art models, the gains confirm that selectively incorporating GNNs is a promising direction. Our contribution lies in showing that lightweight GNN refinements, when carefully integrated, can boost segmentation quality without requiring full 3D architectures or massive computational resources.

However, several limitations remain. First, our study is restricted to the BraTS2020 dataset and 2D slice processing; extending the framework to multi-dataset and volumetric 3D experiments is a natural next step. Second, heuristic design choices for confidence masking and supervoxel generation may limit adaptability and could be replaced by learnable mechanisms. Finally, comparisons with state-of-the-art architectures remain a crucial benchmark for future evaluation.

References

1. Abidin, Z.U., Naqvi, R.A., Haider, A., Kim, H.S., Jeong, D., Lee, S.W.: Recent deep learning-based brain tumor segmentation models using multi-modality magnetic resonance imaging: a prospective survey. Front. Bioeng. Biotechnol. **12**, 1392807 (2024)
2. Achanta, R., Shaji, A., Smith, K., Lucchi, A., Fua, P., Süsstrunk, S.: Slic superpixels. technical report, EPFL (2010)
3. Bakas, S., et al.: Identifying the best machine learning algorithms for brain tumor segmentation, progression assessment, and overall survival prediction in the BraTS challenge. In: Brainlesion: Glioma, Multiple Sclerosis, Stroke and Traumatic Brain Injuries, pp. 20–38. Springer (2020)
4. Gammoudi, I., Ghozi, R., Mahjoub, M.A.: Hybrid architecture for 3d brain tumor image segmentation based on graph neural network pooling. In: International Conference on Computational Collective Intelligence, pp. 337–351. Springer (2022)
5. Li, J., Chen, J., Tang, Y., Wang, C., Landman, B.A., Zhou, S.K.: Transforming medical imaging with transformers? a comparative review of key properties, current progresses, and future perspectives. Med. Image Anal. **85**, 102762 (2023)
6. Menze, B.H., et al.: The multimodal brain tumor image segmentation benchmark (brats). IEEE Trans. Med. Imaging **34**(10), 1993–2024 (2015)
7. Mohammadi, S., Allali, M.: Advancing brain tumor segmentation with spectral-spatial graph neural networks. Appl. Sci. **14**(8), 3424 (2024)
8. Müller, D., Soto-Rey, I., Kramer, F.: Towards a guideline for evaluation metrics in medical image segmentation. BMC. Res. Notes **15**(1), 210 (2022)
9. Patel, D., Patel, D., Saxena, R., Akilan, T.: Multi-class brain tumor segmentation using graph attention network. In: 2023 8th International Conference on Signal and Image Processing (ICSIP), pp. 196–201. IEEE (2023)
10. Pratap Joshi, K., Gowda, V.B., Bidare Divakarachari, P., Siddappa Parameshwarappa, P., Patra, R.K.: VSA-GCNN: attention guided graph neural networks for brain tumor segmentation and classification. Big Data Cogn. Comput. **9**(2), 29 (2025)
11. Saueressig, C., Berkley, A., Kang, E., Munbodh, R., Singh, R.: Exploring graph-based neural networks for automatic brain tumor segmentation. In: International Symposium: From Data to Models and Back, pp. 18–37. Springer (2020)
12. Saueressig, C., Berkley, A., Munbodh, R., Singh, R.: A joint graph and image convolution network for automatic brain tumor segmentation. In: International MICCAI Brainlesion Workshop, pp. 356–365. Springer (2021)

13. Umarani, C.M., Gollagi, S., Allagi, S., Sambrekar, K., Ankali, S.B.: Advancements in deep learning techniques for brain tumor segmentation: a survey. Inf. Med. Unlocked **50**, 101576 (2024)
14. Vatanpour, M., Haddadnia, J., Bajestani, S.S.: Combination of graph and convolutional networks for brain tumor segmentation from multi-modal MR images in clinical applications. Front. Biomed. Tech. **12**(3), 547–556 (2025)
15. Ziaeetabar, F.: Efficientgformer: multimodal brain tumor segmentation via pruned graph-augmented transformer. arXiv preprint arXiv:2508.01465 (2025)

HeteroGAT: A Heterogeneous Graph Attention Network for Multimodal Fake News Detection in Chinese Social Media

Soufiane Khedairia[1(✉)], Mohamed Aimen Ouacel[2], and Aida Chefrour[2]

[1] LiM Laboratory, Department of Computer Science, Faculty of Science and Technology, University of Souk Ahras, 41000 Souk Ahras, Algeria
soufiane.khedairia@univ-soukahras.dz

[2] Department of Computer Science, Faculty of Science and Technology, University of Souk Ahras, 41000 Souk Ahras, Algeria
{m.ouacel,aida.chefrour}@univ-soukahras.dz

Abstract. The widespread dissemination of misinformation poses significant challenges to public health, safety, and information integrity, particularly on social media platforms. To address this, we propose the HeteroGAT model, a novel heterogeneous graph attention network designed for multimodal Chinese fake news detection integrated with genetic algorithm-based hyperparameter optimization. The proposed model combines textual features extracted via BERT-base-chinese and visual features obtained through ResNet-50 in a unified heterogeneous graph structure. Hyperparameters are optimized with a genetic algorithm, resulting in a compact and efficient architecture. Experimental evaluation demonstrates that HeteroGAT achieves state-of-the-art performance with an accuracy of 99.41% and an F1-score of 99.23%, significantly outperforming strong baselines including deep learning and transformer-based models. These results highlight the critical role of integrating multimodal features and heterogeneous social graph structures for the robust identification of fake news.

Keywords: Fake news detection · Heterogeneous graph attention network · Multimodal fake news detection

1 Introduction

The spread of fake news in social media platforms is a is a major worldwide issue that have a serious societal implications, ranging from political manipulation to public health misinformation [6]. For example, a false claim that chloroquine could treat COVID-19 prompted many people to take chloroquine phosphate in the hopes that it would protect them from the deadly virus [14]. Unfortunately, this tragedy led to hospitalisations and deaths. In recent years, the rapid dissemination of fake news on social media involves diverse data forms, including textual content, images, videos, and social context, where news articles that combine images and text are gaining more popularity in social media [10]. Instead

T. Ensari et al. (Eds.): ISPR 2025, CCIS 2859, pp. 194–206, 2026.
https://doi.org/10.1007/978-3-032-21585-7_14

of concentrating on just one modality of content which often fail to grasp the nuanced correlations and interactions across modalities, the multimodal nature of fake news necessitates detection methods that can jointly model and integrate heterogeneous data sources to capture rich semantic and relational information.

The multimodal fake news detection mechanism involves feature extraction and feature fusion. In feature extraction, CNNs are typically used for visual features, while Bi-LSTM, textCNN, or BERT can be used for textual features, ensuring a comprehensive analysis of multimodal data [20]. Heterogeneous GNNs provide a powerful framework to meet this challenge by representing different modalities as distinct node types in a graph, enabling the model to learn both intra-modal and inter-modal relationships effectively. Due to its ability to capture more significant information, the heterogeneous graph was judged to be better than the homogeneous graph. Consequently, it achieves superior outcomes compared to homogeneous graphs [17]. For instance, heterogeneous GNNs, which contain different types of edges and nodes, can model dependencies such as semantic relations in text, spatial features in images, and user interactions in social networks, all within a unified graph structure, which conventional models cannot easily achieve.

Furthermore, the success of heterogeneous GNNs in this domain significantly depends on the effective tuning of hyperparameters, including learning rates, number of layers, attention heads, and fusion weights across modalities. Due to the complexity and high dimensionality of this parameter space, conventional grid or random search methods are inefficient and may lead to suboptimal performance. A good hyperparameter configuration for a GNN will result in efficient training and precise predictions, but a bad configuration will provide different outcomes. These hyperparameters will impact the training process and learning outcomes [25]. However, most deep learning models, including GNNs, have more complex architectures and require more time to train than typical machine learning techniques. This indicates that hyperparameter optimisation (HPO) for GNN is in fact a very costly task [25].

The existing HPO methods widely used across machine learning models, including GNNs: random search, grid search [3], Gaussian, Bayesian methods [18], and many other methods. Nevertheless, the majority of these are plagued by high processing costs. Using Genetic Algorithms (GA) for HPO introduces an evolutionary strategy inspired by natural selection that iteratively explores and exploits the search space to find near-optimal hyperparameter sets. By optimising the graph architecture, attention mechanisms, and feature fusion, this method improves model performance and accelerates convergence and generalisability.

The primary purpose of this work is to develop a heterogeneous GNN-based multimodal fake news detection approach with hyperparameter tuning via genetic algorithms to effectively utilise the synergy between several data modalities. Enhancing detection accuracy, preserving scalability for real-time applications, and tackling the complexities of fake news dessimination on social media platforms. The proposed approach addresses the following issues:

- Semantic and structural heterogeneity: By modeling fake news in social media, which involves diverse data forms, including textual content, images, videos, and social context within a unified heterogenous graph, capturing key indicators that represent fake news.
- Robust and optimal model configuration: By automatically identifying hyper-parameters, improving convergence, reducing overfitting, and enhancing scalability for deployment in real-world social platforms.

2 Related Work

2.1 Multimodal Fake News Detection

Images contain richer information and are more effective at expressing and disseminating content than text-only articles, which draws in more readers. According to Jin et al. [9] study, articles with images receive 11 times as many retweets on average as articles without images. Consequently, Combining text and visual features may improve the performance of fake news detection, by fully learning the semantic features of the data. To overcome the single model limitation, an increasing number of studies incorporate multimodal information, thereby enhancing detection performance [10,20,26]. Giachanou et al. [7] combined semantic, visual, and textual information to leverage the complementary strengths of different modalities. Kumari et al. [12] proposed a multimodal framework for detecting fake news that improves multimodal information representation by maximizing the correlation between text and image features. Zhang et al. [26] proposed a multimodal approach for fake news detection named GBCA by creating a model for a graph classification task based on the propagation structure of fake news. Then, the second model is created by extracting semantic features from fake news using BERT model. Finally, the co-attention mechanism is employed to integrate the propagation structure and semantic features, improving the accuracy of fake news detection.

A common practice in the production of multimodal fake news is the inclusion of irrelevant images with textual posts. Hence, the coherence the coherence between textual content and accompanying images plays a crucial role in multimodal fake news detection, as inconsistencies or mismatches often serve as strong indicators of misinformation [2]. Zhou et al. [27] proposed a SAFE model that examined similarity measurement in assessing multimodal information consistency. Alternatively, Xue et al. [21] presented the Multimodal Consistency Neural Network (MCNN), which evaluated the correlation between text and images by incorporating a similarity measurement technique.

2.2 Heterogeneous Graph Neural Networks

Most existing GNN-based models for detecting fake news create homogeneous graphs. Nevertheless, it is challenging to display all of the news text, pictures, and videos on these graphs at once. The heterogeneous graph has richer semantics and more comprehensive information because it permits different kinds of edges or nodes to have different feature or attribute dimensions. Let G a graph represented as:

$$G = (V, E) \tag{1}$$

where V stands for the graph's nodes (vertices) and E for its edges. In heterogeneous graph $V = V_1 \cup V_2 \cup ... \cup V_k$, where for $i \neq j$, $V_i \cap V_j = \emptyset$.

Fake news typically involves intricate relationships across different modalities such as text, images, users, and knowledge entities. Ke et al. [11] proposed a rumor detection approach called KZWANG based on heterogeneous graph by collecting the local and global connections between users, reposts, and sources on Weibo. A two-layer heterogeneous graph attention mechanism was proposed by Sun et al. [19] to address the issues of low efficacy in detecting fake news based on heterogeneous maps and the challenge of extracting useful features from brief texts. The Decision-based Heterogeneous Graph Attention Network (DHGAT) represents a significant advancement in this domain, introducing dynamic neighborhood optimization for enhanced fake news detection [13]. DHGAT addresses limitations of traditional GNNs by dynamically selecting optimal neighborhood types for each node, resulting in approximately 4% accuracy improvements over existing methods on the LIAR dataset.

3 Methodologie

The overall structure of the proposed model is shown in Fig. 1. The proposed HeteroGat multimodal fake news detection system comprises four primary components: data preprocessing and feature extraction, heterogeneous graph construction, HeteroGAT model implementation, and genetic algorithm-based hyperparameter optimization. The HeteroGat model processes the news posts with binary classification labels, achieving comprehensive multimodal analysis through the integration of textual and visual modalities.

3.1 Textual Feature Extraction

Text and image information are the two primary modalities that are checked by multimodal fake news detection techniques. News events are primarily expressed in the text, which offers crucial hints for determining the news's credibility. In order to capture the superficial features of the text, most methods model the input textual information using recurrent neural networks. However, these methods struggle to effectively capture the complex semantic features of fake news [10]. The transformer (BERT) provides pretrained bidirectional encoder representations, which effectively model rich contextual semantics, enhancing the accuracy of fake news detection. In this work, we employs the BERT-base-chinese [4], a pre-trained transformer model specifically designed for Chinese text understanding. The implementation utilizes BertTokenizer with truncation at 128 tokens and padding to maximum length, generating 768-dimensional embeddings from the token. This method extracts contextual dependencies and semantic relationships within news text, providing rich textual representations

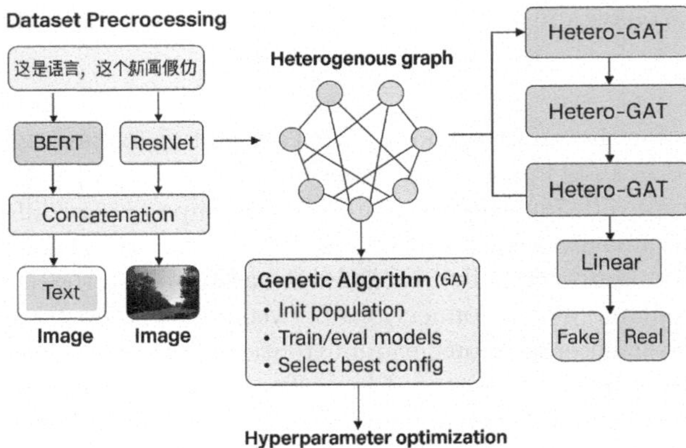

Fig. 1. The HeteroGat multimodal approach for fake news detection consists of four modules: data preprocessing and feature extraction, heterogeneous graph construction, HeteroGAT model implementation, and genetic algorithm-based hyperparameter optimization.

that traditional bag-of-words methods cannot achieve. Tokenized text is processed by the model using a number of transformer layers using self-attention mechanisms to capture contextual information and long-range dependencies that are crucial for differentiating authentic content from fake [5].

3.2 Visual Feature Extraction

The authenticity of an article can also be determined by its images, and articles with images that do not correspond with the text are frequently not authentic. In this work, visual features are extracted using ResNet-50 [8], a deep convolutional neural network pre-trained on ImageNet. For each news item, the associated image is loaded, resized to 224×224 pixels, normalized according to ImageNet standards, and processed through the penultimate layer of a pre-trained ResNet-50 model to produce a 2048-dimensional visual feature vector. If an image is missing or unreadable, a zero vector replaces it to maintain data consistency. The robust feature extraction powers of ResNet-50 extracts spatial patterns, textures, and high-level visual concepts crucial for identifying manipulated or misleading images commonly associated with fake news. Resnet50 model can learn hierarchical representations from low-level edge detection to high-level semantic understanding, making it particularly useful for detecting visual inconsistencies and manipulations in news-related imagery.

3.3 Heterogeneous Graph Construction

The HeteroGat approach constructs heterogeneous graphs representing news posts as multi-node structures with distinct node types and edge relationships.

Each news post is modeled as a graph containing post nodes with concatenated text and image features (768 + 2048 = 2816 dimensions) and user (representing the news creator) nodes initialized with 64-dimensional features as shown in Table 1. The preprocessing pipeline also establishes edge connections between post and user nodes to reflect social relationships relevant for graph-based modeling. This stratified, multimodal feature extraction approach ensures balanced, robust input for downstream graph neural network learning, accommodating both content and context in the detection pipeline. In summary, the heterogeneous graph construction pipeline includes the following steps:

Table 1. The node's type of the constructed heterogeneous graph

Node	Feature Type	Dimensionality	Role
post	BERT+ResNet-50 fusion	2816	Repr. multimodal article
user	Characterics vector	64	Repr. Creator/user

- **Data integration:** News items from authentic and fake datasets are combined. Each is associated with the respective user and the image if available
- **Feature extraction:** Textual and visual characteristics are extracted and concatenated for each post node. Missing or unreadable images are handled robustly by replacing them with zero vectors to maintain feature consistency.
- **Node and edge creation:** The post node acquires the pre-fused multimedia feature and its binary labeling (fake or real). The user node is associated to each node post, and edges are formed to establish the user-post relationship in both directions.
- **Graph serialization:** Each constructed heterogeneous graph (one per news post) is saved for further use in model training, validation, and testing.

The heterogeneous graph structure includes multiple edge types: 'posts' and 'rev_posts' relationships, enabling bidirectional information flow between users and content.

3.4 HeteroGAT Model Architecture

The Heterogeneous Graph Attention Network (HeteroGAT) model implements a sophisticated architecture designed to handle heterogeneous data that contain multiple types of nodes and/or edges. This architecture is particularly well-suited for multimodal tasks such as fake news detection, where textual, visual, and relational information are integrated to improve the detection of fake news. The proposed architecture includes configurable attention heads (1–14 heads based on genetic algorithm optimization), edge-type specific attention mechanisms, and sum aggregation across multiple edge types. By leveraging attention mechanisms across different relation types, HeteroGAT can effectively model both content and context within complex social information networks.

a. Graph Attention and HeteroConv Layers

The HeteroGAT model extends the graph attention network paradigm to heterogeneous graphs using PyTorch Geometric's HeteroConv and GATConv modules. Basically, the model uses a series of HeteroConv layers, each containing a distinct set of GATConv operations, one per edge type. Each GATConv applies self-attention to aggregate messages from node's neighbors by learning dynamic and context-dependent weights for each edge. The use of attention enables selective information flow and helps focus learning on more relevant relationships, which is essential in multimodal graphs where noise and irrelevant connections are common. For each layer l, every node and edge type, the following operations are executed:

- **Node Feature Aggregation:** Updates each node by aggregating features from neighbors through attention weights adapted to each edge type.
- **Multi-head Attention:** Multiple independent attention mechanisms are used (several number of heads); their outputs are concatenated for richer representation.
- **Activation and Dropout:** After each aggregation, non-linear activation (ELU) and dropout are applied to promote model expressiveness and avoid overfitting.

b. Output and Prediction

The output of the final attention layer is passed through a fully connected layer, projecting from the concatenated hidden representations (hidden_channels × num_heads) to the target dimension (typically, the number of classes for classification). For fake news detection task, only the "post" node output is required for prediction; other node types are included for relational context, enabling binary classification of news content as authentic or fake.

3.5 Hyperparameter Optimization

The different hyperparameters to be optimized for the HeteroGat model are described below:

- **Hidden Channels:** Controls the feature dimensionality in each hidden layer.
- **Num Layers:** The number of stacked attention layers.
- **Num Heads:** The number of attention heads, supporting learning of multiple, independent attention patterns.
- **Dropout Rate:** Regularization applied after each layer.
- **Learning Rate and Weight Decay:** Optimizer-specific hyperparameters.

These parameters are optimized through a genetic algorithm in the experimental setup, where candidate hyperparameter sets are evolved based on validation performance.

4 Experimental Results and Analysis

In this section, we begin by describing the dataset used to benchmark our approach. The empirical evaluation of the proposed approach is then discussed, with an emphasis on the fake news detection issue.

4.1 Dataset Description

For model evaluation, we have used a benchmark dataset called "the Chinese COVID-19 Fake News Dataset" (CHECKED) [22] developed to facilitate research on COVID-19 misinformation detection within Chinese social media environments. The first dataset of its kind in Chinese, CHECKED offers a comprehensive and multifaceted corpus of microblogs on the COVID-19 epidemic that are richly annotated for information credibility. CHECKED gathers 1,760 authentic posts with comments and reposts on Weibo between December 2019 and August 2020, along with 344 fraudulent posts. Since fake news is far less prevalent than true news, it has a class-unbalanced issue. The detailed data content description of the CHECKED dataset is presented in Table 2.

Table 2. The detailed data content description of the CHECKED dataset

Data Content Component	Description
Microblog Posts	- Full textual content in Chinese (hashtags, mentions,...). - Authentic social media language about COVID-19-related news.
Image Content	Associated images per post(photos, charts, screenshots, memes).
Post Metadata	- Publication timestamp (date and time). - Post identifiers and source information.
User Profiles	- Anonymized user identifiers linked to posts and interactions. - Includes profile features like user activity level for social context.
Engagement Metrics	- Numerical data: number of reposts, comments, likes.
Propagation and Network	- Repost chains showing how content spreads across user networks. - Comment threads forming discussion hierarchies. - User-post edge relations capturing interactions.
Labels	Binary labels (Fake or Real) assigned by expert fact-checking.

4.2 Experimental Settings

a. Implementation Details and Hyperparameters:

All experiments are carried out on a Linux machine, while PyTorch and PyTorch-Geometric are used to create all models. The CHECKED dataset is divided into training (72%, 1,514 samples), validation (8%, 169 samples), and test (20%, 421 samples) sets using a stratified sampling approach to maintain balanced class distributions of fake and real news instances across all splits. For training the HeteroGat model on CHECKED dataset, a small batch size of 8 is typically used to handle the complexity of graph data effectively while maintaining stable optimization. The HeteroGat model variants are trained using the Adam optimizer and cross-entropy loss as the objective function for the binary classification task of fake news detection. Hyperparameters are optimized using the genetic algorithm. The hidden_channel values include 32, 64, and 128, balancing model expressiveness with computational cost. The num_layers hyperparameter is set to 1, 2, or 3, specifying how many graph attention convolutional layers the model contains. The number of parallel attention heads per layer (num_heads), is selected from a range including 1, 2, 4, 6, 8, 10, 12, or 14, allowing flexibility in capturing diverse relational patterns. Additionally, the dropout_rate values are set within the range of 0.0001 to 0.5, controlling the strength of regularization during training to prevent overfitting. Finally, for the HeteroGAT experiments, the learning rate is typically chosen within the range of 0.0001 to 0.5, with 0.0001 being a common optimal setting found during genetic algorithm hyperparameter tuning. The HeteroGAT hyperparameter values optimized using the genetic algorithm in this work are described in Table 3.

Table 3. Key Hyperparameters

Hyperparameter	Description	Typical Value/Range
hidden_channels	Latent feature dimensions in each layer	{32, 64, 128}
num_layers	Number of GAT-based convolution layers	{1, 2, 3}
num_heads	Attention heads per GAT operation	{1, 2, 4, 6, 8, 10, 12, 14}
dropout_rate	Dropout probability per layer	{0.0001 – 0.5}
learning_rate	The learning rate controls the step size	{0.0001 – 0.5}

b. Baselines:

To validate the performance of our proposed model, we evaluate the proposed HeteoroGat model and baselines in terms of Accuracy and F1-score on the CHECKED dataset.

- **SVM-TS** Ma et al. [16] presents a linear SVM classifier that models the variation of social environment characteristics using time-series structures.
- **CNN** is the acronym for a convolutional neural network that groups similar posts into fixed-length convolutional sequences for the purpose of classifying and representing rumours [24].
- **GRU** A multilayer generalised recurrent neural network with gates, known as GRU [15], uses microblogs represented as variable time series to train a rumour classifier.

- **CSN-BERT** Yang and Pan [23] post-trained the original BERT to the COVID-19 Social Network BERT model,specifically tailored for COVID-19 user posts on social networks.
- **CNFRD** A Few-Shot Rumor Detection Framework via Capsule Network for COVID-19 developed by chen et al. [1] (Table 4).

Table 4. Evaluation metrics of baselines and the HeteroGat model on the CHECKED dataset for Chinese COVID-19 rumor detection.

Methods	Accuracy (%)	F1-score
SVM-TS	65.22	64.88
CNN	75.56	77.06
GRU	80.52	79.60
CNFRD	88.92	88.62
CSN-BERT	98.89	98.18
HeteroGat(ours)	99.41	99.23

c. Evaluation Metrics:

Considering that the fake news detection task is binary, we evaluate our model's performance using accuracy (Acc.) and F1-score metrics.

4.3 Results and Analysis

The HeteroGAT model achieves outstanding performance in detecting COVID-19 rumors within the Chinese checked dataset, reaching an accuracy of 99.41% and an F1-score of 99.23%. This remarkable result reflects the ability of the model to efficiently combined multimodal information, such as textual embeddings captured using BERT and visual features from ResNet, and relational social context encoded by heterogeneous graph attention networks. Compared to several baseline methods, including traditional machine learning models like SVM with text features, which achieved around 65% accuracy, and deep learning models such as CNN and GRU, which achieved moderate improvements with accuracies of roughly 75.5% and 80.5%, respectively, HeteroGAT's superiority is clear. More advanced baselines designed for fake news detection, such as CNFRD, achieved an accuracy of nearly 88.9%, while transformer-based models like CSN-BERT, which leverage pre-trained language representations, improved significantly to approximately 98.9%. The HeteroGAT model outperforms these robust benchmarks by explicitly incorporating a heterogeneous graph structure and multimodal content through attention mechanisms, enabling it to adaptively weight different types of information on nodes and edges. This allows for better distinction between real and fake information, particularly in complex and noisy social media environments. The results highlight the importance of combining content characteristics with social propagation and relational context, aspects that are often overlooked by traditional sequence-based or unimodal approaches. Overall, these results highlight that heterogeneous graph attention networks, when combined with robust multimodal feature fusion and optimized hyperparameters, offer a powerful and state-of-the-art solution for detecting COVID-19 misinformation on Chinese social media platforms.

5 Conclusion

This study presents a comprehensive multimodal fake news detection framework that successfully integrates heterogeneous Graph Neural Networks with genetic algorithm-based hyperparameter optimization. The integration of textual features derived from BERT and visual embeddings extracted by ResNet-50 enables deep multimodal representation at the node level, while the graph attention layers capture complex propagation patterns and social dependencies critical for misinformation identification.

Experimental results on the CHECKED Chinese COVID -19 Fake News Dataset validate the efficiency of the proposed approach, where the HeteroGAT model has achieved a state-of-the-art accuracy of 99.41% and an F1-score of 99.23%, outperforming a comprehensive set of baseline methods including traditional machine learning, deep learning, and transformer-based models. The genetic algorithm-based hyperparameter optimization further contributed to robust and efficient model configurations, balancing expressiveness and regularization for optimal performance.

These findings highlight the importance of combining content and context via heterogeneous graph modeling and attentive fusion for tackling complex, real-world misinformation challenges. The scalability and modularity of HeteroGAT make it a promising framework for further research and deployment in dynamic social media environments. Future work may explore extending the model to incorporate richer user profiling, temporal dynamics, and cross-lingual capabilities.

References

1. Chen, D., Chen, X., Lu, P., Wang, X., Lan, X.: Cnfrd: A few-shot rumor detection framework via capsule network for covid-19. Int. J. Intell. Syst. **2023**(1), 2467539 (2023)
2. Chen, Y., Li, D., Zhang, P., Sui, J., Lv, Q., Tun, L., Shang, L.: Cross-modal ambiguity learning for multimodal fake news detection. In: Proceedings of the ACM Web Conference 2022, pp. 2897–2905 (2022)
3. Claesen, M., De Moor, B.: Hyperparameter search in machine learning. arXiv preprint arXiv:1502.02127 (2015)

4. Devlin, J., Chang, M.W., Lee, K., Toutanova, K.: Bert: Pre-training of deep bidirectional transformers for language understanding. In: Proceedings of the 2019 conference of the North American chapter of the association for computational linguistics: human language technologies, volume 1 (long and short papers), pp. 4171–4186 (2019)
5. Dong, Z., Tang, T., Li, L., Zhao, W.X.: A survey on long text modeling with transformers. arXiv preprint arXiv:2302.14502 (2023)
6. Ferrara, E., Cresci, S., Luceri, L.: Misinformation, manipulation, and abuse on social media in the era of covid-19. J. Comput. Soc. Sci. **3**(2), 271–277 (2020)
7. Giachanou, A., Zhang, G., Rosso, P.: Multimodal fake news detection with textual, visual and semantic information. In: International Conference on Text, Speech, and Dialogue, pp. 30–38. Springer (2020)
8. He, K., Zhang, X., Ren, S., Sun, J.: Deep residual learning for image recognition. In: Proceedings of the IEEE Conference on Computer Vision and Pattern Recognition, pp. 770–778 (2016)
9. Jin, Z., Cao, J., Zhang, Y., Zhou, J., Tian, Q.: Novel visual and statistical image features for microblogs news verification. IEEE Trans. Multimedia **19**(3), 598–608 (2016)
10. Jing, J., Wu, H., Sun, J., Fang, X., Zhang, H.: Multimodal fake news detection via progressive fusion networks. Inf. Process. Manage. **60**(1), 103120 (2023)
11. Ke, Z., Li, Z., Zhou, C., Sheng, J., Silamu, W., Guo, Q.: Rumor detection on social media via fused semantic information and a propagation heterogeneous graph. Symmetry **12**(11), 1806 (2020)
12. Kumari, R., Ekbal, A.: Amfb: attention based multimodal factorized bilinear pooling for multimodal fake news detection. Expert Syst. Appl. **184**, 115412 (2021)
13. Lakzaei, B., Chehreghani, M.H., Bagheri, A.: A decision-based heterogenous graph attention network for multi-class fake news detection. arXiv preprint arXiv:2501.03290 (2025)
14. Luvembe, A.M., Li, W., Li, S., Liu, F., Wu, X.: Caf-odnn: complementary attention fusion with optimized deep neural network for multimodal fake news detection. Inf. Process. Manage. **61**(3), 103653 (2024)
15. Ma, J., Gao, W., Mitra, P., Kwon, S., Jansen, B.J., Wong, K.F., Cha, M.: Detecting rumors from microblogs with recurrent neural networks (2016)
16. Ma, J., Gao, W., Wei, Z., Lu, Y., Wong, K.F.: Detect rumors using time series of social context information on microblogging websites. In: Proceedings of the 24th ACM International on Conference on Information and Knowledge Management, pp. 1751–1754 (2015)
17. Phan, H.T., Nguyen, N.T., Hwang, D.: Fake news detection: a survey of graph neural network methods. Appl. Soft Comput. **139**, 110235 (2023)
18. Snoek, J., Larochelle, H., Adams, R.P.: Practical bayesian optimization of machine learning algorithms. Advances in neural information processing systems **25** (2012)
19. Sun, L., Wang, H.: Topic-aware fake news detection based on heterogeneous graph. IEEE Access **11**, 103743–103752 (2023)
20. Wang, J., Mao, H., Li, H.: Fmfn: fine-grained multimodal fusion networks for fake news detection. Appl. Sci. **12**(3), 1093 (2022)
21. Xue, J., Wang, Y., Tian, Y., Li, Y., Shi, L., Wei, L.: Detecting fake news by exploring the consistency of multimodal data. Inf. Process. Manage. **58**(5), 102610 (2021)
22. Yang, C., Zhou, X., Zafarani, R.: Checked: Chinese covid-19 fake news dataset. Soc. Netw. Anal. Min. **11**(1), 58 (2021)

23. Yang, J., Pan, Y.: Covid-19 rumor detection on social networks based on content information and user response. Front. Phys. **9**, 763081 (2021)
24. Yu, F., Liu, Q., Wu, S., Wang, L., Tan, T., et al.: A convolutional approach for misinformation identification. In: IJCAI, vol. 2017, pp. 3901–3907 (2017)
25. Yuan, Y., Wang, W., Pang, W.: A genetic algorithm with tree-structured mutation for hyperparameter optimisation of graph neural networks. In: 2021 IEEE Congress on Evolutionary Computation (CEC), pp. 482–489. IEEE (2021)
26. Zhang, Z., Lv, Q., Jia, X., Yun, W., Miao, G., Mao, Z., Wu, G.: Gbca: Graph convolution network and bert combined with co-attention for fake news detection. Pattern Recogn. Lett. **180**, 26–32 (2024)
27. Zhou, X., Wu, J., Zafarani, R.: Safe: similarity-aware multi-modal fake news detection. arXiv preprint arXiv:2003.04981 (2020)

An Event-Based Data Generation Framework for Autonomous Driving Scenarios

Abdessamad El Kaouri[1,2(✉)], Mohamed Kas[1], Yassine Ruichek[1], and Youssef El Merabet[2]

[1] Université Marie et Louis Pasteur, UTBM, CIAD UR 7533, 90010 Belfort, France
abdessamad.elkaouri@uit.ac.ma, {mohamed.kas, yassine.ruichek}@utbm.fr

[2] Faculty of Science (FSK), SETIME, Ibn Tofail University (UIT), 14000 Kenitra, Morocco
youssef.elmerabet@uit.ac.ma

Abstract. The development of autonomous driving systems increasingly relies on rich, diverse, and well-annotated datasets, particularly for training deep learning models. However, collecting real-world data, especially from event cameras, remains a costly and labor-intensive task. In this work, we introduce a synthetic data generation tool built upon the CARLA simulator and extended with advanced event simulation capabilities via the V2E (Video to Events) framework. Our system enables the simulation of driving scenarios with fine-grained control over weather, lighting, traffic, and sensor configurations. It supports multi-modal sensor data capture, including RGB, depth, semantic segmentation, and instance segmentation, as well as realistic DVS event streams. By combining CARLA's diverse environments and rich sensor suite with high-fidelity event simulation, the tool facilitates the creation of scalable, customizable datasets that are well-suited for training and evaluating both frame-based and event-based perception algorithms.

Code: https://github.com/A-KAOURI/Carla_DVSV2E

Keywords: Autonomous Driving · CARLA · Simulation · Event-camera · Supervised Learning · Datasets · Depth · Segmentation

1 Introduction

Computer vision is crucial for autonomous driving, as it enables vehicles to interpret and understand their environment, significantly impacting decision-making and the overall performance of the vehicle's automation system. Vision for autonomous vehicles is based on various sensors, such as RGB cameras, LiDAR, and sonar, which provide raw data that are used for a variety of tasks, including low-level operations such as optical flow and depth estimation, as well as higher-level tasks such as semantic segmentation, object detection and recognition, road and lane detection, object tracking, and 3D reconstruction.

Though vision with a conventional captor like the RGB camera has achieved tremendous results, event cameras have emerged as a promising alternative or complimentary sensor [1, 4, 10]. These novel bio-inspired sensors work in a completely different

T. Ensari et al. (Eds.): ISPR 2025, CCIS 2859, pp. 207–218, 2026.
https://doi.org/10.1007/978-3-032-21585-7_15

paradigm: instead of capturing static frames at fixed time intervals, they asynchronously report brightness changes at each pixel at their time of occurrence. The resulting output is a stream of events that describe all the changes that have occurred in the scene by indicating the coordinates, the timestamp, and the polarity of each change (increase or decrease of brightness). The asynchronous nature of this data representation enables several key advantages: a high temporal resolution (on the order of microseconds), a high dynamic range (> 120 dB), immunity to motion blur, and low power consumption. These unique features make event-based vision very well suited for challenging scenarios that include high-speed motion, vast luminance variations, low light, or resource-constrained applications, where conventional cameras often suffer.

On the other hand, the advent of Deep Learning has revolutionized computer vision and brought significant enhancements in robustness, precision, and processing time. Therefore, remarkably impacting both frame-based and event-based vision. Nonetheless, the training and evaluation of deep neural networks remain heavily reliant on the availability of data, which is often limited — especially for event-based approaches. In response to this demand, a few realistic datasets [5, 13] have been gathered and annotated to provide sensor data alongside corresponding ground truths. However, the gathered data is still considered scarce. In fact, the collection of a substantial amount of annotated data is a challenging task, especially for applications like depth or semantic segmentation, which demand annotations at the pixel level. Moreover, many steps of the data preparation process can only be done manually, which makes it time-consuming and labor-intensive, further complicating dataset preparation. Compounded by the high cost of event-based cameras, these challenges limit accessibility and scalability for many researchers and developers, hence emphasizing the growing need for synthetic datasets which can be generated more efficiently through simulation.

In this work, we propose a new synthetic data collection tool that is based on CARLA simulator [3], taking full advantage of its comprehensive sensor suite to generate diverse high-quality synthetic data, including RGB images, depth maps, and semantic segmentation. While CARLA includes a basic event simulation system based on [12], its capabilities are limited in terms of accuracy and realism. To address this limitation, we leverage CARLA's capability of producing high frame rate image streams, by integrating V2E [7], an advanced event simulation tool that allows our system to capture intricate scene dynamics and to provide superior event-based annotations in addition to conventional visual data.

2 The CARLA Simulator

2.1 Overview

CARLA (Car Learning to Act) [3] is a widely used open-source simulation platform designed for the development, training, and validation of autonomous driving systems. Developed from ground up, it addresses the complexities of urban environments, such as multi-agent interactions, adherence to traffic rules, and the presence of pedestrians as well as dynamic obstacles. By providing a safe, controllable, and richly featured simulation environment, CARLA offers a compelling alternative to real-world testing, which is often constrained by high infrastructure costs, safety concerns, and logistical complexities. In doing so, CARLA helps to democratize autonomous driving research, making it accessible to a broader range of institutions and researchers who may lack the resources for large-scale physical experimentation.

Built on Unreal Engine, CARLA provides high-fidelity rendering quality, realistic physics, and a robust environment for simulation. It offers rich urban environments composed of meticulously designed 3D models for static objects like buildings, vegetation, and traffic infrastructure, as well as a wide range of vehicles. Additionally, the simulator supports a wide range of weather and lighting conditions, such as rain, fog, and sunset scenes. Moreover, it incorporates sophisticated non-player character (NPC) behavior. Non-player vehicles follow lanes, respect traffic lights and speed limits, and make decisions at intersections. On the other hand, pedestrians navigate according to a cost map that encourages sidewalk use and marked crossings but also allows for realistic jaywalking behavior.

2.2 Maps and Weather Conditions

CARLA offers a diverse set of 12 maps (virtual towns). These maps vary significantly in layout, scale, and environmental context, ranging from small, simple towns to larger, complex urban landscapes with roundabouts and multiple lanes. The collection includes specialized environments like a mountain town with a unique highway, long multi-lane highways with various entrances and exits, and even rural settings with narrow roads and minimal traffic lights. Furthermore, there are two large, expansive maps for broad environmental testing. Notably, two of these maps are secret unseen towns, specifically reserved for the CARLA leader-board challenge and not publicly accessible. These maps illustrate the environmental diversity CARLA offers, allowing developers to evaluate autonomous driving performance in different scenarios.

Beyond the physical road geometry, CARLA provides extensive control over weather and lighting conditions. The simulator includes a set of preset weather profiles (e.g. ClearNoon, CloudyNoon, HardRainNoon, ClearSunset, SoftRainSunset, etc.), spanning different times of day and precipitation levels. These presets adjust parameters like sky cloudiness, rain intensity, wetness, fog, wind, and sun angle to simulate various conditions. Recent versions of CARLA have also introduced nighttime settings (e.g. clear or cloudy night conditions) to test driving systems in low-light scenarios. Additionally, the weather system is fully configurable via the API: users may fine-tune continuous weather parameters (cloud cover percentage, sun altitude, rain magnitude, fog density, etc.) or interpolate conditions over time. This ability to vary environmental conditions in combination with diverse maps helps ensure that autonomous driving algorithms are robust to a wide range of scenarios and do not overfit to a single environment.

2.3 The Sensor Suite

A core feature of CARLA is its comprehensive flexible sensor suite, which models the real sensors and data streams used by autonomous vehicles. The simulator supports over a dozen sensor types, covering cameras, ranging sensors, and various special detectors.

CARLA includes multiple camera modalities such as standard RGB cameras, depth cameras, semantic segmentation cameras for ground-truth object labeling, instance segmentation cameras, optical flow camera, and a DVS event camera that simulates bio-inspired dynamic vision. Additionally, it supports 3D range sensors like LIDARs and radars and provides sensors for localization and state estimation (such as a GNSS receiver for GPS coordinates and an IMU for acceleration/gyroscope data), and utility sensors including collision detectors, lane invasion detectors, and obstacle detectors. This rich variety of sensors allows researchers to reconstruct a full autonomous vehicle sensor suite and obtain highly realistic data.

Importantly, CARLA's sensors are configurable and customizable to simulate different hardware models or experiment with sensor setups. Each sensor in CARLA is an actor with adjustable attributes that the user can specify through the API (by modifying the sensor's blueprint parameters before spawning it). For example, a camera's resolution, field of view, orientation, and capture rate can be tailored to the needs of the experiment. LIDAR sensors allow configuration of key properties such as number of laser channels, maximum range, rotation frequency, points generated per second, vertical and horizontal field of view, noise level, etc... Radar sensors similarly have tunable range and horizontal/vertical field-of-view angles, among other parameters.

3 Event-Camera Simulation

The development of effective event camera simulators is not trivial due to the fundamental differences in how event cameras operate compared to frame-based cameras. Early efforts in DVS simulation [9, 11] attempted accurate event camera simulation by generating events through images rendering from a 3D scene at a very high frame rate, using either Blender or a custom rendering engine. However, the first event simulator that could generate a large amount of dependable event data was ESIM [12]. Instead of rendering frames at a fixed high frequency to compute intensity changes, ESIM introduced an adaptive sampling strategy by triggering the rendering engine only when needed to produce an event. By establishing a tight coupling with the rendering engine, this approach allows ESIM to markedly reduce redundant computation while still capturing fine temporal changes, accurately mimicking the asynchronous nature of event cameras, and producing high-quality synthetic DVS events.

A limitation, however, is that ESIM on its own is an open-loop simulator – the camera trajectory must be specified (or pre-recorded) since ESIM does not natively include a physics world with moving objects. In practice, researchers either replay logged camera motions or integrate ESIM with a simulator for interactive scenarios. Moreover, while ESIM provided a significant leap forward, it often relied on an idealistic model of DVS pixels which assume infinite bandwidth, no temporal noise or leak events, and a uniformly distributed threshold mismatch. Such assumptions do not fully represent the complexities of real-world cameras, especially under challenging lighting conditions.

In contrast, our tool leverages the V2E (Video-to-Events) simulator [7], which focuses on modelling crucial real-world conditions, and instead of depending on the rendering engine, it can convert any standard video frames into synthetic event streams that simulate real DVS behavior. V2E incorporates realistic models of the sensor's behavior: it simulates the limited bandwidth of photoreceptors, per-pixel threshold variability, leak events, and intensity-dependent noise. By modelling these characteristics, the generated event stream more closely mimics the statistics of a real event camera, avoiding overly idealized output. Given that the simulator works better when given a high framerate image stream, its pipeline usually involves generating intermediate frames using SuperSloMo [8], then computing the difference in logarithmic brightness to generate events at microsecond intervals. Fig. 1 shows that V2E's events are more accurate, more danse and rich with details as opposed to ESIM's overly clean event streams that lack realistic noise and variability.

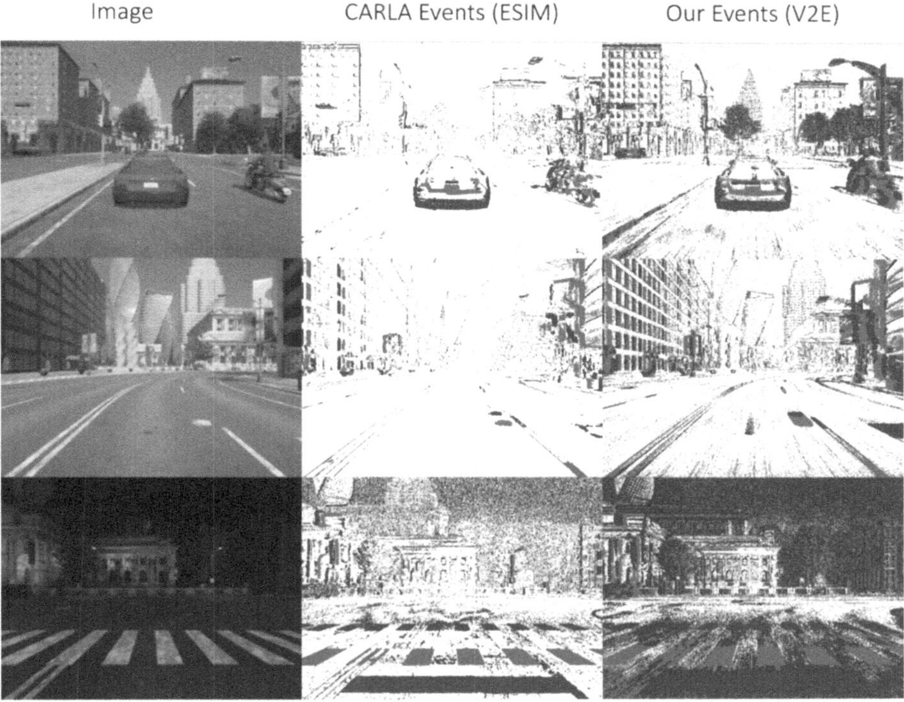

Fig. 1. Comparison between CARLA's built-in DVS output and our V2E-based events, over a 100 ms time window. Positive polarity events are shown in blue, and negative polarity events in red.

However, a key limitation of V2E is its computational expense. Achieving highly precise timestamp resolution, starting from a low frequency RGB stream, means interpolating a huge number of frames and processing them, which is far slower than real time – often hundreds of times slower. Thus, V2E is typically used as an offline tool to pre-generate event data rather than for live simulation. Moreover, the quality of the event output ultimately depends on the input video quality and frame rate; extremely high-speed motions that were not captured in the source frames can only be approximated via interpolation. Despite these caveats, V2E have become important for quickly producing event camera datasets and for research in event-based learning, as it allows leveraging abundant existing vision data.

4 Our Framework

4.1 Overview

We introduce a novel synthetic data generation framework, that is built on the CARLA simulator, and that enables the generation of time-aligned multi-modal datasets across a wide range of sensor modalities. To leverage CARLA's diversity, our framework allows for a detailed configuration of scenarios and sensors through CARLA's client API. These include simulation parameters such as sampling rate, and output resolution; environment variables such as the map, weather settings, and number of vehicles; and sensor settings such as the sensors' relative position, motion blur, and shutter speed. Additionally, our framework preserves control over all agents, including the ego vehicle, by leveraging the native CARLA API and its built-in traffic manager. This allows for automatic traffic orchestration, realistic multi-agent interactions, and reproducible driving scenarios without the need for manual scripting. Moreover, for advanced event-based vision emulation, we combine CARLA's power with the V2E simulator to produce our event data. The configuration process is fully scriptable, making it easy to automate large-scale data generation over varying environmental conditions or sensor configurations.

4.2 CARLA'S DVS Limitations

Though CARLA has already extended its visual sensors suite with a DVS camera by integrating ESIM, it doesn't take full advantage of the simulator as it simplifies the temporal sampling by using fixed-rate frame captures (frame differencing) as opposed to ESIM's continuous, demand-driven rendering. CARLA essentially computes events between each pair of rendered frames, interpolating their timestamps, rather than continuously sampling the scene in an event-driven manner. Even though increasing the update frequency of the DVS sensor could improve the temporal resolution of the generated event stream – by reducing the interval between consecutive rendered frames – this approach introduces significant challenges when combined with other modalities operating at lower frequencies. To maintain synchronization across sensors, it becomes necessary to accumulate events over fixed time windows. However, in CARLA's implementation, each event stream is generated independently at each sensor tick, based solely on the intensity change from the immediately preceding frame. There is no continuous accumulation of intensity change across frames, and past frame history is not preserved. As a result, when high-frequency DVS streams are post-processed into temporally aligned event sets, the generated events may originate from distinct rendering intervals, raising concerns about the temporal and visual consistency of the final event representation.

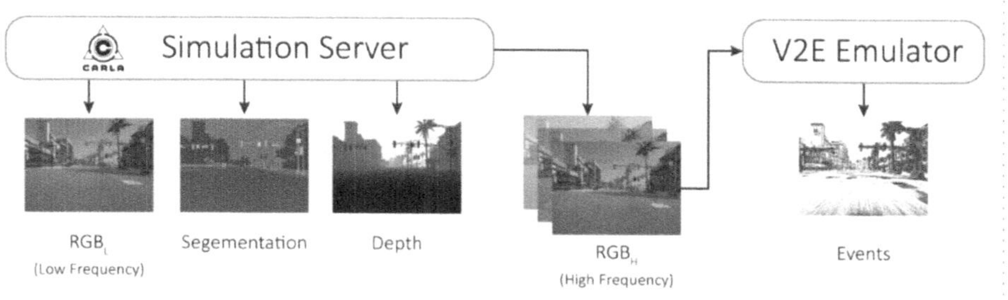

Fig. 2. The overall architecture of our data generation and event simulation strategy.

4.3 Our Strategy for Event Simulation

Given ESIM's simplified and limited implementation in CARLA, we favored using V2E, which has a more accurate model for DVS simulation. In fact, our approach takes advantage of CARLA's ability to operate at high simulated temporal frequencies by assigning a dedicated high framerate RGB sensor specifically for event generation (Fig. 2). This sensor is configured with motion blur disabled to ensure sharp, temporally precise intensity changes – crucial for accurate event triggering. In parallel, other visual sensors – including a separate RGB sensor – operate at a lower framerate while maintaining temporal alignment and synchronization with the simulated events. Furthermore, our framework leverages the full range of V2E configuration options, enabling detailed control over the event generation process. These include adjustable positive and negative contrast thresholds, leak rate, refractory time, temporal noise characteristics, and per-pixel threshold variability. Such configurability makes it possible to emulate a variety of real-world event camera behaviors or to systematically explore sensor response under different conditions. Furthermore, V2E offers seamless integration with any RGB stream in CARLA, allowing easy adaptation to various driving conditions or custom camera models without modifying the core simulation logic. This makes the pipeline highly reusable and adaptable across tasks.

5 Experiments

5.1 Data Collection and Simulation Configuration

To compare our V2E-based event generation framework with the built-in CARLA DVS, we conducted multiple simulation runs under three distinct lighting conditions: daytime, evening, and nighttime (with active road and vehicle lights). In each run, the vehicle was assigned the same trajectory to ensure consistency across conditions. Sensor data were recorded at a resolution of 640×480 pixels and a base frequency of 10 Hz over 120 s of simulated driving on an NVIDIA RTX 3090 GPU. For the CARLA DVS, events are generated internally by the simulator at each sensor tick, operating at the baseline frequency of 10 Hz. For the V2E-based simulation, we generated events from RGB feeds with higher frequencies, corresponding to $1\times$, $20\mathbf{x}$, $50\mathbf{x}$, and $100\mathbf{x}$ the baseline rate, yielding effective input rates up to 1000 Hz at the highest setting. In both pipelines, the sensor period of 100 ms was used as the fundamental time unit. For each period, the corresponding events that capture the changes that occurred between two consecutive frames, were stored in a separate file. All reported metrics were first computed over these 100 ms event segments and then averaged across the full 120 s sequence.

5.2 Evaluation Metrics

Evaluating simulated event data requires balancing two often conflicting aspects: the quality of the generated events and the efficiency of the simulation. High temporal fidelity and realistic noise characteristics improve the realism of event streams, but they also increase computational cost. To capture this trade-off, we employ complementary metrics: entropy-based measures that quantify the irregularity and information content of the events, and performance metrics that assess the runtime required to produce them. Together, these allow us to compare the realism of the generated streams with their computational demands, and to highlight differences between our V2E-based framework, the Carla DVS, and real data.

Spatiotemporal Entropy. To assess the statistical realism of the generated event streams, we employed a spatiotemporal entropy metric based on Shannon entropy [2]. Let each event be $e_i = (x_i, y_i, t_i, p_i)$ with pixel (x_i, y_i), timestamp t_i, and polarity $p_i \in \{-1, +1\}$. For a fixed temporal window of duration $\Delta > 0$ starting at s_k, each half open interval $\mathcal{W}_k = [s_k, s_k + \Delta)$ is divided into B equal temporal bins of width $\delta > 0$.

For any event with $t_i \in \mathcal{W}_k$, its bin index is:

$$b_i^{(k)} = \left\lfloor \frac{t_i - s_k}{\delta} \right\rfloor \in \{0, \ldots, B-1\} \tag{1}$$

The 4D count tensor over space-time-polarity is defined as:

$$C_k(x, y, b, p) = \sum\nolimits_{i:t_i \in \mathcal{W}_k} 1[x_i = x] 1[y_i = y] 1\left[b_i^{(k)} = b\right] 1[p_i = p] \tag{2}$$

With total events N_k, the empirical Probability Mass Function (PMF) is written:

$$P_k(x, y, b, p) = \frac{C_k(x, y, b, p)}{N_k} \tag{3}$$

Using log base $\beta = 2$ (for bits), the joint spatiotemporal-polarity Shannon entropy for the window k is:

$$H_k = -\sum\nolimits_{x,y,b,p} P_k(x, y, b, p) \log_\beta P_k(x, y, b, p) \tag{4}$$

In our experiments, we use non-overlapping windows of $\Delta = 100$ms and temporal bins of $\delta = 1$ *ms*; we report the average Shannon entropy over all windows in each sequence, which is expressed as:

$$\overline{H} = \frac{1}{K} \sum\nolimits_{k=0}^{K-1} H_k \tag{5}$$

Where K is the total number of temporal windows.

For comparison with the realistic data, we selected sequences with driving patterns similar to our CARLA runs from the DSEC dataset [5] and applied the same evaluation settings. For each lighting condition, we report the average spatiotemporal entropy of over 1,200 non-overlapping 100 ms event windows ($\approx$120 s of data).

Runtime and Event Density. To capture both the output characteristics of the event streams and the computational demands of the simulation, we evaluated two complementary metrics across the different V2E sampling rates, the CARLA DVS, and the DSEC dataset. The first metric is the simulation performance cost, measured as the wall-clock time required to generate one time-unit of simulated events, measured in seconds per time-unit (s/t-u). This value was averaged over each 120-s sequence to ensure stability and provide a direct indicator of runtime efficiency, particularly as sampling frequency increases. The second metric is the average event rate, expressed in millions of events per time-unit (ME/t-u). This metric yields a normalized measure of event density and allows us to compare how different sampling rates affect event volume and to benchmark against typical rates reported for real event cameras.

6 Results

Fig. 3. Sample outputs generated by our framework. These examples illustrate the diversity and synchronization of the generated multi-modal data.

As illustrated in Fig. 3, our synthetic data generation framework effectively demonstrates the ability to produce high-quality, multi-modal datasets across a wide range of simulated driving conditions. By integrating a high framerate RGB sensor with the V2E simulator, we enable realistic emulation of dynamic vision sensor outputs, capturing key characteristics such as pixel-level noise, polarity thresholds, and spatial variability. This results in temporally accurate and visually consistent event streams, which remain synchronized with ground-truth annotations—including semantic segmentation, depth, and instance masks—provided by the CARLA simulation engine. In addition to event-based outputs, the tool supports seamless configuration of camera parameters such as resolution, shutter speed, and lens distortion, allowing researchers to emulate different sensor models with a high degree of control.

Table 1. Shannon spatiotemporal entropy measured for each 100 ms event segment using a 1 ms temporal bin and averaged for each lighting condition.

	Average Spatiotemporal Entropy (bits)		
	Day	Evening	Night
DSEC	20.23	20.19	20.68
Ours (100×)	18.25	18.75	21.26
Ours (50×)	18.18	18.62	21.01
Ours (20×)	18.14	18.46	20.75
Ours (1×)	16.55	16.57	18.15
Carla DVS	15.75	15.90	18.50

The entropy analysis shown on Table 1 reveals clear differences between the event generation pipelines. CARLA's built-in DVS consistently produces lower entropy than the DSEC dataset, with an average gap of about 4.5 bits under day and evening conditions and 2.18 bits at night. In contrast, our V2E-based pipeline shows a steady increase in entropy as the internal sampling rate rises, reaching values within 2 bits of DSEC across all lighting conditions when using 100× the base frequency. This indicates that finer temporal sampling enables our framework to more faithfully reproduce the spatiotemporal dynamics of physical event cameras. Notably, even in the 1× configuration—where events are generated only from consecutive frames, like CARLA's DVS—our pipeline yields slightly higher entropy, reflecting the inherent advantages of V2E for event simulation.

Across lighting conditions, the simulated entropy values for day and evening remain closely aligned, whereas the night setting exhibits a marked increase. By comparison, DSEC events maintain relatively consistent entropy across all lighting scenarios, with only a modest rise of approximately 0.5 bits at night. This discrepancy highlights the persistent challenge of accurately simulating event data under harsh illumination, where the noise-to-signal ratio plays a more dominant role than in physical sensors.

In the performance metrics reported in Table 2, CARLA's DVS requires on average 0.15 s per 100 ms of simulated time, with little sensitivity to lighting changes. In comparison, our method at the 1× setting is about three times faster under day and evening conditions but shows a stronger slowdown at night, where the runtime nearly doubles, reducing the speed advantage by half. At higher sampling rates, the simulation cost increases steadily, reaching 2.57 and 3.01 s per 100 ms in day and evening, and up to 5.64 s at night. This runtime growth correlates with the event rate, which rises with input frequency and provides richer detail, peaking at 0.38 and 0.50 million events per 100 ms in the 100× setting. Even at 1×, our method yields a denser event stream than CARLA's DVS, except at night where it produces about 10,000 fewer events per unit.

Table 2. Comparison of runtime and event density for 100 ms event volumes across sensor configurations and lighting conditions.

	Performance (s/t-u)			Event Rate (ME/t-u)		
	Day	Evening	Night	Day	Evening	Night
DSEC	–	–	–	1.471	1.535	1.992
V2E (x100)	2.57	3.01	5.64	0.379	0.492	2.582
V2E (x50)	1.37	1.50	2.60	0.037	0.045	2.181
V2E (x20)	0.64	0.68	1.06	0.036	0.042	0.183
V2E (x1)	**0.04**	**0.05**	**0.09**	0.011	0.011	0.03
Carla DVS	0.14	0.14	0.16	0.006	0.007	0.04

Nevertheless, both pipelines remain below the DSEC dataset in terms of event rate, with a gap of roughly one million events in day and evening and about 0.6 million at night, where our method unexpectedly surpasses DSEC in event production. This convergence is likely explained by the increase in noise inherent to low-light simulation, which elevates the event counts and partially bridges the difference to real data. This noise increase is consistent with the behavior of physical DVS cameras and other event camera simulators. In practice, low-light environments are known to increase background activity due to sensor-level noise and reduced signal-to-noise ratios [6]. Thus, rather than indicating a shortcoming, the higher nighttime event counts reinforce the realism of our simulation by reproducing a well-documented property of actual event-based sensors.

While our tool supports the simulation of all CARLA's visual sensors, we chose not to include optical flow in our experiments due to unresolved limitations in its simulation. In particular, the ground-truth flow currently exhibits visual artifacts—such as unnatural layering effects and quantization—especially under high-speed motion or challenging lighting conditions. These artifacts can reduce the reliability of the flow data for training or evaluation. While this remains an active area of discussion within the CARLA community, we opted to exclude optical flow until a more stable and consistent simulation method is available.

7 Conclusion and Perspectives

We presented a synthetic data generation framework that extends CARLA's capabilities with realistic event-based vision simulation through the V2E pipeline. Our tool enables flexible scenario design and synchronized multi-sensor data collection, supporting both conventional frame-based and event-driven modalities. It addresses a critical gap in event camera datasets by offering a scalable and cost-effective alternative to real-world data acquisition, while preserving a high level of realism, configurability, and reproducibility. With this tool, researchers can generate customized datasets tailored to specific perception tasks in autonomous driving, thereby facilitating the development and benchmarking of models designed to operate reliably under complex, dynamic, and low-visibility conditions.

Future work may include addressing the current limitation of optical flow simulation by contributing directly to CARLA's open-source development or by integrating an external module capable of producing higher-quality flow data. This extension would broaden the applicability of our tool and support the pretraining of a wider range of deep learning models, including those requiring fine-grained motion supervision. Additionally, we aim to enhance control over agent behavior by enabling user-defined trajectories for all actors, including the ego-vehicle. This would move beyond the constraints imposed by CARLA's traffic manager and allow for the generation of more structured, repeatable, and interactive driving scenarios.

References

1. Brandli, C. et al.: A 240 x 180 130 dB 3 μs latency global shutter spatiotemporal vision sensor (2014).
2. Claramunt, C.: Towards a Spatio-temporal form of entropy. In: Castano, S., et al. (eds.) Advances in Conceptual Modeling, pp. 221–230. Springer, Berlin Heidelberg, Berlin, Heidelberg (2012)
3. Dosovitskiy, A., et al.: CARLA: An Open Urban Driving Simulator. In: Proceedings of the 1st Annual Conference on Robot Learning, pp. 1–16 (2017)
4. Gallego, G. et al.: Event-Based Vision: A Survey (2022).
5. Gehrig, M. et al.: DSEC: A Stereo Event Camera Dataset for Driving Scenarios (2021).
6. Graca, R., Delbruck, T.: Unraveling the paradox of intensity-dependent DVS pixel noise (2021). https://arxiv.org/abs/2109.08640.
7. Hu, Y., et al.: v2e: from video frames to realistic DVS events. In: 2021 IEEE/CVF Conference on Computer Vision and Pattern Recognition Workshops (CVPRW). IEEE (2021)
8. Jiang, H. et al.: Super SloMo: High Quality Estimation of Multiple Intermediate Frames for Video Interpolation (2018).
9. Li, W. et al.: Interiornet: Mega-scale multi-sensor photo-realistic indoor scenes dataset (2019).
10. Lichtsteiner, P., et al.: A 128 x 128 120 dB 15 μs latency asynchronous temporal contrast vision sensor. IEEE J. Solid State Circ. **43**(2), 566–576 (2008)
11. Mueggler, E. et al.: The event-camera dataset and simulator: Event-based data for pose estimation, visual odometry, and SLAM (2017).
12. Rebecq, H. et al.: ESIM: an Open Event Camera Simulator (2018).
13. Zhu, A.Z., et al.: The multivehicle stereo event camera dataset: an event camera dataset for 3D perception. IEEE Rob. Autom. Lett. **3**(3), 2032–2039 (2018)

Improving Low-Resource Dialect Identification through Audio Data Augmentation: A Case Study on Algerian Language

Kawthar Yasmine Zergat[1](✉), Sid Ahmed Selouani[2], Abderraouf Zouaoui[1], and Thouraya Merazi-Meksen[1]

[1] Speech Com. and Signal Proc. Lab, University of Science and Technology Houari Boumediene, P.O. Box 32, El Alia, Bab Ezzouar 16111, Algeria
{kzergat,tmeksen}@usthb.dz, abderraoufzouaoui77@gmail.com
[2] Information Management Department, Université de Moncton, Campus of Shippagan, Shippagan, NB E8S 1P6, Canada
selouani@umcs.ca

Abstract. This work addresses the challenge of improving Algerian dialect identification through the use of audio Data Augmentation (DA). Due to the scarcity of labeled speech data for Algerian dialects, we applied four augmentation techniques-Speed Change, Echo Addition, Time Stretching, and Noise Injection to expand our dataset from just 7.8 h to over 60 h of training data. We then employed Convolutional Neural Networks (CNNs) to perform the dialect classification task. Rather than focusing solely on the overall impact of augmentation, this study provides a comparative analysis of individual and combined augmentation strategies. The results demonstrate that specific combinations, most notably Echo+Speed+Stretch significantly enhance model performance, with accuracy improving from 50% on the original dataset to 98.5% after augmentation. These findings highlight the importance of carefully selecting and tuning augmentation methods to improve generalization in low-resource dialectal speech recognition. Furthermore, the study confirms that data augmentation is not uniformly effective; its impact varies significantly depending on the type and combination of transformations applied

Keywords: Algerian Dialect identification · Low-Resource Languages · CNN · Spectrogram · Data Augmentation

1 Introduction

Automatic Speech Recognition (ASR) is a central domain in speech processing, powering a wide range of applications such as voice assistants, translation systems, and human-computer interaction. In the Arabic-speaking world, dialectal variation presents a considerable challenge. These dialects often differ significantly across regions, and their lack of standardization is compounded by the scarcity of large, labeled corpora [1] necessary for effective model training.

T. Ensari et al. (Eds.): ISPR 2025, CCIS 2859, pp. 219–229, 2026.
https://doi.org/10.1007/978-3-032-21585-7_16

In Algeria, this issue is particularly pronounced due to its rich linguistic diversity. Developing accurate dialect identification systems is complex, as it requires distinguishing fine-grained phonetic, lexical, and prosodic differences between dialects [2]. Furthermore, the limited availability of annotated speech data [3] significantly restricts the performance of deep learning models, especially Convolutional Neural Networks (CNNs), which rely heavily on large-scale training datasets.

To mitigate this data scarcity, one widely adopted strategy is audio data augmentation (DA) [4], a technique used to artificially increase the size and diversity of a dataset by applying systematic modifications to the existing data. In speech processing, it helps improve model robustness and generalization, particularly in low-resource settings where annotated data is limited [5].

In this work, we explore the impact of audio DA on Algerian dialect identification under low-resource conditions. We use a portion of the Sawt El-Djazaïr corpus [3], which we developed in earlier work to represent the linguistic diversity of four major Algerian regions: North, East, West, and South. The selected subset includes only 7.8 h of annotated audio, which is insufficient for training a robust CNN-based classifier. To overcome this limitation, we applied four classical audio DA techniques: speed change, time stretching, echo addition, and background noise injection, to expand the dataset to over 60 h. We conducted a comparative study to systematically evaluate the impact of individual and combined augmentation strategies on dialect classification performance, aiming to identify the most effective configurations for enhancing model accuracy and generalization.

The remainder of this paper is structured as follows: Sect. 2 reviews related work on audio data augmentation in speech processing. Section 3 presents the proposed system, including corpus augmentation and the complete CNN pipeline. Section 4 describes the experimental setup and analyzes the results. Finally, Sect. 5 concludes the study and outlines directions for future work

2 Related Work

Data augmentation has emerged as a fundamental technique for enhancing the performance of audio recognition systems by addressing challenges such as data imbalance and limited training samples. Numerous studies have investigated various augmentation strategies, ranging from classical signal processing methods to advanced deep learning approaches.

Anastasios et al. [6] showed that several studies have explored DA techniques to improve environmental sound recognition by increasing dataset variability. Methods such as time stretching, pitch shifting, and noise addition have been widely used and frequently applied to audio datasets. Deep learning architectures, especially CNNs and self-supervised models like wav2vec 2.0, have demonstrated improved classification performance with augmented data. Additionally, recent works have proposed web-based tools to streamline the augmentation process for researchers.

In [7], the authors addressed data imbalance in Speech Emotion Recognition using a conditional Generative Adversarial Networks (GAN) to generate synthetic spectrograms. Their method improved performance by 10% on the IEMOCAP dataset and by 5% on FEEL-25 k, compared to standard augmentation techniques.

The authors in [8] explored DA techniques, including pitch shifting, time stretching, and inverse filtering, to enhance whispered speech recognition. Their results showed that combining traditional augmentation with inverse filtering significantly improves the performance of both HMM- and CNN-based ASR systems, with statistical tests confirming the improvements.

A Random Audio Region Patch Masking (RARPM) technique was introduced in [9], a DA method that enhances audio recognition by applying random patches to log Mel-spectrograms. This method outperforms SpecAugment, achieving a mean Average Precision (mAP) of 0.385 compared to 0.366. The results confirm its effectiveness and potential for integration into various neural networks.

In [10], the authors proposed a multimodal GAN for data augmentation in the field of audio-visual emotion recognition. This method, tested on different datasets, achieved a highest recorded accuracy of up to 71.13%. Three data augmentation techniques were applied in [11], combined with CNNs using Mel spectrograms to improve Amazigh speech recognition. Experiments with 42 participants showed that the VGG19 model achieved 95.66% accuracy, with a 4.67% improvement attributed to data augmentation. These findings demonstrate the impact of augmentation on model accuracy. The study in [12] focused on the problem of limited data in offensive speech detection by using data augmentation techniques such as noise injection. This method effectively increased the training dataset and improved the model's performance on the Vera Am Mittag (VAM) corpus.

In [13], the authors studied the impact of normalization and data augmentation on named entity recognition (NER) for Algerian text using Arabic pre-trained models. They evaluated four scenarios: without normalization or augmentation (ARBERT, F1 score of 84.4%), with normalization (Dziribert, F1 score of 81.9%), with DA (AraBERT v0.2, F1 score of 85.1%), and with both techniques combined (AraBERT v0.2, F1 score of 86.2%).

3 Proposed System

3.1 Data Collection

The database used in this study is part of the Sawt El-Djazaïr corpus [3]. It consists of audio recordings from various regions of Algeria, each representing a specific dialect. These recordings are available in multiple formats, including WAV and MP3, to ensure both compatibility and diversity of audio sources. The audio files were collected in controlled environments to ensure optimal sound quality. However, the dataset also includes recordings made under varied conditions to reflect the diversity of real-world usage contexts. For classification purposes, the dialects were grouped into four main categories: North, East, West, and South.

To standardize the dataset, all audio files were segmented into 20-s clips, with a uniform sampling rate and consistent file sizes. This process resulted in a total of 1,404 audio files in WAV format, representing approximately 7.8 h of audio.

3.2 Data Augmentation Methodology

Audio Data Augmentation specifically involves modifying the raw audio signal (waveform) through various transformation [8]. Commonly used techniques include:

- **Noise Injection**

Noise injection is a data augmentation technique that involves artificially adding background noise to speech recordings to simulate diverse acoustic conditions. It enhances the robustness and generalization of speech recognition models by training them to handle distorted signals [6].

$$x'(t) = x(t) + n(t) \tag{1}$$

where x(t) is the original signal and n(t)is the added noise, typically modeled as Gaussian white noise or environmental noise.

- **Time Stretching**

Time stretching alters the duration of a speech signal while preserving its pitch [8]. This method changes the rhythm and timing of the audio without affecting its intonation [14]. It helps speech recognition models better handle variations in speaking speed while maintaining the spectral characteristics of the original signal. The process is mathematically expressed as:

$$x\ (t) = x(\Phi(t)) \tag{2}$$

where () is a time-scaling function used to stretch or compress the signal while preserving phase coherence, often through techniques such as overlap-add.

- **Echo and Reverberation**

Adding echo simulates reverberant environments to improve the robustness of speech recognition systems in acoustically challenging conditions [6]. It exposes models to real-world scenarios like large rooms or open spaces, helping them distinguish speech from unwanted reflections. The process is represented by the equation:

$$x'(t) = x(t) + \gamma x(t - \tau) \tag{3}$$

where () is the original signal, is the echo amplitude coefficient, and is the echo delay in seconds.

- **Changing Speed**

Speed modification means changing how fast the audio is played. This affects how long the audio lasts and how high or low it sounds. It helps speech recognition systems understand different speaking speeds better [15].

$$y(t) = x\left(\frac{t}{\alpha}\right) \tag{4}$$

where x(t) is the original signal, y(t) is the modified signal, and α is the speed factor.

- If $\alpha > 1$, the speech is faster (shorter duration).
- If $\alpha < 1$, the speech is slower (longer duration).

3.3 Data Augmentation of Sawt El-Djazaïr

Before applying the augmentation techniques, all audio files are converted to a uniform format. It is important to note that the files are converted to mono in order to standardize the processing format. The following augmentation techniques are applied to audio data:

- **Speed Change**: This technique changes the playback speed of audio recordings. Using a speed factor of 1.5, recordings are transformed to increase or decrease their tempo without affecting the pitch of the sound.
- **Adding Echo**: Adding Echo simulates reverberant acoustic environments. A delay factor of 0.2 s and a decay of 0.5 were used to create a realistic echo effect.

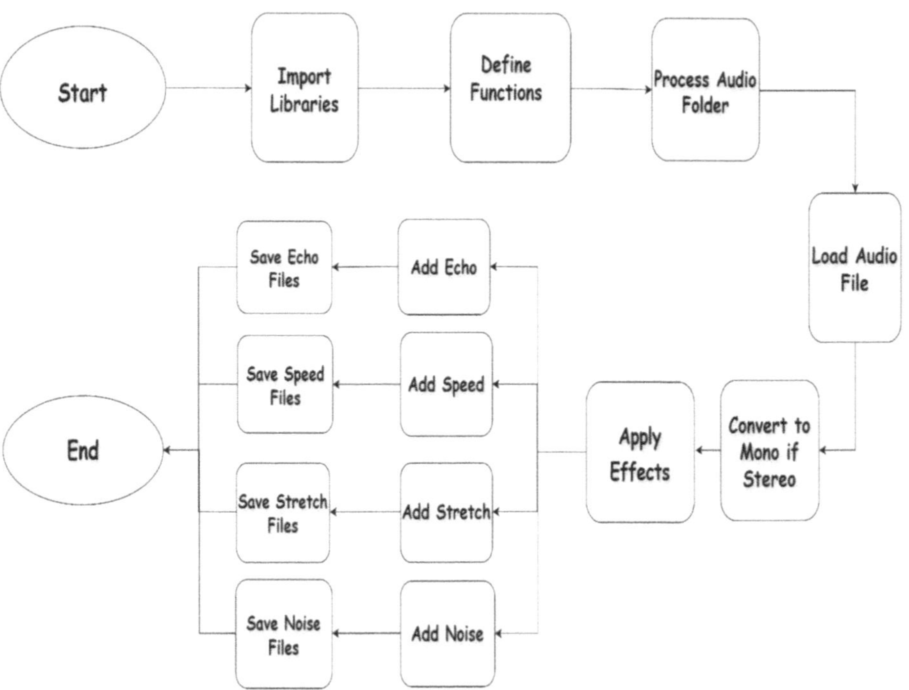

Fig. 1. Audio File Processing and Augmentation

- **Time Stretching**: This technique consists of stretching or compressing audio recordings in time without changing their pitch. A factor of 1.5 was used for this transformation.
- **Added Noise**: To increase the robustness of the models to interference, a Gaussian noise with a level of 0.01 has been added to the audio recordings. This simulates recording conditions in a noisy environment.

After applying the effects, the audio files are saved under new names to indicate the modifications applied. Fig. 1 illustrates the process of handling audio files for augmentation. Since annotated audio data is costly and time-consuming to collect, DA offers an effective way to artificially expand the dataset without the need for additional labeled samples. To achieve this, the four DA techniques mentioned above were applied to the original corpus. The resulting augmented dataset is detailed in Table 1.

Table 1. Number of Samples and Duration of Each Class before and after Data Augmentation.

		Before DA		After DA	
Region	**Speakers (M/F)**	**Samples**	**Duration (min)**	**Samples**	Duration (min)
North	2/1	615	205	7380	1594.44
East	2/2	212	70.67	2544	549.63
West	2/2	243	81	2916	630
South	2/3	334	111.33	4008	865.92
Total		1404	468	16 848	3639.99

3.4 CNN Pipeline Design for Algerian Dialect Identification

The audio preprocessing pipeline begins by converting raw audio waveforms into log-scaled spectrograms, typically using Mel-scaled filters, which provide a compact time-frequency representation of the audio signal. Once the spectrogram is generated, a Z-score normalization is applied across the dataset to center the features around a mean of zero and scale them to unit variance. This step helps the neural network converge more efficiently during training and reduces sensitivity to variations in amplitude across recordings. Finally, the class labels associated with each audio sample are converted into one-hot encoded vectors, transforming categorical labels into a binary matrix format where each class is represented by a unique vector.

The model starts with an input layer shaped as (250, 128), which represents 250 time steps with 128 features each. From there, the data passes through three convolutional blocks. Each block includes a Conv1D layer to extract features, followed by batch normalization to improve training stability, max pooling to reduce the data size, and dropout to help prevent overfitting. We gradually increase the number of filters in each block so the model can learn more complex patterns at deeper levels. After these convolutional layers, the data is flattened and passed through a dense layer with 256 units, and finally through an output layer with 4 units, each corresponding to a different class. This step-by-step pipeline helps the model learn meaningful patterns in the audio and make accurate predictions.

In our implementation, all convolutional and dense layers use the ReLU activation function, as it is widely adopted in deep learning research due to its computational efficiency, while the output layer applies softmax for multi-class classification. Each Conv1D layer employs a kernel size of 3 with the default stride of 1, and max pooling is performed with a pool size of 2. A dropout rate of 0.3 is applied after each convolutional block to reduce overfitting. Fig. 2 illustrates the architecture of the proposed CNN used in this study. It visually outlines the sequential flow of data through the input layer, convolutional blocks, and fully connected layers, leading to the final output.

Fig. 2. Architecture of the Proposed CNN for Audio Classification

4 Simulation and Results

To evaluate the effectiveness of the proposed approach for Algerian dialect identification, a series of experiments were conducted using a CNN trained on the augmented Sawt El-Djazaïr speech corpus. The simulations aimed to assess how different combinations of DA techniques impact the model's classification performance. Metrics such as accuracy, precision, recall, and F1-score were used to provide a comprehensive evaluation.

This section presents the experimental setup, describes the augmentation strategies applied, and discusses the results obtained for each configuration. We first trained and evaluated the CNN model on the original dataset without applying any DA. To do so, the complete Algerian dialect audio dataset was split into training (60%), validation (20%), and test (20%) sets.

During training, the model achieved a final training accuracy of approximately 93%, indicating effective pattern learning. However, validation accuracy stabilized early, remaining between 75% and 78%. This performance gap suggests overfitting. This is further supported by the loss curves: while the training loss continuously decreased to around 0.7, the validation loss remained much higher, oscillating between 1.3 and 1.5. These observations indicate that the model had difficulty generalizing to unseen data, motivating the need for DA to improve robustness and generalization. Fig. 3 presents the model accuracy and loss curves.

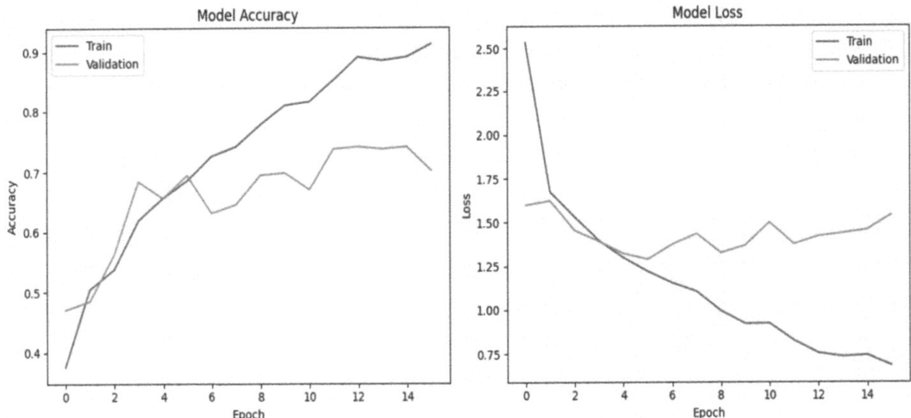

Fig. 3. Model Accuracy and Loss Curves Before Data Augmentation.

After applying various DA techniques, the CNN model exhibited notable improvements in both training and validation performance. Fig. 4 shows the updated accuracy and loss curves, illustrating a more stable and consistent learning process. Unlike the previous setup, the gap between training and validation performance has significantly narrowed, indicating enhanced generalization and reduced overfitting.

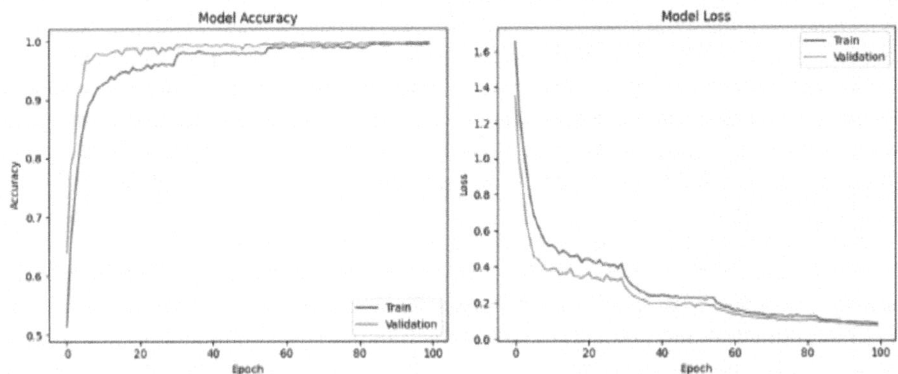

Fig. 4. Model Accuracy and Loss Curves After Data Augmentation.

Table 2 presents a comprehensive comparison of the CNN model's performance metrics before and after implementing DA techniques.

Table 2. Performance Metrics on Test Set

	Accuracy	Precision	Recall	F1-Score	Support
Before Augmentation	0.5	0.48	0.5	0.47	282
After Augmentation	0.985	0.985	0.985	0.984	3371

To further evaluate the classification performance, confusion matrices were generated for both the baseline and augmented CNN models. Fig. 5a and b present the confusion matrices before and after applying traditional DA techniques, respectively. Before augmentation, predictions were notably imbalanced for instance, only 13 samples were correctly classified as West, and 34 as South, highlighting the model's struggle to distinguish between dialects. In contrast, after augmentation, the classification results improved: 505 East samples were correctly identified, and North achieved 1455 correct predictions, indicating very interesting performance.

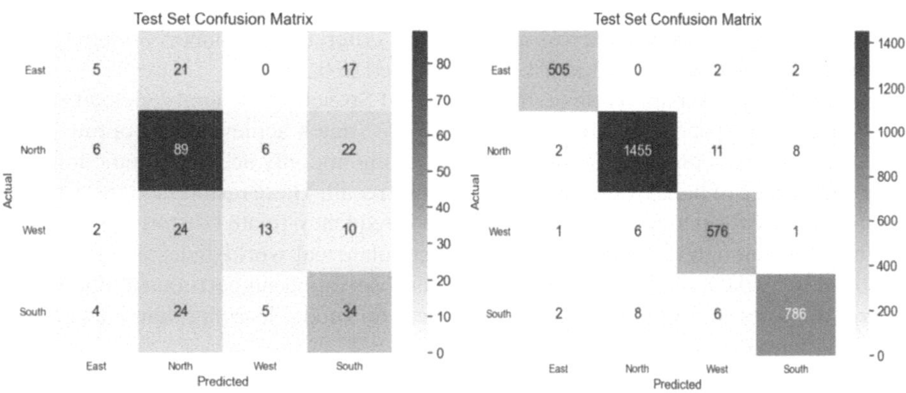

Fig. 5. Confusion matrix (a) Before Data Augmentation (b) After Data Augmentation.

Table 3 summarizes the CNN model's performance before and after applying data augmentation techniques, comparing different combinations to assess their impact.

Table 3. Performance Comparison across Different Combinations of data augmentation

Combination	Accuracy	Precision	Recall	F1-Score
No Data Augmentation	0.50	0.48	0.5	0.47
Stretch+Noise	0.49	0.68	0.49	0.39
Echo+Noise	0.66	0.79	0.66	0.63
Echo+Speed+Stretch+Noise	0.88	0.90	0.88	0.88
Echo+Speed	0.89	0.89	0.89	0.88

(continued)

Table 3. (*continued*)

Combination	Accuracy	Precision	Recall	F1-Score
Echo+Stretch+Noise	0.89	0.89	0.89	0.89
Echo+Speed+Noise	0.91	0.92	0.91	0.91
Echo+Stretch	0.91	0.92	0.91	0.91
Speed+Noise	0.95	0.95	0.95	0.95
Speed+Stretch+Noise	0.95	0.95	0.95	0.95
Speed+Stretch	0.97	0.97	0.97	0.97
Echo+Speed+Stretch	0.98	0.98	0.98	0.98

Prior to augmentation, the model achieved a test accuracy of 0.50 and an F1-score of 0.47 on a relatively small test set of 282 samples, reflecting limited learning capacity and generalization. Applying basic augmentation combinations such as Stretch+Noise or Echo+Noise led to minor improvements in individual metrics like precision but failed to deliver balanced performance across all evaluation criteria. For instance, Stretch+Noise improved precision to 0.68, but its F1-score dropped to 0.39.

In contrast, combinations involving Speed and Stretch consistently yielded strong results. Notably, Speed+Stretch, and Echo+Speed+Stretch achieved near-optimal performance with F1-scores of 0.97 and 0.98, indicating not only accurate predictions but also a high degree of balance between precision and recall. These results demonstrate that time-based transformations (speed changes) and frequency-related distortions (stretching) introduce enough variability to effectively simulate real-world dialectal differences, thereby helping the model generalize better. This observation suggests that temporal and spectral diversity in training data is more critical for model learning than background noise simulation alone.

5 Conclusion

The main objective of this study is the automatic identification of Algerian dialects. To achieve this, we focused on enriching Sawt El-Djazaïr database with augmented audio recordings, aiming to increase the diversity and robustness of the training data.

The experimental results clearly demonstrate that applying DA significantly improves model performance compared to using no augmentation. Among all tested combinations, Echo+Speed+Stretch achieved the highest overall performance, with an accuracy, precision, recall, and F1-score of 0.98, indicating optimal classification results. In contrast, the model trained without any DA performed the worst, with an accuracy of 0.50, suggesting poor generalization and reliability. Simple augmentation combinations like Stretch+Noise or Echo+Noise showed modest improvements, but remained below acceptable thresholds for robust classification.

The CNN model demonstrated promising accuracy and generalization capabilities, effectively leveraging the enriched dataset. Overall, our approach confirms the importance of DA in dialect identification tasks and represents a meaningful step toward building more reliable and dialect aware speech recognition systems for the Algerian linguistic landscape.

In future research, we will involve testing the system on real-world, noise variance data to evaluate its robustness in practical applications, and potentially developing a dialect-aware speech recognition system tailored to the Algerian linguistic landscape.

References

1. Besdouri, F.Z., Zribi, I., Hadrich Belguith, L.: Arabic automatic speech recognition: challenges and Progress. Speech Comm. **163**, 103110 (2024). https://doi.org/10.1016/j.specom.2024.103110
2. Biadsy, F., Hirschberg, J., Habash, N.: Spoken Arabic dialect identification using phonotactic modeling. In: Proc. EACL Workshop on Computational Approaches to Semitic Languages, pp. 53–61. Association for Computational Linguistics (2009)
3. Zergat, K.Y., Selouani, S.A., Amrouche, A., Kahil, Y., Merazi-Meksen, T.: The voice as a material clue: a new forensic Algerian corpus. Multimed. Tools Appl. **82**, 29095–29113 (2023). https://doi.org/10.1007/s11042-023-14412-2
4. Ding, K., Li, R., Xu, Y., et al.: Adaptive data augmentation for mandarin automatic speech recognition. Appl. Intell. **54**, 5674–5687 (2024). https://doi.org/10.1007/s10489-024-05381-6
5. Ko, T., Peddinti, V., Povey, D., Khudanpur, S.: Audio augmentation for speech recognition. In: Proceedings Interspeech, vol. 2015, pp. 3586–3589 (2015)
6. Sarris, A.L., Vryzas, N., Vrysis, L., Dimoulas, C.: Investigation of data augmentation techniques in environmental sound recognition. Electronics. **13**(11) (2024). https://doi.org/10.3390/electronics13234719
7. Chatziagapi, A., Paraskevopoulos, G., Sgouropoulos, D., Pantazopoulos, G., Nikandrou, M., Giannakopoulos, T.: Data augmentation using GANs for speech emotion recognition. In: Proceedings Interspeech, vol. 2019, pp. 171–175 (2019). https://doi.org/10.21437/Interspeech.2019-2561
8. Galić, J., Marković, B., Grozdić, Đ., Popović, B., Šajić, S.: Whispered speech recognition based on audio data augmentation and inverse filtering. Appl. Sci. **14** (2024). https://doi.org/10.3390/app14188223
9. Wong, W., Li, Y., Li, S.: Improving audio recognition with randomized area ratio patch masking: a data augmentation perspective. IEEE Access. **12**, 3475508 (2024). https://doi.org/10.1109/ACCESS.2024.3475508
10. Ma, F., Li, Y., Ni, S., Huang, S.-L., Zhang, L.: Data augmentation for audio visual emotion recognition with an efficient multimodal conditional GAN. Appl. Sci. **12**(1), 527 (2022). https://doi.org/10.3390/app12010527
11. Boulal, H., Bouroumane, F., Hamidi, M., Barkani, J., Abarkan, M.: Exploring data augmentation for Amazigh speech recognition with convolutional neural networks. Int. J. Speech Technol. **28**(1), 53–65 (2024). https://doi.org/10.1007/s10772-024-10164-y
12. Sekkate, S., Chebbi, S., Adib, A., et al.: Handling data scarcity through data augmentation for detecting offensive speech. Ann. Telecommun. **80**, 417–426 (2025). https://doi.org/10.1007/s12243-025-01072-6
13. Dahou, A.H.H., Cheragui, M.A.: Impact of normalization and data augmentation in NER for Algerian Arabic dialect. In: Modelling and Implementation of Complex Systems (MIC 2022), vol. 593, pp. 249–262. LNNS (2022). https://doi.org/10.1007/978-3-031-18516-8_18
14. Cauli, N., Reforgiato Recupero, D.: Survey on videos data augmentation for deep learning models. Future Internet. **14**(3), 93 (2022). https://doi.org/10.3390/fi14030093
15. Herrera, J.M., Barrios, M.: Pangeanic Voice: Generation of a Large Speech Corpus for the Languages of Spain Using Data Augmentation, p. 22. Pangeanic, Technical Report (2023)

From Farm to Warehouse: Toward a Semantic and AI-Based Architecture for Wheat Lifecycle

Djakhdjakha Lynda[1,2](✉), Bouacida Imane[1,2], Farou Brahim[1,2], and Harizi Rana[2]

[1] LabSTIC Laboratory, Computer Science Department, 8 Mai 1945 Guelma University, 24000 Guelma, Algeria
{bouacida.imane,farou.brahim}@univ-guelma.dz

[2] Computer Science Department, 8 Mai 1945 Guelma University, 24000 Guelma, Algeria
djakhdjakha.lynda@univ-guelma.dz

Abstract. The wheat lifecycle involves multiple stages, from crop cultivation and disease monitoring to harvesting, transportation, and storage, each generating vast and heterogeneous data from various devices, platforms, and users. The lack of interoperability between these systems poses a significant challenge to achieving seamless supervision, data integration, and intelligent decision-making across the entire chain. This paper provides a theoretical overview of a conceptual architecture for an ontology-based system intended to supervise and integrate all phases of the wheat production chain, from farm to warehouse. The proposed architecture is conceptually structured into five main layers: data acquisition, semantic integration, artificial intelligence, reasoning and decision support, and user interface. It is designed to resolve semantic and structural heterogeneity in order to achieve interoperability among diverse data sources. It supports real-time data analysis, semantic reasoning, and full traceability through the integration and linking of distributed information. This conceptual work lays the theoretical foundation for the future development of an intelligent, modular, and interoperable platform for sustainable wheat production management. The proposed framework was validated through a dedicated use case on wheat grain quality classification. This validation involved applying deep learning models integrated with semantic ontology support to ensure both accurate prediction and meaningful interpretation of results.

Keywords: Smart Agriculture · Wheat Production Chain · Semantic Integration · Ontology · Deep Learning · Data Heterogeneity · Intelligent Systems

---

B. Imane, F. Brahim and H. Rana—These authors contributed equally to this work.

T. Ensari et al. (Eds.): ISPR 2025, CCIS 2859, pp. 230–243, 2026.
https://doi.org/10.1007/978-3-032-21585-7_17

1 Introduction

Wheat is an essential crop that plays an important role in ensuring global food security [1]. Its cultivation, harvesting, storage, and distribution involve complex and multi-stage processes that must be carefully managed to ensure quality and meet growing market and regulatory demands [2]. Across the wheat production lifecycle, from sowing and growth to harvesting, transport, and warehousing, there is a critical need for accurate data monitoring and traceability to ensure efficiency in the supply chain [3].

Recent advances in precision agriculture have introduced sveral technologies such as the Internet of Things (IoT), deep learning, remote sensing, and cloud computing [4]. These technologies enable real-time monitoring of crop health, environmental conditions, and grain quality. However, despite these technological advancements, many existing systems operate in silos, use incompatible data formats, and lack semantic understanding of the information they generate [5].

A major challenge in implementing an integrated and smart agricultural systems is data heterogeneity [6, 7]. The wheat production chain involves a variety of data sources: (i) heterogeneous sensors with different protocols and units, (ii) AI models with different outputs and metadata, and (iii) software platforms with different terminologies and data structures. This lack of interoperability prevents seamless integration and reasoning across different stages of the wheat cycle [8].

Moreover, current systems often focus on isolated components of the agricultural process, disease detection only [10], or warehouse management only [10], without providing an end-to-end, unified view of the entire wheat lifecycle.

In this paper, we aim to design a comprehensive ontology-based system to supervise the entire wheat production cycle, from cultivation on the farm to storage in the warehouse. The proposed system will model key concepts involved throughout the wheat lifecycle, such as stage of growth of the crops, diseases, sensor data, quality metrics, logistics processes, and storage conditions, using domain-specific ontologies. It seeks to resolve both the semantic and structural heterogeneity inherent in data collected from various devices, platforms, and users. The system will also integrate deep learning models for disease detection and grain quality assessment into the semantic framework, allowing AI outputs to be understood and utilized in a standardized and interoperable manner. An additional objective is to ensure full traceability across all stages of production by semantically linking data generated from field activities, transportation, and storage operations within a unified framework.

The structure of the paper is designed to present the theoretical foundations, motivations, and design considerations of the proposed AI and ontology-based system. Section 2 reviews related work, highlighting existing research and systems in smart agriculture, disease detection using deep learning, and ontology-based integration in agriculture. Section 3 introduces the conceptual architecture of the proposed system, detailing its core components. Section 4 presents the case study on wheat grain quality classification, serving as a first validation of the framework. Finally, the conclusion summarizes the main contributions and outlines directions for future work.

2 Related Work

In recent years, the integration of new technologies in agricultural domain has gained significant attention as a means to ensure sustainable farming practices [11]. Several studies have focused on applying deep learning techniques for crop and grain disease detection [12]. For instance, convolutional neural networks (CNNs) have been employed to classify wheat leaf diseases based on visual symptoms captured through mobile or drone imaging [13]. Models such as ResNet, VGGNet, and YOLO have demonstrated high accuracy in recognizing diseases like rust, blight, and mildew in field conditions [14–21]. However, the cited approaches are often limited to disease detection or disease classification tasks and lack integration with the entire wheat production cycle.

Parallel to this, the rise of smart agriculture systems using IoT devices has facilitated the continuous monitoring of environmental and crop conditions [22,23]. IoT-based platforms have been used to track soil moisture, temperature, humidity, and other variables that influence wheat growth and grain storage [24]. Despite the growing number of such solutions, most of them operate as standalone applications with limited interoperability [25]. The absence of a unified semantic framework prevents efficient data sharing and holistic process supervision.

To address this challenge, researchers have explored the use of ontologies and semantic web technologies in the agricultural domain [26]. Initiatives like OAK [26], AGROVOC [27], and the FoodOn [28] have provided reusable vocabularies for standardizing agricultural concepts and promoting interoperability. Some studies have proposed ontology-based platforms for traceability [28]; however, these are lack integration with AI models outputs.

To the best of our knowledge, only a few works have attempted to combine semantic modeling with AI-driven analysis within a comprehensive and unified system [29]. In addition, existing semantic systems rarely cover the entire wheat production lifecycle, from field-level data collection and disease diagnosis to storage management and quality control. This gap highlights the need for a flexible, ontology-driven architecture capable of supervising all stages of the wheat supply chain while enabling the integration of heterogeneous data sources, including AI model outputs and sensor streams.

3 Proposed System Architecture

The present architecture is designed to provide a semantic and end-to-end supervision of the wheat lifecycle, from field cultivation to grain storage in warehouses. It is conceptually structured into five main layers: data acquisition, semantic integration, artificial intelligence, reasoning and decision support, and user interface.

3.1 The Data Acquisition Layer

This layer plays a fundamental role in the proposed system by serving as the entry point for all raw data collected throughout the wheat production cycle. It aggregates information from a variety of heterogeneous sources, each capturing specific aspects of the production environment. On the farm, IoT-based sensors measure essential parameters such as soil moisture, ambient temperature, soil pH, and rainfall, providing real-time environmental data that directly impacts crop growth. Complementing these ground-based sensors, drones and satellite imaging systems capture high-resolution visual and spectral data for large-scale crop monitoring, enabling early detection of diseases, nutrient deficiencies, and water stress. Field operators also contribute through mobile applications and handheld devices, which are used to input observations, annotate crop conditions, or respond to system alerts during planting, inspection, and harvesting. In the post-harvest phase, storage facilities are equipped with specialized sensors that monitor warehouse conditions such as humidity, temperature, carbon dioxide concentration, and pest presence, all of which are critical for maintaining grain quality. Additionally, agricultural machinery such as harvesters and transport vehicles generate operational data including GPS coordinates, yield maps, and transport logs.

This multi-source data is typically diverse in structure and format, ranging from CSV tables and JSON sensor outputs to satellite images and XML-based metadata; making it necessary to apply semantic normalization in the next layer of the system to ensure interoperability and integration within the ontology-based framework.

3.2 The Semantic Integration Layer

The Semantic Integration Layer forms the core of the proposed architecture. It addresses the critical issue of data heterogeneity through the application of semantic web technologies. Once data is collected from the diverse sources in the Data Acquisition Layer, this layer is responsible for semantically annotating and structuring the information into a unified and interoperable structure.

The central process in this layer is the use of carefully designed domain ontology that models the key concepts and relationships relevant to the wheat lifecycle including crop growth stages, disease types, sensor measurements, quality assessment parameters, logistics events, storage conditions and AI models outputs. This ontology serves as a common vocabulary for mapping and aligning disparate data formats, enabling the system to interpret and link diverse datasets meaningfully. OWL (Web Ontology Language) [30] is used to represent knowledge. Data is then stored in a knowledge base, which acts as a centralized repository accessible by components of the other layers of the system.

3.3 The AI Layer

The AI Layer augments the system's analytical capabilities. It integrates deep learning models to automate disease detection and grain quality assessment throughout the wheat production lifecycle. This layer serves as a bridge between captured data and high-level semantic interpretation. It employs Convolutional Neural Networks (CNNs) [31] based on wheat datasets available in literature to analyze collected images. In the context of crop health monitoring, CNN models will be trained to detect different plant stages and visual symptoms of prevalent wheat diseases and insects. These models process field imagery to produce annotated classifications and detections. During post-harvest operations, deep learning will be applied to assess grain quality by identifying warehouses insects and indicators such as damaged grain, broken, or fungal contamination, thus enabling automated quality control.

The outputs generated by these AI models will be semantically enriched through a dedicated ontology.

3.4 The Reasoning Layer

This layer is responsible for transforming semantically integrated data and AI model outputs into actionable knowledge that supports proactive and informed decision making throughout the wheat production chain. Building upon the structured knowledge base populated by the previous layers, this component applies rule based reasoning, semantic inference, and query mechanisms to detect risks and generate context specific recommendations. Semantic rules expressed in SWRL (Semantic Web Rule Language) [32] are used to encode expert knowledge and domain specific logic.

The outputs of this layer include real time alerts, system notifications and automated recommendations which are then relayed to the user interface.

3.5 The User Interaction Layer

The User Interaction Layer is the main point where users communicate with the system. It allows farmers, warehouse managers, and agricultural decision-makers to access and use the information provided by the system in the right time. Its main goal is to interpret complex, semantically processed data and AI outputs into simple visuals that help users make good decisions. This layer includes different types of interfaces, designed to meet the needs of each user and the devices they use, including interactive dashboards which show stage updates and reports on crop health, disease alerts, grain quality, and conditions inside the storage warehouses.

All user interfaces will be connected to the ontology. This means users always see the latest and most accurate information. In addition, the system uses role-based access control, so each user sees only the data and tools that match their permissions.

4 Experimental Validation on Wheat Grain Storage Step

To strengthen the proposed architecture, it is essential to demonstrate its applicability through a concrete validation. While the architecture provides a conceptual framework integrating artificial intelligence with semantic technologies across the wheat value chain, its effectiveness must be assessed on a representative and critical use case.

In this section, Wheat grain storage step was selected as a validation scenario because it directly influences storage decisions, market pricing, and food safety. Traditional quality inspection methods are labor-intensive and subjective, creating a pressing need for intelligent, scalable, and explainable solutions.

This use case demonstrates how the system integrates IoT-based sensing, deep learning–driven image analysis, and ontology-based reasoning to ensure reliable and intelligent grain storage management.

4.1 Image Dataset Description

In this case study, we used the GrainSet dataset proposed in [33] to conduct experiments on wheat grain disease and defect classification. GrainSet is a large-scale, high-resolution image collection originally consisting of over 350,000 samples covering multiple cereals, including wheat, maize, sorghum, and rice. For our experiments, we focused on the subset of wheat grains, where each sample was captured from both top and bottom views using a dedicated optical imaging system. Expert inspectors manually annotated the images to ensure reliability and consistency of the ground truth labels.

The wheat subset used in our case study experiments contains 200,000 images distributed across eight quality categories. The majority of samples belong to the Normal (NOR) class, with 120,000 images representing healthy grains. The remaining categories correspond to different types of defects: Fusarium & Shriveled (F&S), Sprouted (SD), Moldy (MY), and Pest Attacked (AP), each containing approximately 13,000 images; Broken (BN) with 14,500 images; Black Point (BP) with 3,500 images; and Impurities (IM) with 10,000 images.

4.2 Deep Learning Classification

For wheat grain classification, we adopt MobileNetV3-Large [34], owing to its strong balance between accuracy, inference speed, and computational efficiency. MobileNet [35] belongs to a family of lightweight architectures designed for resource-constrained environments, where depthwise separable convolutions, inverted residuals, and linear bottlenecks reduce computational cost without sacrificing accuracy. The latest version, MobileNetV3, further integrates Squeeze-and-Excitation (SE) modules and the h-swish activation function, offering an optimized trade-off between performance and efficiency.

The architecture of MobileNetV3-Large (Fig. 1) consists of three main components: an initial standard convolution layer, followed by a sequence of bottleneck blocks with SE and h-swish enhancements, and a final classification head. This design ensures efficient feature extraction and accurate classification with minimal resource consumption.

As a preprocessing step, all images are resized to $224 \times 224 \times 3$, which matches the standard input size expected by MobileNetV3-Large. This ensures uniform processing across samples of varying original dimensions while preserving the proportions of grain structures, thereby enabling consistent and robust model training.

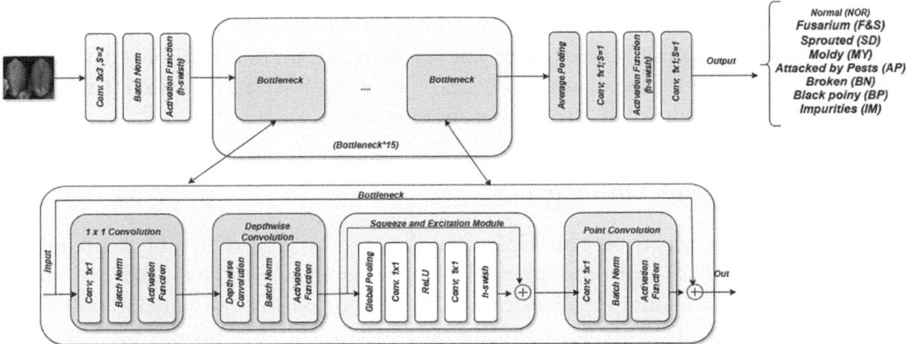

Fig. 1. The Architecture of MobileNetV3-Large.

4.3 Experimental Setup

All experiments were executed on an HP Z8 G4 workstation equipped with dual Intel(R) Xeon(R) Gold 6138 CPUs (40 cores, 80 threads), 128 GB of DDR4 RAM, and an NVIDIA Quadro RTX 6000 GPU with 24 GB of dedicated VRAM. The dataset was divided into two subsets: Training set: 80% (180,000 images), and Testing & validation set: 20% (20,000 images)

The model was trained from scratch, and hyperparameters were tuned empirically through extensive trials to balance generalization and avoid overfitting. The Adam optimizer [36] was employed with default parameters due to its robustness in handling adaptive learning rates and momentum.

A summary of the tested hyperparameters and the final chosen values is presented in Table 1.

Table 1. Hyperparameters configuration

Hyperparameter	Values Explored	Selected values
Batch size	8, 16, 32, 64, 128, 256	64
Epochs	1,2, .. .,1000	301
Learning rate	0.000001, 0.00005, 0.00002, 0.00001, 0.0005, 0.0002, 0.0001, 0.005, 0.002, 0.001, 0.01	0.00005

4.4 Evaluation and Results

To evaluate the performance of the proposed system, we employed several classification metrics. Table 2 summarizes the accuracy and loss values for the training and testing sets, while Fig. 2 depicts the progression of the training loss and accuracy over the epochs. As shown in Fig. 2, the model's accuracy rises quickly in the early epochs and stabilizes after about 301 epochs, reaching an optimal performance of 99.27%. The loss function follows the opposite trend, decreasing consistently during the early epochs before stabilizing at a minimum value of 0.0226.

For the testing set, the model achieves an accuracy of 98.33% with a corresponding loss of 0.0574. The distribution of accuracy across the classes is illustrated by the confusion matrix in Fig. 3. The individual class accuracies are well aligned, as observed, demonstrating consistent performance across different classes. To analyze the classification errors of the model, we conducted a detailed examination of the misclassified images across all eight classes. The highest misclassification error was observed in the AP class. Images in this class often contain small pest-induced holes that are not easily visible. This lack of clarity causes the model to misclassify them as NOR or, in some instances, as SP, due to the visual resemblance between pest damage and the early stages of sprouting. Another notable misclassification error was identified in the MY class. Wheat images with small powdery spots were often misclassified as F&S, likely due to the visual resemblance between the grain curling patterns and those present in F&S samples. Misclassification errors in the F&S class arise from images of grains with pronounced curling, which resemble the pest-induced holes observed in the AP class. In the NOR class, misclassification arises from images of grains exhibiting an indented lower edge, which closely resembles pest-induced edge damage in the AP class. This similarity frequently leads to their misclassification as AP. In the SD class, misclassification errors arise in images of grains displaying early sprouting stages. These visual features resemble grains from the AP class, which are characterized by small pest-induced holes. Some misclassifications were detected in the BP, BN, and IM classes. Despite this, the model shows almost perfect classification performance for these classes, with the limited errors largely attributed to images without clear or distinctive features.

Table 2. Performance on training and testing sets

	Accuracy	Loss
Train	0.9927	0.0226
Test	0.9833	0.0574

As discussed in subsection Image Dataset Description, the dataset exhibits a certain level of imbalance. To evaluate its impact, we performed a more detailed analysis of the model's performance. In addition to the confusion matrix, we used three core metrics: precision, recall, and F1-score. Table 3 summarizes the results of these metrics. The recall value of 0.9833 indicates that the model correctly identifies approximately 98.33% of all actual positive cases. A precision score of 0.9832 shows that when the model predicts a positive class, it is correct about 98.32% of the time. The F1-score, recorded at 0.9831, demonstrates that the model maintains a strong balance between precision and recall. The close alignment of precision, recall, and F1-score further reflects the balanced nature of the classification performance.

Figure 3 shows that the Normal (NOR) class, despite being the majority, does not achieve the highest accuracy, and minority classes such as Black Point (BP) and Insect Damage (IM) are not the least accurate. This demonstrates the model's ability to recognize minority classes without strong bias toward the majority. Furthermore, we examined the false positives (FP) (last ligne) and the false negatives (FN) (last column) across all classes. The results indicate that the majority class, Normal (NOR), produces only a small number of false positives, suggesting that the model rarely misclassifies other classes as Normal. This demonstrates that the model is not biased toward the majority class and confirms its ability to clearly differentiate NOR from other classes, even under conditions of class imbalance. Conversely, the minority classes, such as Black Point (BP) and Impurities (IM), show only a small number of false negatives. This suggests that the model successfully classifies most samples from these classes, demonstrating its capability to accurately identify and distinguish minority class instances despite their limited occurrence in the dataset.

4.5 Ontology Development for the Case Study

The ontology forms the semantic backbone of the proposed architecture. In this case study, our aim is the integration of heterogeneous data sources-including

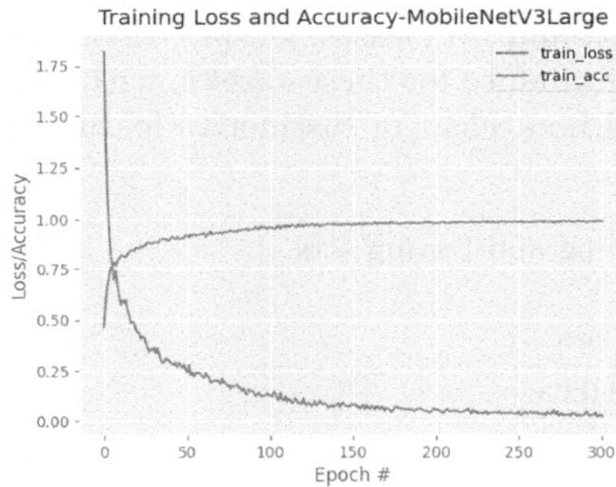

Fig. 2. Training set loss and accuracy progression.

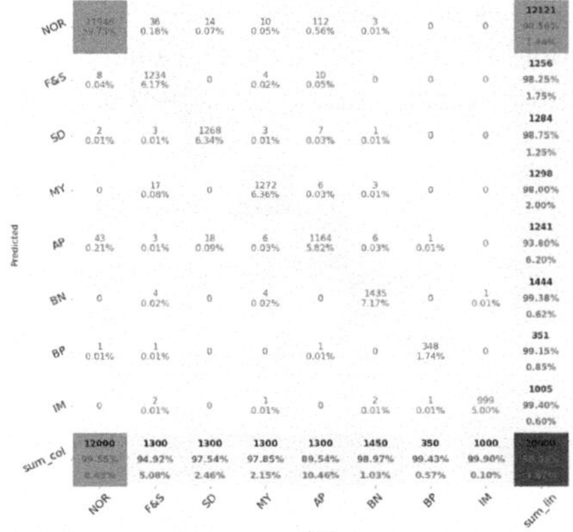

Fig. 3. Confusion matrix.

Table 3. Performance on testing set

	F1-score	Precision	Recall
Test	0.9831	0.9832	0.9833

deep learning-based grain classification results and IoT sensor readings-into a unified reasoning environment. The first ontology was developed following the Methontology methodology [37]. This ontology provides formal representations of the key entities involved in wheat storage and their relationships, thereby supporting intelligent decision-making through SWRL rules. The main objective of the ontology is to prevent quality degradation in storage facilities by combining MobileNetV3-Large defect classifications with environmental parameters such as humidity and temperature.

To ensure both scientific rigor and operational relevance, its development was based on two complementary sources: (i) a review of academic literature on grain storage, post-harvest quality assessment, and pest management, and (ii) practical observations from an internship at the Directorate of Cereals and Pulses in Guelma, Algeria, where real-world challenges in wheat storage management were directly studied. This dual approach ensured that the modeled knowledge reflects both theoretical insights and field practices. A structured set of competency questions guided the design process, focusing on five main dimensions: storage conditions, infrastructure, pest management, monitoring practices; and operational strategies. These questions helped ensure that the ontology addresses practical decision-making needs in warehouse supervision. During the conceptualization stage, core concepts were identified such as:

- GrainState, representing different physical states of wheat grains including Normal, BlackPoint, Broken, Fusarium, Shriveled, Sprouted, Impurities, Moldy and PestAtteacked.
- Problem, describing possible storage or crop issues such as FungalInfection, GrainPest, MoistureDamage, Comtamination, SproutingRisk and PestInfestation.
- Sensor, listing types of devices like Camera, CO2Sensor, HumiditySensor, O2Sensor, and TemperatureSensor used for monitoring silos.
- Silo, representing the physical grain storage units.

Concepts were modeled with their properties and relationships and implemented using OWLReady2 in Python, with validation and visualization performed in Protégé.

To illustrate the operation of the proposed system, we present a simulation scenario. Consider a batch of wheat stored in a warehouse. The system begins by continuously collecting environmental data and images: IoT sensors record a rise in humidity, a slight increase in temperature and CO2. At the same time, a camera captures images of wheat grains from the silo. Then, the images are processed by the MobileNetV3-Large model. Both the sensor readings and the

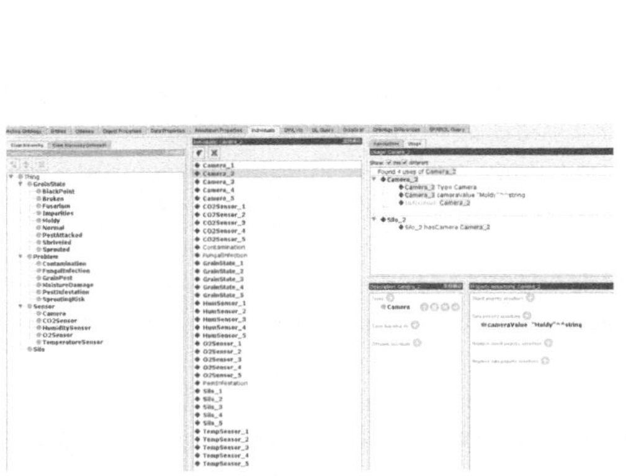

Fig. 4. A part of Ontology classes hierarchy and instances.

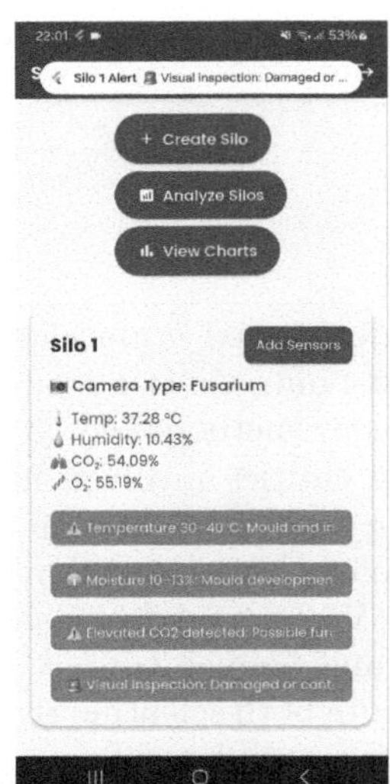

Fig. 5. Alerts Generated by the System.

classification results are semantically annotated and injected into the ontology (see Fig. 4). Using predefined reasoning rules, the system infers a potential risk to grain quality. As a result, alerts are automatically generated and communicated to the warehouse manager, recommending preventive actions to preserve the silo (see Fig. 5).

Through this semantic structure, the framework bridges the gap between AI-based perception and semantic reasoning, enabling automated interventions that improve intelligence in wheat storage management step.

5 Conclusion and Future Work

In the present paper, we provided the theoretical foundations for a conceptual framework designed to semantically integrate and supervise the entire wheat production cycle. The proposed architecture leverages ontologies and artificial intelligence to overcome the challenges of data heterogeneity, lack of interoperability, and fragmented decision-making. The integration of diverse data sources including IoT sensors, remote sensing images, and AI model outputs into a unified semantic layer enables the system to support real-time monitoring and traceability. The use of ontologies allows for the alignment of terminology and the modeling of domain knowledge, while deep learning models contribute to automated disease detection and grain quality evaluation.

To demonstrate the feasibility of the proposed system, we validated the conceptual framework through a use case in smart wheat storage. This case study showed how the integration of MobileNetV3-based grain quality classification,

IoT sensor data, and ontology-driven reasoning can support early risk detection and decision-making in real storage conditions.

Despite these conceptual advancements, the implementation of such a system requires further development, particularly in the area of knowledge representation. As future work, we aim to continue the implementation of the semantic structure for the wheat production cycle. The ontology will cover major components such as crop phenology, diseases, environmental conditions, logistics, sensor data, and storage parameters. A modular approach will promote flexibility, reuse, and interoperability with existing ontologies, including AGROVOC [27], the Crop Ontology [38], SSN/SOSA [39] for sensor data, FOODON [28] for post-harvest representation, and WTO [40] for wheat traits and phenotypes. Additionally, we plan to extend the ontology with metadata standards based on W3C recommendations such as DCAT [41], enabling the description, sharing, and integration of agricultural data across heterogeneous platforms. This work will ultimately pave the way for building a fully interoperable, AI-enhanced, and semantically consistent smart agriculture system, capable of improving traceability in the wheat value chain.

References

1. Giraldo, P., Benavente, E., Manzano-Agugliaro, F., Gimenez, E.: Worldwide research trends on wheat and barley: a bibliometric comparative analysis. Agronomy **9**(7), 352 (2019)
2. Karmakar, P., Teng, S.W., Murshed, M., Pang, P., Van Bui, C.: A guide to employ hyperspectral imaging for assessing wheat quality at different stages of supply chain in Australia: a review. IEEE Trans. AgriFood Electron. **1**(1), 29–40 (2023)
3. Lucena, P., Binotto, A.P., Momo, F.d.S., Kim, H.: A case study for grain quality assurance tracking based on a blockchain business network. arXiv preprint arXiv:1803.07877 (2018)
4. Fuentes-Peñailillo, F., Gutter, K., Vega, R., Silva, G.C.: Transformative technologies in digital agriculture: leveraging internet of things, remote sensing, and artificial intelligence for smart crop management. J. Sens. Actuator Netw. **13**(4), 39 (2024)
5. Aydin, S., Aydin, M.N.: Semantic and syntactic interoperability for agricultural open-data platforms in the context of IoT using crop-specific trait ontologies. Appl. Sci. **10**(13), 4460 (2020)
6. Akanbi, A.K., Masinde, M.: Semantic interoperability middleware architecture for heterogeneous environmental data sources. In: 2018 IST-Africa Week Conference (IST-Africa), p. 1. IEEE (2018)
7. Yeumo, E.D., et al.: Developing data interoperability using standards: a wheat community use case. F1000Research **6**, 1843 (2017)
8. Zeginis, D., Kalampokis, E., Palma, R., Atkinson, R., Tarabanis, K.: A semantic meta-model for data integration and exploitation in precision agriculture and livestock farming. Semantic Web **15**(4), 1165–1193 (2024)
9. Bouacida, I., Farou, B., Djakhdjakha, L., Seridi, H., Kurulay, M.: Innovative deep learning approach for cross-crop plant disease detection: a generalized method for identifying unhealthy leaves. Inf. Process. Agric. **12**(1), 54–67 (2025)

10. Lynda, D., Brahim, F., Imane, B., Aya, M.: A semantic framework for monitoring wheat grain warehouses (2024)
11. Bertoglio, R., Corbo, C., Renga, F.M., Matteucci, M.: The digital agricultural revolution: a bibliometric analysis literature review. IEEE Access **9**, 134762–134782 (2021)
12. Kamilaris, A., Prenafeta-Boldú, F.X.: Deep learning in agriculture: a survey. Comput. Electron. Agric. **147**, 70–90 (2018)
13. Lu, J., Hu, J., Zhao, G., Mei, F., Zhang, C.: An in-field automatic wheat disease diagnosis system. Comput. Electron. Agric. **142**, 369–379 (2017)
14. Xu, L., et al.: Wheat leaf disease identification based on deep learning algorithms. Physiol. Mol. Plant Pathol. **123**, 101940 (2023)
15. Mumtaz, R., Maqsood, M.H., Haq, I., Shafi, U., Mahmood, Z., Mumtaz, M.: Integrated digital image processing techniques and deep learning approaches for wheat stripe rust disease detection and grading. Decis. Anal. J. **8**, 100305 (2023)
16. Goyal, L., Sharma, C.M., Singh, A., Singh, P.K.: Leaf and spike wheat disease detection & classification using an improved deep convolutional architecture. Inform. Med. Unlocked **25**, 100642 (2021)
17. Long, M., Hartley, M., Morris, R.J., Brown, J.K.: Classification of wheat diseases using deep learning networks with field and glasshouse images. Plant. Pathol. **72**(3), 536–547 (2023)
18. Bukhari, H.R., et al.: Assessing the impact of segmentation on wheat stripe rust disease classification using computer vision and deep learning. IEEE Access **9**, 164986–165004 (2021)
19. Aboneh, T., Rorissa, A., Srinivasagan, R., Gemechu, A.: Computer vision framework for wheat disease identification and classification using jetson GPU infrastructure. Technologies **9**(3), 47 (2021)
20. Shafi, U., et al.: Embedded ai for wheat yellow rust infection type classification. IEEE Access **11**, 23726–23738 (2023)
21. Genaev, M.A., Skolotneva, E.S., Gultyaeva, E.I., Orlova, E.A., Bechtold, N.P., Afonnikov, D.A.: Image-based wheat fungi diseases identification by deep learning. Plants **10**(8), 1500 (2021)
22. Navarro, E., Costa, N., Pereira, A.: A systematic review of IoT solutions for smart farming. Sensors **20**(15), 4231 (2020)
23. Dong, M., Yu, H., Sun, Z., Zhang, L., Sui, Y., Zhao, R.: Research on agricultural environmental monitoring internet of things based on edge computing and deep learning. J. Intell. Syst. **33**(1), 20230114 (2024)
24. Lynda, D., Brahim, F., Hamid, S., Hamadoun, C.: Towards a semantic structure for classifying IoT agriculture sensor datasets: an approach based on machine learning and web semantic technologies. J. King Saud Univ.-Comput. Inf. Sci. **35**(8), 101700 (2023)
25. Ahoa, E., Kassahun, A., Verdouw, C., Tekinerdogan, B.: Challenges and solution directions for the integration of smart information systems in the agri-food sector. Sensors **25**(8), 2362 (2025)
26. Ngo, Q.H., Kechadi, T., Le-Khac, N.-A.: Oak: ontology-based knowledge map model for digital agriculture. In: International Conference on Future Data and Security Engineering, pp. 245–259. Springer (2020)
27. Subirats-Coll, I., et al.: Agrovoc: the linked data concept hub for food and agriculture. Comput. Electron. Agric. **196**, 105965 (2022)
28. Dooley, D.M., et al.: Foodon: a harmonized food ontology to increase global food traceability, quality control and data integration. NPJ Sci. Food **2**(1), 23 (2018)

29. Skobelev, P., Laryukhin, V., Simonova, E., Goryanin, O., Yalovenko, V., Yalovenko, O.: Multi-agent approach for developing a digital twin of wheat. In: 2020 IEEE International Conference on Smart Computing (SMARTCOMP), pp. 268–273. IEEE (2020)
30. McGuinness, D.L., Van Harmelen, F., et al.: Owl web ontology language overview. W3C Recommendation **10**(10), 2004 (2004)
31. Alzubaidi, L., et al.: Review of deep learning: concepts, CNN architectures, challenges, applications, future directions. J. Big Data **8**(1), 53 (2021)
32. Horrocks, I., et al.: SWRL: a semantic web rule language combining owl and ruleml. W3C Member Submission **21**(79), 1–31 (2004)
33. Fan, L., et al.: An annotated grain kernel image database for visual quality inspection. Sci. Data **10**(1), 778 (2023). https://doi.org/10.1038/s41597-023-02529-2
34. Howard, A., et al.: Searching for mobilenetv3. In: Proceedings of the IEEE/CVF International Conference on Computer Vision (ICCV), pp. 1314–1324 (2019)
35. Howard, A.G., et al.: Mobilenets: efficient convolutional neural networks for mobile vision applications. arXiv preprint arXiv:1704.04861 (2017)
36. Tato, A., Nkambou, R.: Improving adam optimizer (2018)
37. Fernández-López, M., Gómez-Pérez, A., Juristo Juzgado, N.: Methontology: from ontological art towards ontological engineering (1997)
38. Matteis, L., et al.: Crop ontology: vocabulary for croprelated concepts. Semantics for Biodiversity (S4BioDiv 2013) **37** (2013)
39. Janowicz, K., Haller, A., Cox, S.J., Le Phuoc, D., Lefrançois, M.: Sosa: a lightweight ontology for sensors, observations, samples, and actuators. J. Web Semantics **56**, 1–10 (2019)
40. Nédellec, C., Ibanescu, L., Bossy, R., Sourdille, P.: WTO, an ontology for wheat traits and phenotypes in scientific publications. Genomics Inform. **18**(2), 14 (2020)
41. Data catalog vocabulary (DCAT) (2014)

Spatio-temporal Transformers for High-Accuracy Detection of Embryo Developmental Transitions

Aissa Benfettoume Souda[1], Mohammed El Amine Bechar[1(✉)], Souaad Hamza-Cherif[2], Jean-Marie Guyader[1], Marwa Elbouz[1], Fréderic Morel[3,4], Aurore Perrin[3,4], and Nesma Settouti[1(✉)]

[1] LabISEN, LSL Teams, Marseille, France
{mohammed-el-amine.bechar,nesma.settouti}@isen-ouest.yncrea.fr
[2] Biomedical Engineering Laboratory, University of Tlemcen, Chetouane, Algeria
[3] Service de Génétique Médicale et Biologie de la Reproduction, CHU de Brest, Brest, France
[4] Faculté de Médecine et des Sciences de la Santé, Université de Brest, EFS, UMR1078, GGB, Brest, France

Abstract. Accurate and objective assessment of embryo development is a cornerstone of successful in vitro fertilization (IVF), yet current practice remains highly dependent on manual and subjective evaluation of morphokinetic events. This work introduces a deep learning framework leveraging spatio-temporal transformers to automatically detect developmental phase transitions in human embryos from time-lapse imaging. We formulate this task as a binary sequence classification problem and benchmark a transformer-based architecture (TimeSformer) against a strong convolutional baseline (frame-stacked ResNet-18). Using an embryo-level split on a curated time-lapse dataset, TimeSformer achieved a substantial performance gain, reaching an F1-score of 90.44% compared to 71.82% for the CNN baseline. These results demonstrate the critical importance of modeling temporal dependencies and highlight transformers as a state-of-the-art solution for morphokinetic analysis. Beyond outperforming conventional CNNs, our approach lays the groundwork for robust, automated, and clinically relevant decision-support tools, advancing the integration of AI into IVF workflows.

Keywords: In Vitro Fertilization · Embryo Phase Transition · Temporal Deep Learning · Vision Transformers · Morphokinetic Analysis

1 Introduction

In vitro fertilization (IVF) has become a cornerstone of reproductive medicine, yet its success remains critically dependent on accurately selecting the most

T. Ensari et al. (Eds.): ISPR 2025, CCIS 2859, pp. 244–255, 2026.
https://doi.org/10.1007/978-3-032-21585-7_18

viable embryo for transfer. Traditionally, embryologists rely on static morphological assessment at fixed time points: evaluating cell number, symmetry, fragmentation, or blastocyst expansion. However, such evaluations are subjective, prone to inter-observer variability, and inconsistent across clinics [13].

Time-lapse imaging (TLI) systems have transformed embryo monitoring by enabling continuous, non-invasive observation throughout development [6,11]. These systems generate rich spatiotemporal data capturing the full morphokinetic trajectory. Despite this, most automated and deep learning (DL)-based approaches either use single static frames or manually annotated events, neglecting the temporal dynamics that are central to embryo viability [3,10,14,15].

For example, Khosravi et al. [10] trained STORK on over 50000 static embryo images to predict implantation outcomes, achieving expert-level performance but entirely disregarding developmental timing. Similarly, Bormann et al. [3] and Tran et al. [14] focused on still frames or handcrafted morphokinetic features, rather than exploiting end-to-end temporal learning. While promising, these methods overlook critical features such as cleavage intervals, morula compaction timing, and blastocyst expansion speed–key indicators linked to implantation success [1,12].

Recent advances in video-based deep learning offer new opportunities. Vision Transformers such as TimeSformer [2] explicitly model spatiotemporal dependencies through divided attention, showing state-of-the-art performance in video understanding tasks. Yet, their application to embryo morphokinetic analysis remains unexplored.

This study investigates transformer-based temporal modeling for the automated detection of developmental phase transitions in human embryos. Unlike static-stage classification, phase transition detection focuses on capturing subtle, transient changes (e.g., morula-to-blastocyst transition) that are crucial but often overlooked in IVF workflows.

Figure 1 provides a conceptual overview contrasting traditional manual embryo assessment with our proposed deep temporal modeling pipeline. While conventional approaches rely on static morphological inspection, our framework leverages time-lapse imaging and transformer-based temporal modeling to automatically detect phase transitions in embryo development. Our main contributions are:

- *Exploration of deep temporal modeling* for embryo phase transition detection from time-lapse imaging.
- *Systematic comparison* between a frame-stacked CNN baseline (ResNet-18) and a transformer-based spatiotemporal model (TimeSformer).
- *Empirical evidence* demonstrating the superiority of temporal attention mechanisms, achieving a significant F1-score improvement over CNNs.
- *Shift from static evaluation to dynamic modeling*, advancing AI-driven decision-support tools for IVF embryo assessment.

The remainder of this paper is structured as follows: Sect. 2 presents the dataset and preprocessing, the proposed modeling approaches and training pro-

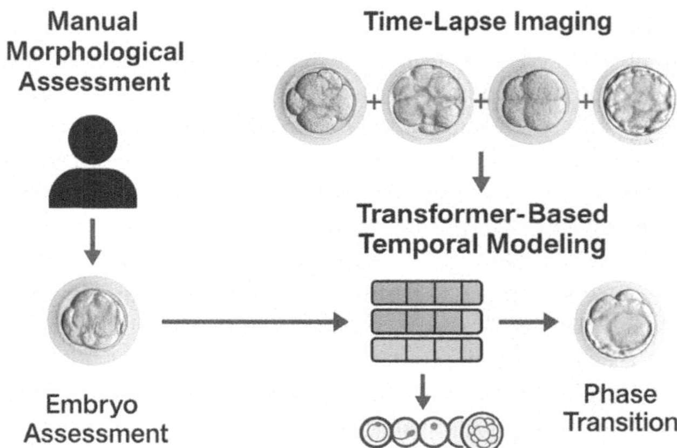

Fig. 1. Conceptual overview contrasting traditional manual embryo assessment (left) with the proposed deep temporal modeling pipeline (right), leveraging time-lapse imaging and transformer-based temporal modeling for phase transition detection.

tocol. Section 3 reports the experimental results and analysis. Finally, Sect. 4 discusses the implications and outlines future work.

2 Methodology

The proposed methodology integrates three core components: (1) a curated time-lapse embryo dataset with fine-grained morphokinetic annotations, (2) a task-specific formulation for phase transition detection, and (3) deep temporal modeling architectures comparing convolutional and transformer-based approaches. Figure 2 provides an overview of the proposed pipeline.

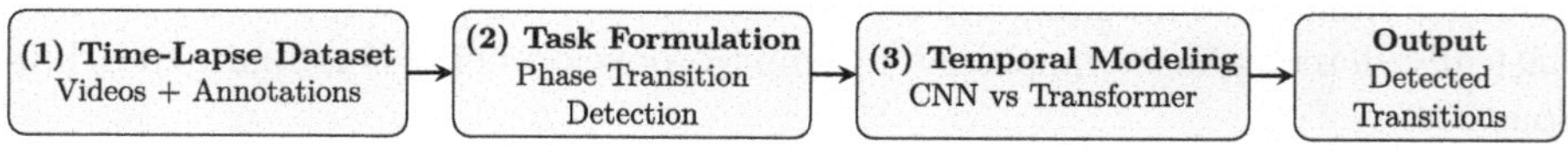

Fig. 2. Proposed pipeline combining curated embryo datasets, temporal modeling (CNN vs Transformer), and automated transition detection.

2.1 Dataset and Annotations

This work utilizes the publicly available dataset introduced by Gomez et al. [7], one of the most comprehensive resources for time-lapse embryo analysis. It contains 704 human embryo videos captured across seven focal planes, comprising approximately 2.4M grayscale images (500 × 500 px).

Each embryo is annotated by certified embryologists following the Ciray morphokinetic nomenclature [5], marking the precise timing of 16 developmental events: from second polar body extrusion (tPB2) and pronuclear dynamics (tPNa, tPNf), through cleavage stages (t2–t9+), compaction (tM), and blastocyst formation and expansion (tSB, tB, tEB, tHB). These annotations provide ground truth for detecting phase transitions. Figure 3 illustrates the annotated developmental timeline.

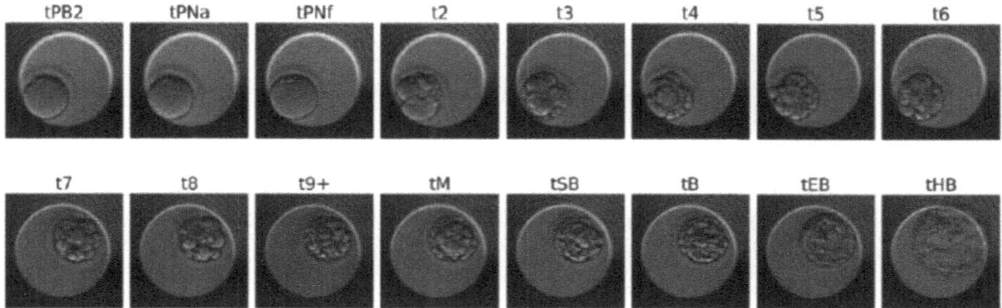

Fig. 3. Morphokinetic events defined in the dataset following Ciray et al. [5], covering the full trajectory from fertilization to hatching blastocyst.

2.2 Task Formulation

The problem is framed as a **binary sequence classification task**: given a temporal window of consecutive frames, the model predicts whether a phase transition occurs within the window (see Fig. 4). This formulation directly targets morphokinetic changes rather than static stage classification, enabling temporal modeling of subtle embryo dynamics (e.g., morula-to-blastocyst transition).

To ensure robust evaluation, an **embryo-level split** was applied, preventing data leakage across training, validation, and test sets.

2.3 Deep Temporal Modeling

To assess the benefit of temporal modeling, two deep learning architectures designed to capture temporal dynamics are compared. The core objective is to move beyond static image analysis and compare how different models interpret information embedded in sequences of frames over time and detect whether embryo morphology is changing (see Fig. 5). The experiment contrasts two distinct approaches, each representing a different strategy for learning from spatio-temporal data: a 2D and ResNet architecture, and a state-of-the-art TimeSformer model [2].

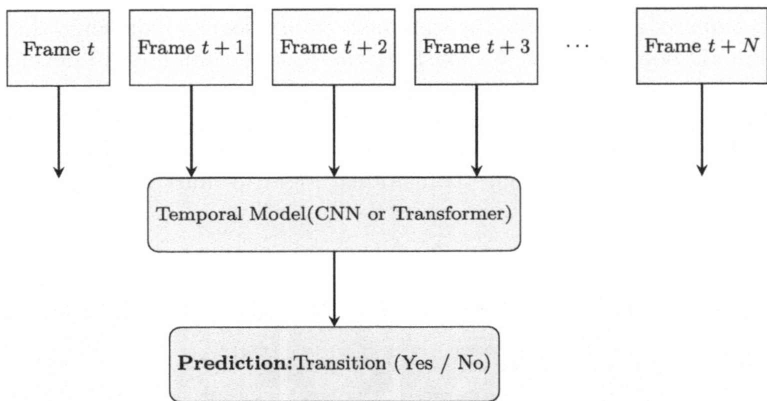

Fig. 4. Task formulation: given a temporal window of N consecutive frames from time-lapse imaging, the model predicts whether a phase transition (e.g., morula-to-blastocyst) occurs within that window.

Frame-Stacked 2D CNN (ResNet-18 Baseline). We adapt a ResNet-18 [8] pre-trained on ImageNet by stacking $N = 8$ consecutive grayscale frames along the channel dimension to form an $8 \times 224 \times 224$ tensor, replacing the usual RGB input. This approach approximates temporal context while retaining a 2D convolutional backbone.

Transformer-Based Temporal Modeling (TimeSformer). TimeSformer [2] is a state-of-the-art video classification model based on the transformer architecture, which replaces convolution with self-attention mechanisms. Unlike traditional CNNs that extract local features, TimeSformer is designed to capture long-range dependencies across both spatial and temporal dimensions, making it particularly effective for video data. In this study, it was selected to move beyond static image analysis and better capture the dynamic information inherent in embryo development sequences.

Architecturally, TimeSformer is a video-adapted variant of the Vision Transformer (ViT) [4]. Each video frame is divided into a grid of patches, which are flattened into a sequence of tokens—similar to words in a sentence. The model then applies a spatio-temporal self-attention mechanism: spatial attention captures relationships within each frame, while temporal attention models dependencies across frames. This decoupling allows for efficient and scalable video processing without the high computational cost typical of full 3D attention.

A pretrained TimeSformer model, available from the Hugging Face repository and originally provided by Meta (Facebook), was used for temporal modeling. Trained on the Kinetics-400 dataset [9]. The input to the model is a tensor of shape $32 \times 3 \times 224 \times 224$, where 32 represents consecutive, non-overlapping frames extracted from embryo development videos. This longer temporal window is crucial for capturing morphokinetic patterns subtle timing and struc-

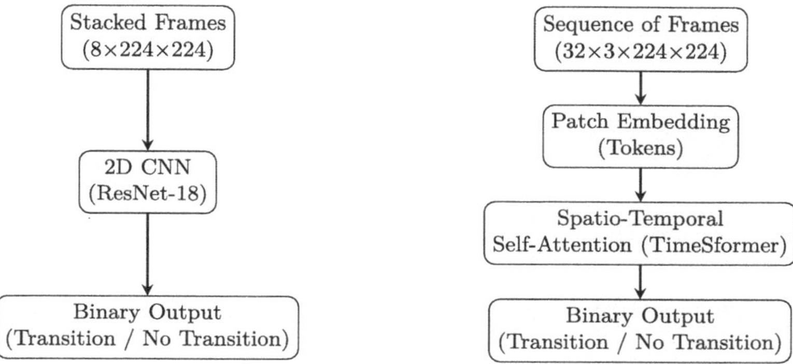

Fig. 5. Temporal modeling architectures: (a) Frame-stacked CNN baseline where temporal context is encoded by stacking frames, vs. (b) Transformer-based TimeSformer with decoupled spatio-temporal attention.

tural changes that characterize different developmental phases. By analyzing an extended sequence, the model can learn to distinguish between phases that may appear visually similar in individual frames. This ability to model temporal dynamics is hypothesized to enhance classification accuracy in tracking embryo development progression.

3 Experiments and Results

3.1 Data Preparation

The dataset Despite its richness, presents multiple challenges that necessitated careful preparation to ensure the reliability of our temporal modeling task. Key issues with the raw data include:

- Severe class imbalance: Certain stages (e.g., tPNa, t9+, t8) are overrepresented, while others (e.g., t3, t5, t7) appear in very few frames as seen in Fig. 6a.
- Heterogeneity across videos: The number of annotated stages per embryo varies significantly, from as few as 6 to as many as 16, leading to inconsistent temporal coverage as illustrated in Fig. 6b.
- Data quality problems: Numerous frames were blurry, under- or overexposed, or miss-annotated. In some cases, entire video segments were corrupted, with all frames labeled as a single incorrect phase or containing empty, dark, or camera-misaligned images as shown in Fig. 7.

To address these issues, a comprehensive data cleaning process was applied:
Every video was examined frame-by-frame. Whenever a quality issue was detected (e.g., blurriness starting at frame 239), the entire segment from that frame to the end was discarded. The rare hatched blastocyst (tHB) phase was excluded entirely due to its extreme scarcity. Miss-annotated images were

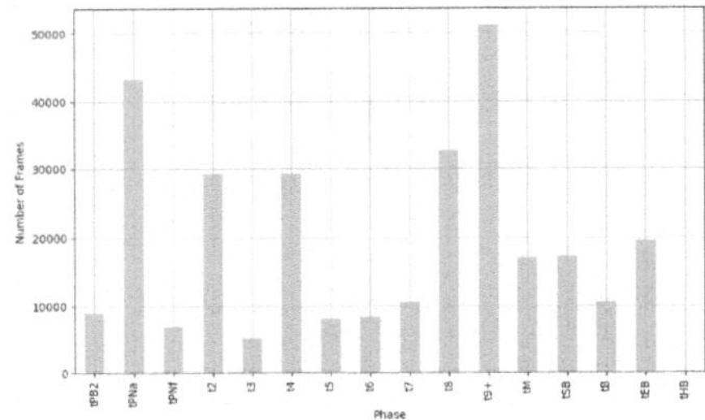

(a) Distribution of frames across phases.

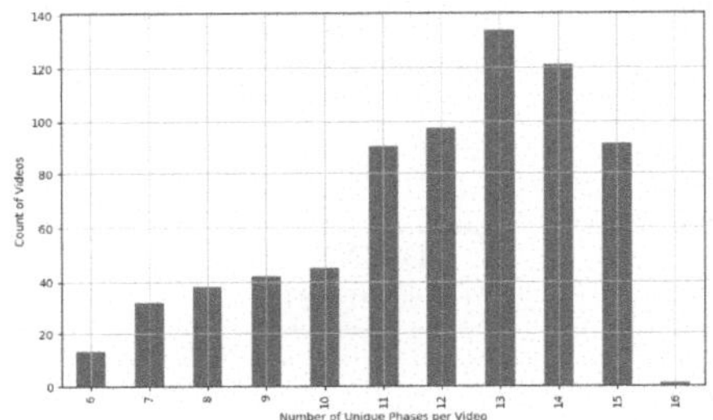

(b) Distribution of phase counts across videos.

Fig. 6. Analysis of the dataset: (a) distribution of frames across morphokinetic phases, and (b) distribution of phase counts across embryo videos.

retained as-is, due to the lack of embryological expertise for manual correction, but were handled cautiously during analysis.

After cleaning, the total number of frames dropped from 297428 to 287147. Table 1 summarizes frame reductions per phase.

For training and evaluation, the dataset was split at the embryo level, ensuring that all frames from a single embryo were assigned exclusively to either the training, validation, or test set. This strategy prevents data leakage and allows for a more realistic assessment of generalization to unseen embryos.

The experimental design exclusively employed an embryo-level data split, with a 70%-15%-15% division across training, validation, and test sets, respectively. This resulted in 704 embryos being partitioned without overlap. Each sample used in training consists of a fixed-length sequence of 8 and 32 consecutive frames centered on the embryo's main focal plane. While multiple focal planes are available, we intentionally restricted our analysis to the central focal plane to simplify modeling and establish robust baselines. Future work will explore multi-focal modeling.

Final preprocessing steps included resizing all frames to 224×224 pixels and normalizing grayscale values. For transformer-based models such as TimeSformer, images were optionally converted to RGB format. This rigorous preparation ensures the quality, fairness, and interpretability of the results, setting a foundation for future exploration of spatiotemporal dynamics in embryo development.

3.2 Hyperparameter Settings

All experiments were conducted using the PyTorch framework (version 2.8) and Python 3.11. The computational workload was managed by an NVIDIA RTX 6000 Ada Generation GPU. To ensure a fair and direct comparison between models, a consistent set of hyperparameters was used across all experiments. This standardized approach is crucial for maintaining the reliability and reproducibility of the results, and the general hyperparameter configurations are summarized

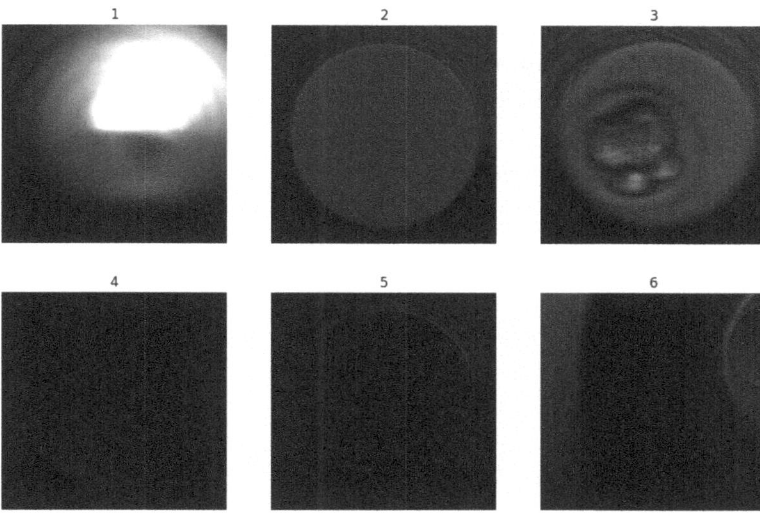

Fig. 7. Examples of corrupted images include: (1) overexposed frames, (2) empty images mislabeled as tEB, (3) blurred frames, (4) underexposed (excessively dark) images, (5) unidentifiable images mislabeled as valid stages, and (6) misaligned frames with the camera.

Table 1. Summary of frame counts before and after data cleaning

Phase	Raw Data	Deleted Frames	New Data
tPB_2	8895	6	8889
tPNa	43244	1	43243
tPNf	6793	10	6783
t2	29189	136	29041
t3	5127	232	4895
t4	29177	641	28536
t5	8045	51	7994
t7	10531	118	10413
t8	32602	1387	31215
$t9^+$	51112	1138	49974
tM	17084	716	16368
tSB	17244	843	16401
tB	10387	1666	8721
tEB	19535	3161	16374
tHB	97	51	46
Total	**297428**	**10281**	**287147**

in Table 2. However, a notable exception was made for the TimeSformer model due to its significant computational cost. With each training epoch requiring approximately 14 h to complete, this specific model was trained with reduced settings, highlighted in blue in Table 2.

Table 2. Hyperparameter settings for model training

Hyperparameter	ResNet-18	TimeSformer
Image Size	224 × 224	
Normalization	Mean & Std	
Optimizer	Adam	
Loss Function	Cross-Entropy	
Learning Rate	1×10^{-3}	1×10^{-4}
Batch Size	16	1
Epochs	100	10

3.3 Results and Discussion

This study evaluated the contribution of temporal information to phase transition detection in embryo development by comparing two deep learning architectures: a frame-stacked 2D convolutional model (ResNet-18) and a transformer-based spatiotemporal model (TimeSformer). The ResNet-18 baseline, pretrained on ImageNet, was adapted to accept 8-channel inputs by stacking consecutive grayscale frames, thereby providing limited temporal context. In contrast, the TimeSformer architecture, derived from Vision Transformers, was specifically designed for video analysis and leverages dual self-attention mechanisms to explicitly capture spatial and temporal dependencies. Both models were trained on non-overlapping frame sequences, using 8-frame windows for ResNet-18 and 32-frame windows for TimeSformer.

The comparative results, summarized in Table 3, demonstrate a clear advantage for transformer-based temporal modeling. TimeSformer achieved an F1-score of **90.44%**, substantially outperforming the ResNet-18 baseline (**71.82%**). Its higher recall (**97.06%**) highlights its sensitivity in identifying subtle morphokinetic transitions, whereas the CNN exhibited limitations in leveraging temporal cues effectively. These findings confirm that explicitly modeling temporal dynamics provides significant benefits over implicit frame-stacking approaches for embryo phase transition detection.

The TimeSformer achieved a substantial improvement over the baseline ResNet-18, reaching an F1-score of **90%** compared to **71%** for the stacked-frame CNN (Table 3). This performance gap underscores the advantage of transformer-based architectures in explicitly modeling temporal dependencies within time-lapse embryo videos.

Table 3. Performance metrics of ResNet18 and TimeSFormer

Model	ResNet18 Metrics (%)				TimeSFormer Metrics (%)			
	Acc	Precision	Recall	F1-score	Acc	Precision	Recall	F1-score
Results	71.04	72.60	71.05	71.82	83.36	84.67	97.06	90.44

While the ResNet-18 baseline can implicitly encode limited temporal context via frame stacking, the TimeSformer leverages self-attention mechanisms to integrate both spatial and temporal information across an extended sequence of **32 frames**, compared to only 8 frames for ResNet-18. This broader temporal window enables the model not only to detect phase transitions but also to contextualize them within preceding developmental patterns.

However, these performance gains come at a cost. The TimeSformer exhibited significant computational demands, requiring GPUs with at least 24 GB of VRAM to support a batch size of 1, which limited training to 10 epochs. In contrast, the ResNet-18 baseline was trained for 100 epochs with a batch size of 16. This highlights a clear trade-off: while transformer-based approaches capture richer spatiotemporal dynamics, their resource-intensive nature currently limits accessibility and practical deployment.

3.4 Limitations

Despite the promising results, this study has several limitations. First, experiments were conducted on a single publicly available dataset and restricted to the central focal plane. This was intentional to establish a reliable baseline, as the central plane provides the clearest view of embryo morphology and such datasets are rare, requiring specialized incubators, expert annotation, and strict ethical approvals. While this ensured controlled evaluation, it may limit generalizability, underscoring the need for multi-focal and multi-center validation. Second, although we tested standard data augmentation (e.g., rotations, flips, brightness adjustments, temporal jittering), it brought no significant improvement and mainly increased training time, likely due to the subtle similarities between consecutive stages. Moreover, the dataset remained imbalanced even after cleaning, with rare developmental transitions being underrepresented (e.g., the pronuclear fading to 2-cell stage *tPNf* → *t2* transition), making the model less sensitive to these events. Alternative strategies such as class-weighted or focal loss, along with larger datasets, may better address this issue. Third, other transformer variants gave only marginal gains and did not match the substantial improvement of TimeSformer, which remains computationally demanding. Finally, interpretability was not directly assessed, and future work should link attention maps to embryological features and validate them with experts to support clinical adoption.

4 Conclusion and Perspectives

This study demonstrates that transformer-based temporal modeling significantly improves the detection of morphokinetic phase transitions in human embryos using time-lapse imaging. The TimeSformer outperformed the ResNet-18 baseline by effectively leveraging self-attention to model extended temporal dependencies and subtle developmental changes. The use of an embryo-level split ensured robust generalization, while focusing on the central focal plane established strong 2D spatiotemporal baselines.

These findings underscore the importance of shifting from static stage classification toward **dynamic transition detection**, a more biologically relevant approach to embryo assessment in IVF. Attention-based models, such as TimeSformer, show strong potential to capture the sequential and progressive nature of embryo development.

Looking forward, the main challenges include improving computational efficiency through model compression, or distillation; enhancing generalizability via multi-focal and multi-center validation; addressing class imbalance through larger datasets and tailored loss functions rather than standard augmentation; and strengthening interpretability by linking attention maps to biologically meaningful features validated by embryologists. Finally, integrating the model into IVF decision-support workflows under regulatory and ethical frameworks will be essential to ensure safe and impactful clinical deployment.

References

1. Basile, N., et al.: The use of morphokinetics as a predictor of implantation: a multicentric study to define and validate an algorithm for embryo selection. Hum. Reprod. **30**(2), 276–283 (2015)
2. Bertasius, G., Wang, H., Torresani, L.: Is space-time attention all you need for video understanding? In: ICML, vol. 2, p. 4 (2021)
3. Bormann, C.L., et al.: Performance of a deep learning based neural network in the selection of human blastocysts for implantation. elife **9**, e55301 (2020)
4. Chen, J., et al.: Transunet: transformers make strong encoders for medical image segmentation. arXiv preprint arXiv:2102.04306 (2021)
5. Ciray, H.N., et al.: Proposed guidelines on the nomenclature and annotation of dynamic human embryo monitoring by a time-lapse user group. Hum. Reprod. **29**(12), 2650–2660 (2014)
6. Conaghan, J., et al.: Improving embryo selection using a computer-automated time-lapse image analysis test plus day 3 morphology: results from a prospective multi-center trial. Fertil. Steril. **100**(2), 412–419 (2013)
7. Gomez, T., et al.: A time-lapse embryo dataset for morphokinetic parameter prediction. Data Brief **42**, 108258 (2022)
8. He, K., Zhang, X., Ren, S., Sun, J.: Deep residual learning for image recognition. In: Proceedings of the IEEE Conference on Computer Vision and Pattern Recognition, pp. 770–778 (2016)
9. Kay, W., et al.: The kinetics human action video dataset. arXiv preprint arXiv:1705.06950 (2017)

10. Khosravi, P., et al.: Deep learning enables robust assessment and selection of human blastocysts after in vitro fertilization. NPJ Digit. Med. **2**(1), 21 (2019)
11. Kirkegaard, K., Agerholm, I.E., Ingerslev, H.J.: Time-lapse monitoring as a tool for clinical embryo assessment. Hum. Reprod. **27**(5), 1277–1285 (2012)
12. Rubio, I., et al.: Clinical validation of embryo culture and selection by morphokinetic analysis: a randomized, controlled trial of the embryoscope. Fertil. Steril. **102**(5), 1287–1294 (2014)
13. Storr, A., Venetis, C., Cooke, S., Kilani, S., Ledger, W.: Time-lapse algorithms and morphological selection of day-5 embryos for transfer: a preclinical validation study. Fertil. Steril. **109**(2), 276–283 (2018)
14. Tran, D., Cooke, S., Illingworth, P.J., Gardner, D.K.: Deep learning as a predictive tool for fetal heart pregnancy following time-lapse incubation and blastocyst transfer. Hum. Reprod. **34**(6), 1011–1018 (2019)
15. VerMilyea, M., et al.: Development of an artificial intelligence-based assessment model for prediction of embryo viability using static images captured by optical light microscopy during IVF. Hum. Reprod. **35**(4), 770–784 (2020)

Prompting Recovery: Filling Missing Diabetes Data with Large Language Models

Lucia Cascone(✉), Simone D'Assisi, Michele Nappi, and Orlando Tomeo

Department of Computer Science, University of Salerno, Via Giovanni Paolo II, 132, 84084 Fisciano, SA, Italy
{lcascone,mnappi}@unisa.it, {s.dassisi,o.tomeo}@studenti.unisa.it

Abstract. Missing data is a persistent challenge in clinical analysis, often undermining the reliability of machine learning models. This study explores the use of large language models (LLMs), both general-purpose and biomedical, for imputing missing values in the Pima Indians Diabetes Dataset. Patient records were reformulated as natural language prompts, with missing features posed as explicit questions to the models. We evaluated GPT-2, Llama-2-7B, BioBERT v1.1, Gemma-2-2B-it, Gemma-2-9B-it, and Bio-Medical-Llama-3-8B under Zero-Shot and Few-Shot prompting. While GPT-2, Llama-2-7B, and BioBERT showed limited performance, Gemma-2-9B-it and Bio-Medical-Llama-3-8B—especially with carefully designed Few-Shot prompts—produced accurate imputations and yielded downstream classification performance comparable to, or better than, traditional (KNN) and generative (CTGAN, TVAE) methods. These results highlight the potential of LLMs as a flexible and effective solution for clinical data imputation, capable of leveraging semantic context to capture complex relationships in medical data without requiring changes to existing analytical pipelines.

Keywords: Large Language Models · Data Imputation · Diabetes Dataset

1 Introduction

Missing data is a pervasive and significant issue in medical datasets, often arising from patient dropouts, measurement errors, or incomplete records. If not properly addressed, these missing values can introduce bias and substantially reduce the accuracy of both clinical analyses and the predictive models built upon them. As a result, a variety of data imputation methods have been developed to estimate plausible values in place of missing entries, thereby enabling more reliable downstream analysis and interpretation.

T. Ensari et al. (Eds.): ISPR 2025, CCIS 2859, pp. 256–270, 2026.
https://doi.org/10.1007/978-3-032-21585-7_19

Recent advances in natural language processing suggest that large language models (LLMs) could offer a novel and flexible approach to data imputation. In this study, we systematically evaluate the effectiveness of several pre-trained LLMs in imputing missing values within the widely used Pima Indians Diabetes Dataset. Our approach reformulates each patient record as a textual prompt, explicitly casting missing features as natural-language questions to the model. We experiment with both Zero-Shot and Few-Shot prompting strategies, providing models with either no example or a small set of representative examples to guide the imputation process. The tested models include `GPT-2`, `Llama-2-7B`, `BioBERT v1.1`, `Gemma-2-2B-it`, `Gemma-2-9B-it`, and `Bio-Medical-Llama`. Model outputs are subsequently parsed and post-processed to yield structured numerical values consistent with the original tabular format.

To maximize imputation accuracy, we design targeted prompts for each language model, carefully tailoring the prompting approach to the unique characteristics and requirements of each architecture, while ensuring a consistent task formulation across experiments. Prompt engineering, therefore, emerges as a critical factor in adapting general-purpose and domain-specific LLMs to the structured imputation setting.

This work is guided by two main research questions, the answers to which represent our primary contributions:

- **[RQ1] Intrinsic Evaluation:** How accurately can large language models impute missing values in structured medical data compared to traditional statistical and generative techniques?
- **[RQ2] Downstream Evaluation:** Do LLM-based imputations yield improved downstream predictive performance when used as input to diagnostic models, compared to standard imputation methods?

The **contributions**[1] of this work extend beyond methodological innovation to provide practical insights for clinical data analysis. By demonstrating that LLMs can effectively impute missing medical values while preserving or enhancing downstream predictive performance, we establish a foundation for integrating advanced language models into clinical data preprocessing pipelines. Our findings suggest that the semantic understanding capabilities of LLMs offer distinct advantages over traditional statistical approaches, especially in capturing complex contextual relationships that characterize medical data. Furthermore, this research provides detailed insights into the practical challenges and solutions associated with applying LLMs to structured data imputation, including prompt engineering strategies, model selection criteria, and output validation approaches. These contributions are essential for practitioners seeking to leverage LLM capabilities for real-world clinical applications where data completeness remains a persistent challenge.

[1] Code for prompt generation, parsing, and the imputation/classification pipeline is available in a private repository and can be shared upon request to the authors.

2 Literature Review

Over the years, a wide variety of imputation techniques were developed, ranging from statistical models to machine learning models. In the present study, in addition to the use of LLMs, we also consider further strategies for the generation of synthetic data. Several artificial intelligence models are specifically designed to produce realistic synthetic datasets, offering high-quality outputs, especially within the medical and healthcare domains. This study reviews the most relevant contributions and identifies models most suitable for clinical data generation to be integrated into the experimental phase.

2.1 Established Imputation Techniques

A traditional imputation approach used in medical domain is the k-Nearest Neighbors (KNN) imputation which estimates missing values based on similar patient records, according to a chosen distance metric, typically using the mean, median, or mode [3]. In addition, the survey presented in *Comprehensive Exploration of Synthetic Data Generation: A Survey* [2] systematically analyzes 417 models for synthetic data generation (SDG) developed over the past decade, providing a comprehensive overview of model categories, functionalities, and emerging trends like Conditional Tabular GANs (CTGAN) that have gained popularity for their ability to model complex dependencies among heterogeneous variables in tabular medical data [18]. Variational Autoencoders (VAEs) and their tabular extensions (TVAEs) offer another powerful framework by learning latent representations that summarize data variability. Recent applications in clinical research have demonstrated the potential of TVAE for generating realistic synthetic datasets that preserve both univariate and multivariate distributions of sensitive medical variables, including laboratory biomarkers and physiological parameters [16]. Another well-established approach for synthetic data generation in imbalanced datasets is SMOTE (Synthetic Minority Oversampling Technique), which interpolates between nearest neighbours to create realistic synthetic samples, improving model robustness [4].

2.2 Large Language Models for Medical Data Imputation

Large Language Models (LLMs) have recently emerged as promising tools for data imputation, owing to their ability to capture semantic relationships and generate contextually accurate values. Unlike traditional statistical or machine learning approaches, LLMs can leverage the full contextual meaning of structured data to infer missing entries with greater nuance. An example is the Pred-LLM framework described in Generating Realistic Tabular Data with Large Language Models [13], which generates realistic tabular datasets while preserving feature–target correlations. The method fine-tunes an LLM by converting tabular rows into natural language sentences, keeping the target variable last to reinforce dependencies, then generates synthetic samples conditioned on given features and predicts missing values through targeted prompts. In evaluations across 20 datasets against 10 baseline models, including CTGAN and TapTap, Pred-LLM consistently reproduced real data structure and class distributions with high fidelity. Similarly, the study Data Imputation using Large Language Model to Accelerate Recommender System [6] demonstrates that fine-tuning LLMs (via LoRA) on partially missing datasets enables them to accurately predict absent values, improving downstream performance in recommendation systems. This approach outperformed classical imputation methods—such as mean, median, or k-Nearest Neighbors—by capturing complex dependencies, a capability particularly relevant for clinical datasets where missingness is often systematic rather than random. Another recent study relevant to medical data imputation is the method **CLAIM** (Contextual Language model for Accurate Imputation Method). This approach transforms tabular records into natural language descriptions and leverages LLMs to impute missing entries, fine-tuning on enriched datasets. CLAIM significantly outperformed traditional techniques including KNN, MICE, and mean imputation especially under complex missingness patterns, demonstrating greater accuracy and robustness across tabular datasets [9]. In general, LLMs represent a promising new approach to address missing data challenges, with the potential to enhance both the precision and resilience of data analysis in diverse domains.

3 Materials and Methods

3.1 Dataset Description

The *Pima Indians Diabetes Dataset*, [11,15] developed by the National Institute of Diabetes and Digestive and Kidney Diseases, is a widely used benchmark for research on Type 2 Diabetes. It collects medical data from 768 women of Pima heritage, aged 21 years or older, residing in Arizona. Each observation includes eight clinical features—number of pregnancies, plasma glucose concentration, diastolic blood pressure, triceps skinfold thickness, two-hour serum insulin level, body mass index (BMI), diabetes pedigree function, and age—along with a binary outcome variable indicating the presence or absence of diabetes.

Although the dataset does not contain explicit `NaN` entries, several features include *zero* values that are not physiologically plausible and were therefore treated as missing. Specifically, we identified 5 missing values in Glucose, 35 in BloodPressure, 227 in SkinThickness, 374 in Insulin, and 11 in BMI. These features were accordingly processed using imputation strategies described in the following sections.

3.2 Imputation Strategies

To address the presence of missing data, we evaluated the performance of multiple imputation strategies, ranging from traditional statistical techniques to advanced generative models and language model-based approaches.

K-Nearest Neighbors (KNN). As a baseline method commonly used in the literature [1], we applied the KNN imputation algorithm, which estimates missing values by averaging the values of the k most similar instances, based on Euclidean distance computed on the observed features. We used $k = 5$ nearest neighbors with uniform weighting, following the setup adopted in the reference study discussed below.

Conditional Tabular GAN (CTGAN). CTGAN [18] is a generative adversarial network specifically designed for tabular data, which often includes a mix of continuous and categorical variables with complex interdependencies. Unlike standard GANs, CTGAN explicitly models the distribution of each variable conditioned on the others, leveraging techniques such as conditional sampling and training-by-sampling to address imbalanced categories and nonlinear relationships. This allows the model to generate synthetic values that preserve the statistical structure of the original data, making it distinctly effective for missing data imputation in structured datasets [10].

Tabular Variational Autoencoder (TVAE). TVAE [17] is a neural network model based on an encoder–decoder architecture. The encoder maps real observations into a lower-dimensional latent space, learning a probabilistic representation—typically a multivariate Gaussian—for each instance. The decoder reconstructs the data by sampling from this latent space, generating new synthetic samples that preserve the statistical properties of the original dataset. Unlike standard VAEs, TVAE introduces customizations in the loss function and decoder structure to handle the mixed data types typically found in tabular datasets. This enables the model to effectively capture nonlinear dependencies and produce both plausible synthetic data and imputations of missing values, while preserving inter-feature relationships.

Large Language Models (LLMs). The large language models employed in this study include both general-purpose and domain-specialized architectures.

GPT-2, developed by OpenAI, is an autoregressive model based on a unidirectional transformer, trained on a large-scale web corpus and available in multiple sizes [14]. It serves as a reference standard for text generation due to its balance between simplicity and expressive power across various NLP tasks.

Llama-2-7B, released by Meta AI, is a 7-billion-parameter open-source model optimized for high performance in text understanding and generation, with manageable computational requirements [12]. *Gemma-2-2B-it* and *Gemma-2-9B-it*, developed by Google [8]. The smaller version prioritizes efficiency, while the larger offers increased contextual understanding, enabling an analysis of the trade-off between imputation quality and resource consumption.

For biomedical applications, we adopted both adapted and domain-specific models. *BioBERT v1.1*, built upon BERT and pretrained on biomedical corpora, enhances the understanding of clinical and scientific texts [7]. *Bio-Medical-Llama-3-8B* (BML), based on the Llama 3 architecture and further tuned on biomedical documents, improves terminological accuracy and reasoning in healthcare contexts [5].

Together, these models provide a diverse experimental foundation, covering both general-purpose and specialized scenarios for missing data imputation.

3.3 Prompt Design for LLMs

To enable imputation via large language models, we reformulated the task as a structured completion problem expressed in natural language. Each patient record was converted into a textual prompt describing a clinical context in which a medical assistant is asked to infer the missing values. Features with missing entries were explicitly masked using a placeholder ("???"), and the model was instructed to return only the missing fields, using a strict attribute=value format.

Zero-Shot vs. Few-Shot. For **Zero-Shot** prompting, models were provided only with the target patient record and explicit instructions to complete missing fields using realistic numerical values. The prompts emphasized strict output formatting requirements to facilitate automated parsing of responses. For **Few-Shot** prompting, we adopted a balanced example selection strategy to provide contextual guidance and mitigate potential class-related biases. Examples were sampled from the training data according to three predefined modes: `only_0` (diabetes-negative cases), `only_1` (diabetes-positive cases), and `balanced` (equal number of cases from both classes). The `balanced` mode was preferred to ensure fair representation across the outcome variable. In cases where the number of provided examples was odd, an `extra_to_class` parameter controlled which class received the additional instance.

All prompts were programmatically generated to ensure consistency across models and test cases. However, minor manual adjustments were made to some of them. While the initial strategy aimed to adopt a single prompt per approach—designed to yield optimal responses across all models—this approach ended up being suboptimal. As a result, model-specific prompts were introduced to better tailor the input to each model's behavior. This decision was driven by the observation that some models failed to generate correct outputs using generic prompts, whereas others successfully returned the expected response.

Example of Zero-Shot prompt.

```
You are a medical assistant. A patient may have type 2 diabetes. Here is the available information:
- Pregnancies: 2
- Glucose: ???
- BloodPressure: 70
- SkinThickness: ???
- Insulin: 120
- BMI: 33.2
- DiabetesPedigreeFunction: 0.65
- Age: 45

Fill in ONLY the missing values (marked with ???) using realistic numeric values.
Respond in the format:
Glucose=..., SkinThickness=...
Do not include explanations or line breaks.
```

Parsing of Model Outputs. The output parsing was automated using regular expressions to extract values from the response of the model. Strict formatting constraints were enforced to minimize the need for post-processing and reduce parsing ambiguity. The predictions were validated to ensure plausibility (e.g., physiological ranges) before being reintegrated into the data set. In cases of a malformed or incomplete output, the prompt was reissued with minor rephrasing to improve model compliance.

4 Experimental Setup

All runs employed fixed random seeds and a standardized preprocessing pipeline to ensure fair comparison.

4.1 Testing Scenarios

We designed three complementary scenarios to rigorously evaluate LLM effectiveness in diabetes data imputation, each targeting a specific research objective.

- ***Masking—Intrinsic Evaluation (RQ1).*** In this setting, we artificially masked a subset of observed values in the test set while keeping the ground-truth values available. This allows for a controlled and quantitative comparison of imputation strategies, based on direct reconstruction error.
- ***Downstream Evaluation (RQ2).*** To measure the practical utility of imputations, we used a downstream classifier trained on complete data and tested it on imputed samples. Improvements in classification performance indicate whether the imputed values preserve or enhance the discriminative structure of the original dataset.

  We adopted two complementary strategies for this purpose:

1. *Strategy 1—Training on Fully Observed Data.* We trained classifiers on a clean subset of the dataset consisting exclusively of rows without missing values. The test set was composed of samples that had undergone imputation through different methods (LLMs, KNN, CTGAN, TVAE). This strategy isolates the effect of imputation on model generalization.
2. *Strategy 2—Training and Testing on Imputed Data.* To simulate more realistic operational conditions, we trained and tested classifiers directly on datasets that were entirely imputed using each method. That is, for each imputation strategy, a model was trained and evaluated using its corresponding completed version of the dataset. This setting measures internal consistency and the ability of the imputed values to support predictive learning. To ensure fair comparison with existing literature [1], we adopted a 70/30 train-test split strategy.

4.2 Evaluation Metrics

We used complementary metrics, each chosen to match the goals of its evaluation scenario, to assess imputation quality and downstream impact.

- *Masking—Intrinsic Evaluation.* For synthetic masking, where ground-truth values are available, we computed the **Mean Absolute Error (MAE)**, for each feature, between the imputed values and the original ground-truth values. This metric measures the numerical accuracy of imputations when the true value is known.
- *Downstream Evaluation.* To evaluate the practical impact of the imputations, we trained several predictive models using the imputed datasets and assessed the **accuracy** as the primary evaluation metric.

4.3 Downstream Classifiers

To evaluate the quality of the imputations, following the approach of the reference study [1], the following classifiers were trained on the imputed datasets:

- **DecisionTreeClassifier (DT)**: A decision tree that splits data based on features to predict classes; limited to a maximum depth of 15.
- **RandomForestClassifier (RF)**: An ensemble of 55 decision trees (random forest) with maximum depth 15, improving robustness and reducing overfitting.
- **Support Vector Classifier (SVC)**: A support vector machine classifier with RBF kernel, using default regularization and gamma parameters, capable of handling non-linear problems.
- **XGBClassifier (XGB)**: A gradient boosting classifier (XGBoost) with 55 estimators and maximum depth 15, optimized for speed and accuracy.
- **ExtraTreesClassifier (ETC)**: An ensemble of 55 extremely randomized trees with maximum depth 30, introducing more randomness compared to random forests.

- **SGDClassifier (SGD)**: A linear classifier using stochastic gradient descent with L2 regularization, suitable for large datasets.
- **GaussianNB (GNB)**: A Naive Bayes classifier assuming Gaussian feature distributions.
- **LogisticRegression (LR)**: A logistic regression model with L2 penalty and lbfgs solver, suitable for binary and multiclass classification.
- **VotingClassifier (VC)**: An ensemble classifier that combines probabilistic predictions (*soft voting*) of three models (`XGB`, `RF`, and `ETC`) to enhance overall performance and robustness.

4.4 Hardware

All experiments ran on a workstation with 56 CPU cores, 512 GB RAM, and two NVIDIA RTX 5000 Ada GPUs (32 GB each), providing the resources needed for LLM training and evaluation.

Table 1. Classification accuracy (%) on test sets completed by different imputation strategies (Strategy 1). Each classifier is trained once on the complete training data and evaluated on test samples imputed by KNN or LLM-based methods. Best results per row are in bold.

Model	KNN	Gemma-2-2B-it	Gemma-2-9B-it	BML
DT	**65.96**	60.98	**65.96**	62.30
ETC	72.61	73.44	**73.67**	73.26
GNB	69.68	69.92	**70.48**	70.05
LR	75.27	**75.61**	74.73	74.60
RF	71.01	71.82	**73.40**	72.19
SGD	65.16	**65.58**	64.63	62.83
SVC	73.40	73.98	73.67	**74.06**
VC	70.21	72.09	**72.87**	70.86
XGB	69.95	69.11	**71.54**	70.59

5 Results and Discussion

We present a detailed analysis of the experimental outcomes to evaluate the effectiveness of large language models for data imputation. Results are discussed with respect to both intrinsic accuracy and downstream utility, with comparative benchmarks against classical (KNN) and generative (TVAE, CTGAN) methods, providing a comprehensive view of the relative performance of each approach.

Masking—Intrinsic Evaluation. This test revealed substantial differences in the intrinsic imputation capabilities of the tested language models. Only a subset of models were able to generate outputs in the required strict format and to provide meaningful numerical imputations. For *GPT-2*, initial attempts were unsuccessful: the model was unable to produce valid numerical values and mostly repeated the prompt. Subsequently, a distilled and fine-tuned version (following the LoRA approach) was adopted, as in previous studies [6]. Even after fine-tuning with a subset of the complete dataset, GPT-2 continued to produce contextually incoherent numeric values, likely due to the limited training set and inherent capacity constraints. *Llama-2-7B* exhibited similar limitations; the majority of outputs simply restated the prompt, and when numerical values were generated, these were highly inconsistent. This performance highlights the model's inadequacy for the specific structured imputation task, leading to its exclusion from further analyses. *BioBERT v1.1* was hindered by excessive verbosity: although prompted to return only numerical values in a strict key-value format, the model often produced full sentences or textual strings (e.g., "Insulin=media") rather than numeric entries. The lack of format and value consistency prevented its use in downstream intrinsic evaluation.

Preliminary tests were conducted using 5, 10, and 20 randomly selected rows from the dataset, ensuring that these rows were excluded from both the training set and in-context examples to prevent any form of data leakage. For each selected instance, between one and three columns were masked to evaluate the model's ability to correctly impute the missing values.

The *Bio-Medical-Llama-3-8B* model showed significant improvements thanks to careful prompt engineering. Initially, its Zero-Shot outputs did not strictly comply with the required format, generating multiline responses instead of single-line, comma-separated values. Explicitly instructing the model in the prompt to provide a single-line response with comma-separated fields greatly enhanced format adherence. Approximately 25% of the Zero-Shot responses were discarded due to empty fields. Through iterative prompt refinements, these issues were largely mitigated for both Zero-Shot and Few-Shot settings. Similarly to *Gemma-2-2B-it*, the Few-Shot configuration reduced the MAE and eliminated anomalies related to missing fields.

When analyzing the larger *Gemma-2-9B-it* model, some formatting difficulties arose. In both Zero-Shot and Few-Shot modes, the model occasionally retained placeholder underscores ("__") for variables not correctly imputed or produced ambiguous formats—for example, reporting blood pressure as a fraction (`BloodPressure=60/120`), which caused parsing errors. Consequently, it was necessary to refine the prompt by explicitly enforcing the generation of a single numerical value for each variable. As observed with previous models, Few-Shot prompting led to a consistent reduction in MAE across all variables compared to Zero-Shot generation.

Overall, across all examined variables, the Few-Shot approach consistently resulted in a reduction of the MAE as shown in Fig. 1.

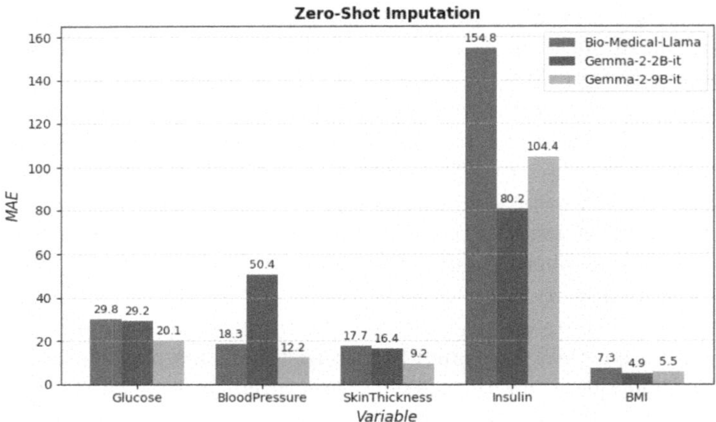

(a) Zero-Shot Performance

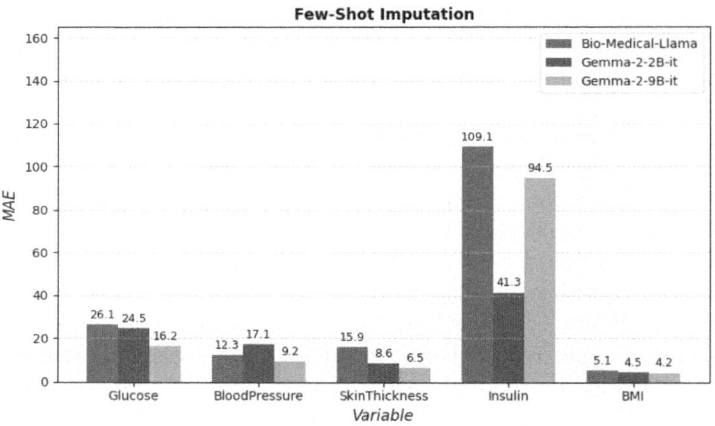

(b) Few-Shot Performance

Fig. 1. Mean Absolute Error (MAE) of LLM imputation performance across five biomedical variables. Panel (a) shows Zero-Shot prompting, and panel (b) shows Few-Shot prompting. Numerical values above each bar indicate the MAE for Bio-Medical-Llama, Gemma-2-2B-it, and Gemma-2-9B-it, highlighting improvements achieved with Few-Shot examples.

The reduction in error is particularly pronounced in more complex tasks, such as insulin prediction. This finding underscores that the incorporation of exemplars is a critical factor in enhancing the accuracy of language models for this task.

Downstream Evaluation. Based on the evidence discussed above, the performance analysis focused on the three most promising models for imputation: *Gemma-2-2B-it*, *Gemma-2-9B-it*, and *Bio-Medical-Llama-3-8B*. For Strategy 1, we compared these LLM-based imputations against the KNN baseline to assess relative improvements, while, for Strategy 2, in order to provide a rigorous benchmark with respect to prior literature, we extended the comparison to include not only the selected LLMs and KNN, but also generative baselines such as CTGAN and TVAE. We report the downstream impact of each imputation method under the two evaluation strategies described in Sect. 4.1. As shown in Table 1, for Strategy 1, LLM-based imputations often yield competitive or superior performance compared to the KNN baseline. Notably, Gemma-2-9B-it consistently outperforms KNN in 6 out of 9 classifiers, including ensemble models like Random Forest (73.40% vs 71.01%) and Voting Classifier (72.87% vs 70.21%). The Bio-Medical-Llama model performs well overall, especially in linear classifiers such as SVC (74.06%) and GNB (70.05%), but shows slightly lower robustness on tree-based models. These results suggest that LLMs, especially larger variants, are capable of learning plausible imputations that preserve class-discriminative structure.

Table 2. Classification accuracy (%) under Strategy 2: comparison across KNN, generative models (CTGAN, TVAE), and LLM-based imputations. Best result per row is highlighted in bold.

Model	KNN	CTGAN	TVAE	Gemma-2-2B-it	Gemma-2-9B-it	BML
DT	**72.29**	70.13	74.46	70.31	70.13	70.43
ETC	74.03	**78.35**	76.19	75.98	75.76	76.09
GNB	74.89	75.76	75.76	75.98	76.19	**77.83**
LR	76.62	76.62	76.62	**79.48**	77.06	77.83
RF	72.29	75.76	76.62	75.55	**79.22**	77.83
SGD	72.73	75.76	72.73	76.42	74.03	**76.96**
SVC	73.59	75.76	74.46	75.98	73.59	**78.26**
XGB	72.29	74.03	74.46	74.24	74.46	**75.65**
VC	73.59	76.19	75.32	76.86	74.03	**76.96**

In Strategy 2, to provide a rigorous benchmark against the existing literature, we replicated the experimental protocol and classifier settings adopted in [1], including training conditions, hyperparameters, and data partitioning strategy. Table 2 reports the classification accuracy across a diverse set of models, highlighting in bold the best performance for each classifier. The results demonstrate that LLM-based imputations, particularly Gemma-2-9B-it and Bio-Medical-Llama, tend to achieve superior performance compared to traditional methods such as KNN and generative models like CTGAN and TVAE across different classifiers. Bio-Medical-Llama achieves the highest accuracy in 5 out of

9 classifier configurations, highlighting strong generalization capability on high-dimensional clinical data. The maximum accuracy was observed with Gemma-2-9B-it paired with Random Forest classifier (79.22%), while BML obtained comparable results with Naïve Bayes, SGD, and Voting Classifier. Generative models showed promising performance in specific cases; for instance, Extra Trees Classifier with CTGAN reached 78.35%, outperforming LLMs in this particular configuration. However, these results proved less consistent across different classifiers compared to LLM-based approaches. The KNN method, while serving as an established baseline, was generally outperformed by all other techniques, both LLM-based and generative. This suggests the limitations of distance-based imputations in capturing the complex and contextual relationships present in the data, which can be better modeled by transformational architectures such as LLMs. In general, LLM-based imputations emerge as a viable and stable solution for tabular clinical data, improving predictive performance without the need for modifications to downstream models or training procedures.

Computational Costs. On the same hardware setup described in Sect. 4.4, we measured the average runtime required to impute 100 test samples. Results indicate that the Gemma-2 models are the most computationally demanding, requiring approximately 15 minutes per 100 samples, with little difference observed between the 2B and 9B parameter variants. By contrast, the Bio-Medical-Llama-3-8B model completed the same task in roughly 3 min, highlighting more efficient runtime characteristics despite its larger parameter size. These findings suggest that, while all LLMs investigated are considerably more resource-intensive than classical imputers such as KNN (which completes the same imputation in less than a second), runtime efficiency is strongly model-dependent. This factor should therefore be taken into account when considering deployment in real-world clinical pipelines.

6 Conclusions

This study investigated the applicability of LLMs for imputing missing values in structured clinical datasets, addressing two main research questions: (i) to what extent LLMs can accurately reconstruct missing entries, and (ii) whether such imputations translate into improved performance in downstream classification tasks. Results show that, with well-designed prompts and Few-Shot examples, models such as Gemma-2-2B-it, Gemma-2-9B-it, and Bio-Medical-Llama-3-8B achieve competitive imputation performance compared to classical (e.g., KNN) and generative (e.g., TVAE, CTGAN) approaches. LLM-based imputations not only produced numerically coherent values but also supported predictive tasks with accuracy comparable to or exceeding that of traditional methods. Despite these encouraging findings, limitations remain. Prompt engineering is task-dependent and non-trivial, and further validation on larger and more diverse datasets is needed to ensure robustness. Future work will explore automated prompt design, uncertainty estimation, and interpretability of LLM-driven imputations. In conclusion, LLMs offer a flexible and effective solution for structured data imputation, with clear potential for integration into clinical data pipelines when accompanied by careful methodological adaptation and validation.

Acknowledgments. D3 4 Health – Digital Driven Diagnostics, Prognostics and Therapeutics for Sustainable Health Care (Project PNC0000001 – CUP: B53C22006090001), funded by the European Union – NextGenerationEU under the National Plan for Complementary Investments to the NRRP.

Disclosure of Interests. The authors declare no competing interests relevant to this article.

References

1. Altamimi, A., et al.: An automated approach to predict diabetic patients using KNN imputation and effective data mining techniques. BMC Med. Res. Methodol. **24**(1), 221 (2024)
2. Bauer, A., et al.: Comprehensive exploration of synthetic data generation: a survey. arXiv preprint arXiv:2401.02524 (2024)
3. Beretta, L., Santaniello, A.: Nearest neighbor imputation algorithms: a critical evaluation. BMC Med. Inform. Decis. Mak. **16**, 197–208 (2016)
4. Chawla, N.V., Bowyer, K.W., Hall, L.O., Kegelmeyer, W.P.: Smote: synthetic minority over-sampling technique. J. Artif. Intell. Res. **16**, 321–357 (2002)
5. ContactDoctor Healthcare Private Ltd: Bio-Medical Llama 3 8B model on Hugging Face. https://huggingface.co/ContactDoctor/Bio-Medical-Llama-3-8B (2024), announcement: ContactDoctor Healthcare, Transforming healthcare with AI: Bio-Medical LLaMA 3 8B, https://medium.com/@mckanth_85918/transforming-healthcare-with-ai-introducing-contact-doctors-bio-medical-multimodal-llama-3-8b-v1-3801f97cd42a
6. Ding, Z., Tian, J., Wang, Z., Zhao, J., Li, S.: Data imputation using large language model to accelerate recommendation system. arXiv preprint arXiv:2407.10078 (2024)
7. DMIS Lab: BioBERT v1.1 model on Hugging Face. https://huggingface.co/dmis-lab/biobert-v1.1 (2023), Official paper: Lee et al., BioBERT: a pre-trained biomedical language representation model for biomedical text mining, Bioinformatics 36(4):1234–1240, 2020, https://academic.oup.com/bioinformatics/article/36/4/1234/5610707
8. Google: Gemma 2B IT model on Hugging Face. https://huggingface.co/google/gemma-2b-it (2024), Official blog post: DeepMind, Gemma: a family of lightweight state-of-the-art open models, https://deepmind.com/blog/article/gemma-a-family-of-lightweight-state-of-the-art-open-models
9. Hayat, A., Hasan, M.R.: Claim your data: enhancing imputation accuracy with contextual large language models. arXiv e-prints pp. arXiv–2405 (2024)
10. Khan, H.S., Panda, J.: Enhancing medical research with synthetic data generation via ctgan on cleveland heart dataset. In: 2025 Second International Conference on Cognitive Robotics and Intelligent Systems (ICC-ROBINS), pp. 1–5. IEEE (2025)
11. Khare, A.D.: Diabetes dataset (2022). https://www.kaggle.com/datasets/akshaydattatraykhare/diabetes-dataset

12. Meta AI: Llama 2 7B model on Hugging Face. https://huggingface.co/meta-llama/Llama-2-7b (2023), Official paper: Touvron et al., LLaMA 2: Open Foundation and Fine-Tuned Chat Models, arXiv:2307.09288 (2023)
13. Nguyen, D., Gupta, S., Do, K., Nguyen, T., Venkatesh, S.: Generating realistic tabular data with large language models. arXiv preprint arXiv:2410.21717 (2024)
14. OpenAI Community: GPT-2 model on Hugging Face. https://huggingface.co/openai-community/gpt2 (2023), official paper: Radford et al., Language Models are Unsupervised Multitask Learners, OpenAI, 2019, https://cdn.openai.com/better-language-models/language_models_are_unsupervised_multitask_learners.pdf
15. Smith, J.W., Everhart, J.E., Dickson, W.C., Knowler, W.C., Johannes, R.S.: Using the pima Indians diabetes dataset for predictive modeling. National Institute of Diabetes and Digestive and Kidney Diseases (1988), original dataset source
16. Titar, R.R., Ramanathan, M.: Variational autoencoders for generative modeling of drug dosing determinants in renal, hepatic, metabolic, and cardiac disease states. Clin. Transl. Sci. **17**(7), e13872 (2024)
17. Tran, V.Q., Byeon, H.: Explainable hybrid tabular variational autoencoder and feature tokenizer transformer for depression prediction. Expert Syst. Appl. **265**, 126084 (2025)
18. Xu, L., Skoularidou, M., Cuesta-Infante, A., Veeramachaneni, K.: Modeling tabular data using conditional GAN. In: Advances in Neural Information Processing Systems, vol. 32 (2019)

Regularized Continual Learning for Generative Crowd Counting

Jianyong Wang, Wenzhe Zhai, and Mingliang Gao(✉)

Shandong University of Technology, Zibo255022, China
{23404020559,20404020495}@stumail.sdut.edu.cn, mlgao@sdut.edu.cn

Abstract. Crowd counting plays a pivotal role in urban management, public safety, and event planning by providing accurate estimates of population density across diverse environments. A significant limitation of conventional crowd counting methods is their typical training and evaluation within a single and homogeneous scenario. Such discrepancy renders the continuous estimation of crowd counts across varying scenarios a significant challenge. This makes the task of continuous estimation of crowd counts in the different scenarios a formidable challenge. To this end, this paper presents a Regularized Continual Learning Network (RCLNet) for crowd counting. The primary objectives are to mitigate catastrophic forgetting and improve model generalization. For a comprehensive assessment, we leveraged the JHU dataset to develop four challenging scene datasets (*e.g.*, fog, snow, stadium, street). The proposed method retains critical knowledge from prior tasks while acquiring new information, thereby improving crowd counting performance under diverse environmental conditions. Experimental results demonstrate that RCLNet outperforms state-of-the-art approaches in terms of MAE and RMSE, and it effectively handles domain shifts while ensuring source-free generalization. This work introduces a new theoretical framework for crowd counting that enhances model robustness and scalability across diverse real-world scenarios. The code will be available at: https://anonymous.4open.science/r/RCLNet-8C2A.

Keywords: Crowd counting · Continual learning · Regularization · Scene generalization

1 Introduction

Crowd counting is a crucial task with significant applications in domains such as public safety [1,2], traffic management [3,4], and emergency response [5]. However, conventional crowd counting methods [6–8] predominantly operate within a single, homogeneous domain during training and evaluation. This paradigm severely constrains their efficacy in real-world environments characterized by inherent complexity and frequent variations, thereby impeding robust generalization across diverse scenarios.

T. Ensari et al. (Eds.): ISPR 2025, CCIS 2859, pp. 271–282, 2026.
https://doi.org/10.1007/978-3-032-21585-7_20

Existing crowd counting methods [6–8] often suffer from catastrophic forgetting, wherein adaptation to unseen data degrades knowledge acquired from prior tasks. This erosion of critical information hinders the model's capacity to generalize and leverage past knowledge when encountering novel data. Compounding this limitation, most models are developed in static environments and exhibit inadequate performance under domain shifts, particularly in dynamic real-world conditions involving variations in crowd density, illumination, and occlusion.

Continual learning provides a viable solution by enabling models [9,10] to acquire new knowledge while preserving previously learned information. This capability is essential for crowd counting systems that must operate in evolving environments without retraining from scratch. The comparison between traditional and continual learning-based crowd counting methods is shown in Fig. 1. As illustrated in Fig. 1 (a), traditional crowd counting methods [6–8] train and evaluate models on a fixed dataset (Dataset A), which does not accommodate the inherent complexities and dynamism of real-world settings. These approaches fail to adapt to unseen environments and often forget previously learned knowledge, limiting their robustness in dynamic scenarios. In contrast, continual learning crowd counting methods (Fig. 1 (b)) sequentially learn from multiple datasets (Datasets A, B, ..., N) while retaining knowledge from prior tasks. The key challenge for continual learning in crowd counting is to mitigate catastrophic forgetting while enabling adaptation to new data. This problem is particularly pressing in dynamic environments where the distribution of crowd data is constantly shifting.

In addition to catastrophic forgetting, scene generalization is a key challenge in crowd counting. In real-world applications, crowd counting systems must work across different environments, such as streets, stadiums, and outdoor settings with varying weather. These environments often differ significantly from the training data, leading to poor performance. Traditional models [6–8], which are trained on homogeneous datasets, struggle when tested in new domains. While domain adaptation (DA) [11] and domain generalization (DG) [12–14] methods aim to address this, DA methods require target-domain data during training, while DG methods still face difficulties in generalizing to highly diverse environments. These issues highlight the need for models that can learn continuously and adapt to new environments while retaining previous knowledge.

To address these limitations, we propose a Regularized Continual Learning Crowd Counting Network (RCLNet). It leverages regularization-based continual learning, which not only addresses catastrophic forgetting but also improves the model's ability to generalize across a variety of scenarios. By balancing knowledge retention from prior tasks with adaptation to new domains, RCLNet achieves robust performance across varied datasets *e.g.,* fog, snow, stadium, street environments.

The primary innovation lies in integrating a regularization term derived from the Fisher Information Matrix (FIM) into the training process. This term constrains parameter updates to prevent forgetting of prior knowledge while enabling adaptation to new data. We evaluate RCLNet using four sub-datasets

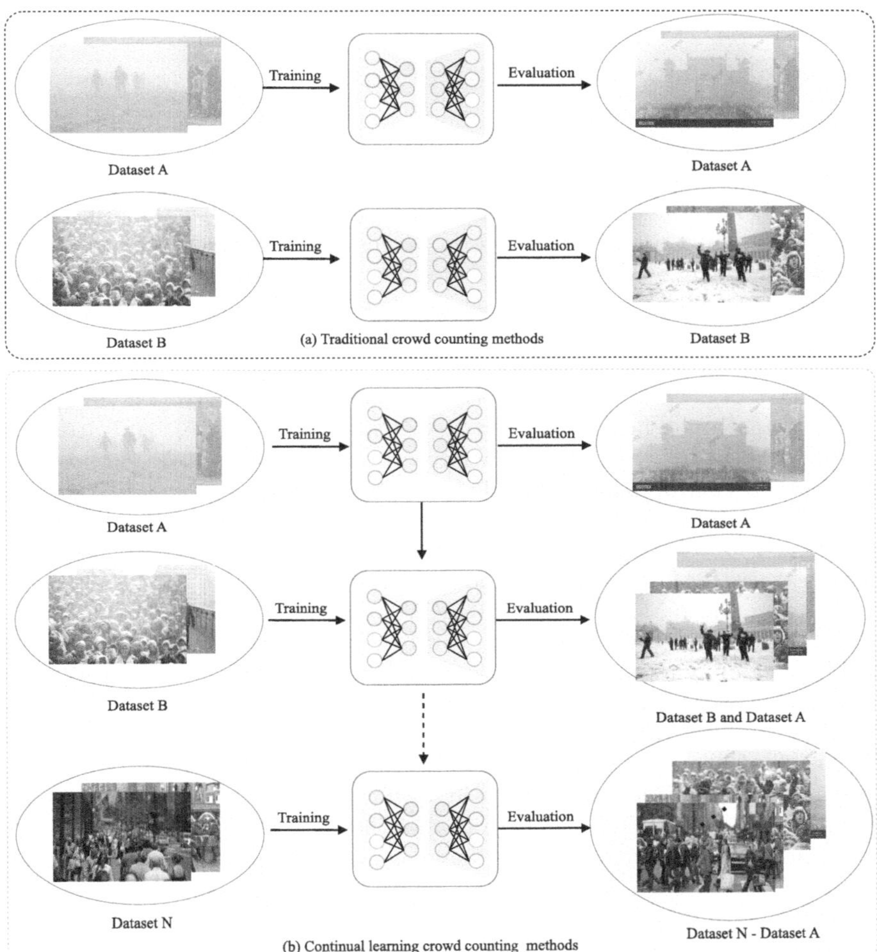

Fig. 1. Comparison of traditional and continual learning-based crowd counting methods. (a) Traditional methods involve training and evaluating the model on a single dataset (Dataset A). (b) Continual learning methods update and evaluate a model sequentially on multiple datasets (Datasets A, B, ..., N), maintaining continuous learning from each dataset.

derived from the JHU-Crowd++ dataset [15] which representing four distinct scenarios, *i.e.*, fog, snow, stadium, street. Results demonstrate its efficacy in mitigating domain shifts and maintaining strong performance across diverse conditions.

2 Related Work

2.1 Crowd Counting

Conventional crowd counting methods consists of detection-based and regression-based methods. They normally leverage handcrafted features to detect individual head instances [16,17] or to directly regress the global crowd count [18]. Nevertheless, these models struggle to model the spatial context of crowds, which results in inaccurate predictions in congested environments. Consequently, state-of-the-art methods have shifted towards deep learning-based frameworks that estimate a density map. However, the diverse and complex scene data pose significant challenges to the task of crowd counting.

To solve the problem of crowd counting in complex and diverse scenarios, a diverse range of deep learning architectures has been proposed. These methods include, but are not limited to, zero/few shot networks [2,5], and attention-based models [19–22]. For zero/few shot networks, Chen *et al.* [2] proposed the Deep Spatial Prior Interaction (DSPI) network, which incorporates spatial priors from a pre-trained Grounding DINO model to enhance spatial awareness in zero-shot object counting. They also introduce a meta adapter to align visual and textual features for improved cross-modality consistency. Zhai *et al.* [5] proposed a Scale-Aware Cross-Modal Adapter (SACMA) framework that leverages a scale-aware module to extract multi-scale features for cross-modal crowd counting. They further employ an adapter-based fusion mechanism to effectively align and integrate visual and textual features. For attention-based models, Zhai *et al.* [21] proposed a Scale-Aware Decoupling Network (SADNet) that separates scale-specific features to better handle variations in crowd density. They also design a decoupled attention mechanism to enhance feature representation across different density levels. The common goal of these approaches is to enhance the model's adaptability across varying domains and conditions, enabling accurate and reliable crowd counting in complex and diverse scenarios.

In addition to these architectures, domain adaptation (DA) and domain generalization (DG) methods have been widely explored to mitigate performance degradation caused by domain shifts. DA approaches [11] typically leverage target-domain data during training to align feature distributions, whereas DG methods [12–14] aim to learn domain-invariant representations without access to target-domain samples. While these strategies improve cross-domain performance, they still struggle when the model must incrementally adapt to multiple evolving domains without retraining from scratch.

To address this challenge, we propose a Regularized Continual Learning Network (RCLNet) for crowd counting wherein a single model is designed to assimilate knowledge from a stream of multiple, sequentially arriving datasets without performance degradation.

2.2 Continual Learning

While deep learning has significantly advanced crowd counting, existing methods typically assume a static, single-domain training scenario. Real-world crowd surveillance systems often operate in dynamic environments where new scenes, crowd densities, or environmental conditions emerge continuously. Training a model from scratch for each new scenario is computationally prohibitive and impractical. Continual learning (or lifelong learning) offers a promising paradigm to incrementally learn new tasks without forgetting previously acquired knowledge.

The primary objective of continual learning is to mitigate catastrophic forgetting, thereby improving model generalization in scenarios characterized by continuously evolving data distributions. The predominant paradigms in continual learning have found extensive application in domains such as image classification [23–25] and numerical prediction [26]. A comprehensive taxonomy of these strategies reveals four principal archetypes: model-growth approaches [27], rehearsal-based techniques [24,28], regularization-based methods [23,24], and knowledge distillation mechanisms [25].

Specifically, model-growth (*e.g.*, PNN [27]) and rehearsal-based methods (*e.g.*, GEM [28]) suffer from high computational and memory costs. This stems from their fundamental mechanisms: model-growth methods add network parameters for new tasks, while rehearsal-based methods must store and reuse old training data. The LwF framework [25] synergistically combines knowledge distillation and fine-tuning, which leverages the former to preserve knowledge from prior tasks while the latter adapts the model to unseen data, thereby improving overall performance.

However, these continual learning techniques designed for classification cannot be straightforwardly transferred to the crowd counting domain. This limitation arises because crowd counting is inherently an open-set problem [29], where the target value (the count) can theoretically range from zero to positive infinity. In dense prediction tasks like crowd counting, both latent feature representations (encoding general visual knowledge) and high-level semantic information at the output layer are critically important.

To address this challenge, we introduce a regularized continual learning network for crowd counting. The proposed method employs the FIM to estimate the importance of each parameter. This FIM-based regularization balances knowledge retention and adaptability without storing past samples or expanding model size and ensures efficiency and scalability in dynamic environments.

3 Proposed Method

3.1 Overall Framework

The architecture of the RCLNet for crowd counting is depicted in Fig. 2. The architecture supports incremental training on heterogeneous domains, including adverse weather conditions and complex urban scenes. A key innovation lies in the integration of a regularization term into the loss function during new-task training. This term constrains parameter updates by penalizing deviations from the optimal parameters learned in prior tasks. This mechanism ensures that critical parameters remain stable and enables the model to accumulate knowledge across sequential tasks while preserving previously learned competencies.

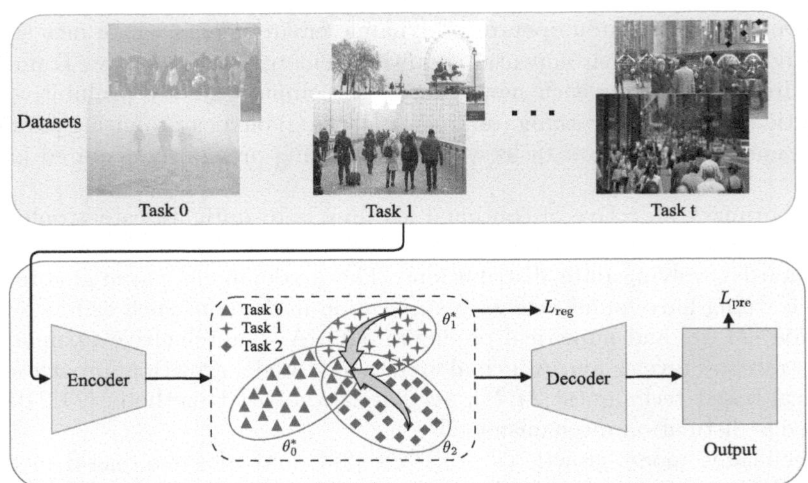

Fig. 2. The architecture of the RCLNet for crowd counting.

3.2 Regularization Training Strategy

The elastic weight consolidation regularization strategy [23] is adopted to address catastrophic forgetting in continual learning settings. This approach involves incorporating a regularization term into the loss function, which discourages large adjustments to parameters crucial for previously learned tasks. The regularization term relies on the Fisher Information Matrix (FIM), which measures the significance of each parameter in retaining knowledge from earlier tasks.

Let θ^* denote the optimal parameters for the previous task and θ the current parameters. The regularization loss is formulated as:

$$L_{\text{Reg}}(\theta) = \sum_i \frac{\lambda}{2} F_i (\theta_i - \theta_i^*)^2 \tag{1}$$

where F_i represents the Fisher information for parameter θ_i, θ_i^* denotes the optimal parameter from the previous task, and λ is a hyperparameter that regulates the strength of the regularization.

The total loss for the model is the sum of the standard crowd counting loss L_{total} and the regularization penalty:

$$L_{\text{total}} = L_{\text{pre}} + \lambda_{\text{Reg}} L_{\text{Reg}} \tag{2}$$

This approach allows the model to maintain the effectiveness of prior tasks while simultaneously adapting to unseen domains. The parameter λ_{Reg} controls the balance between stability and flexibility, which allows the model to learn effectively from both new and old domains without forgetting previously learned knowledge.

3.3 Fisher Information Matrix

The FIM serves as a fundamental tool for quantifying parameter importance within the context of previous tasks. Specifically, parameters deemed critical to a task exhibit larger corresponding values within the FIM. The FIM is computed by taking the second derivative of the log-likelihood of the model concerning its parameters, and it provides a measure of how sensitive the model's predictions are to changes in each parameter.

The FIM is computed by taking the second derivative of the log-likelihood of the model concerning its parameters, and it provides a measure of how sensitive the model's predictions are to changes in each parameter.

$$F_{ij} = \mathbb{E}\left[-\frac{\partial^2 \log p(\mathbf{x}|\theta)}{\partial\theta_i \partial\theta_j}\right] \tag{3}$$

where $p(\mathbf{x}|\theta)$ represents the likelihood function, which is the probability of observing the data $\mathbf{x}$ given the model parameters, θ_i and θ_j are the different parameters of the model. Specifically, θ_i refers to the i-th parameter, and θ_j refers to the j-th parameter. $\frac{\partial^2 \log p(\mathbf{x}|\theta)}{\partial\theta_i \partial\theta_j}$ represents the second-order partial derivative of the log-likelihood function with respect to the parameters θ_i and θ_j. This measures how the likelihood function changes with respect to the two parameters.

The elements F_{ij} in the FIM measure the relationship and sensitivity between the i-th and j-th parameters in terms of their effect on the likelihood function. This matrix quantifies the influence of each parameter on the model's predictions and identifies the parameters that are crucial for retaining knowledge from prior tasks.

4 Experiments

4.1 Training Details

This study employed the computational capabilities of the NVIDIA RTX 3090 Ti GPU, with the PyTorch framework serving as the primary platform for both model training and testing phases. The feature extraction component of our model is based on the VGG16-BN network [30]. Batch normalization is consistently employed after all intermediate convolutional layers, but is omitted from the final layers of each head to preserve the raw output values. For model optimization, we employ the AdamW optimizer [31]. To manage the learning rate, we adopt the OneCycleLR scheduling policy [32]. The maximum learning rate is set to 1e−3, and training is conducted for a total of 500 epochs.

4.2 Datasets

We conduct the experiments on the JHU-Crowd++ dataset [15], a large-scale benchmark for crowd counting. The dataset comprises a total of 4,372 images, which are partitioned into 2,722 for training, 500 for validation, and 1,600 for testing. Beyond crowd count annotations, the JHU-Crowd dataset is also equipped with rich image-level metadata, which encompasses different distinct scene classes and weather conditions.

For the domain adaptation study, the dataset is first categorized by weather conditions into snow (SN, 201 images) and fog/haze (FH, 168 images). Based on scene annotations, it is further divided into two sub-datasets: the stadium (SD) domain with 879 images and the street (ST) domain with 573 images. For each domain, we partition the data into an 80% training set and a 20% test set. Notably, images sharing the same label within this dataset exhibit a narrower data distribution compared to mainstream benchmarks. This characteristic makes the task of Source Domain Generalization (SDG) particularly challenging, as it requires the model to generalize from a more homogeneous source.

4.3 Evaluation Metrics

The performance of the proposed model is quantitatively assessed using two widely-adopted evaluation metrics: Mean Absolute Error (MAE) and Root Mean Square Error (RMSE). The mathematical formulations for these metrics are provided below:

$$\mathrm{MAE} = \frac{1}{N}\sum_{1}^{N}|x_i - \hat{x}_i|, \tag{4}$$

$$\mathrm{RMSE} = \sqrt{\frac{1}{N}\sum_{1}^{N}|x_i - \hat{x}_i|^2}, \tag{5}$$

where N corresponds to the total number of images in the evaluation set. For each image i, x_i signifies its ground-truth count, whereas $\hat{x}_i$ signifies the count estimated by the model.

4.4 Comparison with SOTA Methods

To evaluate the effectiveness of the proposed RCLNet, we compared it against several state-of-the-art (SOTA) methods on the challenging JHU-Crowd++ [15] dataset. It encompasses four subsets (SN, FH, SD, SR) that test domain adaptation (DA), domain generalization (DG), and continual learning (CL) capabilities.

Comparative results are depicted in Table 1. It proves that the proposed RCLNet ranks first place in three subsets (SN, SD and SR) and achieves a competitive result in FH. Specifically, traditional methods such as BL [33] and MAN [34], which lack explicit CL or DG strategies, demonstrated limited generalization. For example, BL achieved high MAE/RMSE scores of 262.7/1063.9 (SD) and 343.8/770.5 (SN), while MAN showed slightly better performance but still faced challenges, especially on complex subsets like SD (MAE 246.1, RMSE 950.8) and SN. Methods incorporating DA or DG techniques, such as

Table 1. Performance comparison on the JHU-Crowd++ dataset (SN, FH, SD, SR subsets). The best performance for each metric is highlighted in **bold**.

Dataset				SN		FH		SD		SR	
Method	DA	DG	CL	MAE	RMSE	MAE	RMSE	MAE	RMSE	MAE	RMSE
BL [33]	✗	✗	✗	343.8	770.5	48.1	129.5	262.7	1,063.9	42.1	79.0
MAN [34]	✗	✗	✗	445.0	979.3	**38.1**	**68.0**	246.1	950.8	45.1	79.0
DAOT [11]	✓	✗	✗	151.6	273.9	42.3	73.0	278.7	1,624.3	45.3	88.0
IBN [12]	✗	✓	✗	491.8	1,110.4	109.7	267.7	318.1	1,420.4	92.2	178.0
SW [13]	✗	✓	✗	381.3	825.0	131.5	306.6	312.6	1,072.4	110.3	202.4
ISW [14]	✗	✓	✗	276.6	439.8	151.6	365.7	385.9	1,464.8	108.1	212.4
RCLNet (Ours)	✗	✗	✓	**97.6**	**333.4**	54.9	115.2	**119.9**	**275.4**	**28.4**	**69.4**

DAOT [11] and IBN [12], showed some improvement. However, they still experienced significant performance degradation on unseen domains and struggled with catastrophic forgetting or robust multi-domain generalization, as reflected in their consistently higher MAE/RMSE scores compared to RCLNet across SR and SN.

In contrast, RCLNet achieves the best MAE and RMSE results in SR, SN, and SD. On SN, RCLNet (97.6 in MAE, 333.4 in RMSE) reduces the MAE by 36% and RMSE by 22% compared to the best baseline DAOT [11] (151.6 in MAE, 273.9 in RMSE). On SD, RCLNet (119.9 in MAE, 275.4 in RMSE) achieves a 51% reduction in MAE and a 74% reduction in RMSE compared to BL [33] (262.7 in MAE, 1063.9 in RMSE), and still outperforms DAOT by 57% in MAE and 83% in RMSE. For SR, RCLNet improves MAE by 37% and RMSE by 21% compared to the best performing baseline SW [13]. Although performance on FH is slightly worse than DAOT, it remains competitive while delivering superior results in all other scenarios.

This performance gain stems from its regularization-based continual learning strategy, which mitigates catastrophic forgetting while enabling incremental adaptation to new domains. Notably, RCLNet achieves this without requiring access to or storage of target domain data during training, which is a crucial advantage in source-free generalization scenarios. This eliminates the need for costly data storage and retraining on historical data, enhancing the model's efficiency and scalability.

The qualitative performance evaluation of the proposed RCLNet framework is shown in Fig. 3. The figure compares the ground truth (GT) data with the predictions (Pred) in four challenging crowd counting scenarios (fog, snow, stadium, and street environments). In fog and snow, where visibility is low, RCLNet estimates crowd counts with high accuracy. The predictions are close to the GT values, even with significant environmental obstructions. In the stadium setting, which has large variations in crowd density and complex layouts, the model performs well. The predictions are also close to the actual crowd numbers. These

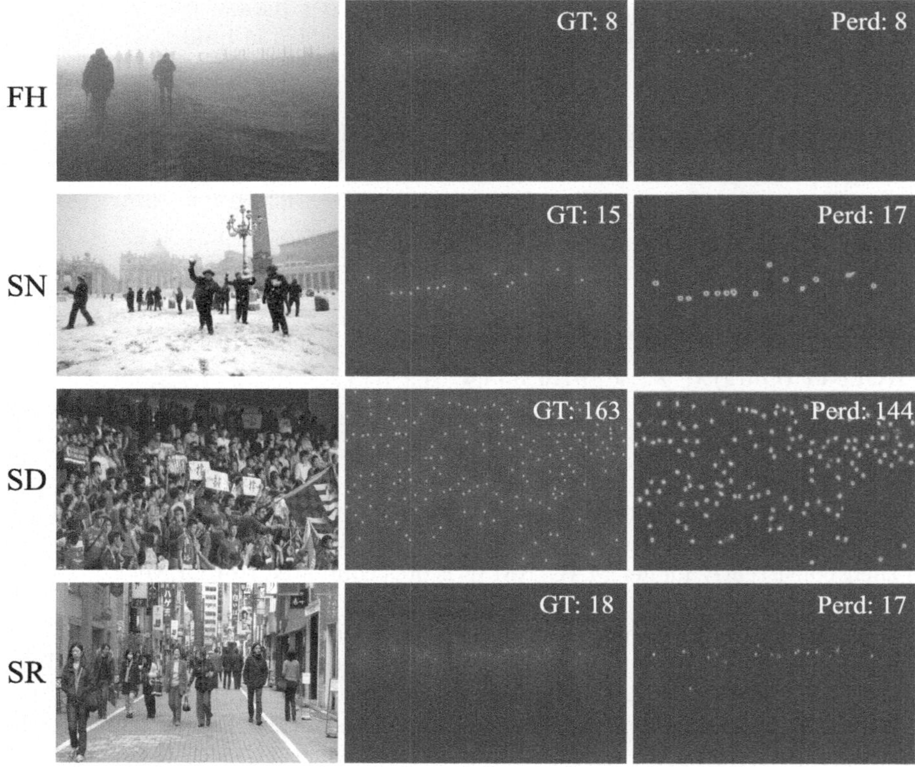

Fig. 3. The qualitative results of the proposed model in different crowd datasets. GT refers to the ground truth, and Pred represents the estimated results generated by RCLNet.

results prove that RCLNet generalizes well across different domains and performs well in dynamic conditions.

5 Conclusion

In this work, we address the key challenges of crowd counting in dynamic and diverse environments. These challenges include catastrophic forgetting during sequential learning and the inability to adapt efficiently without retraining from scratch. We propose a regularization-based continual learning framework (RCLNet) that uses the Fisher Information Matrix to identify and preserve parameters critical for spatial density estimation. The framework also allows the model to adapt to new domains without losing important knowledge. Experimental results demonstrate consistent improvements in MAE and RMSE over representative baselines. They also confirm the effectiveness of our method for source-free generalization. Future work will focus on reducing sensitivity in parameter-importance estimation, enhancing adaptability to more diverse scene structures, and extending evaluation to more complex scenarios. Integrating multi-modal data sources could further improve real-world applicability and performance.

References

1. Zhai, W., Gao, M., Li, Q., Jeon, G., Anisetti, M.: Fpanet: feature pyramid attention network for crowd counting. Appl. Intell. **53**(16), 19:199–19:216 (2023)
2. Chen, J., Li, Q., Gao, M., Zhai, W., Jeon, G., Camacho, D.: Towards zero-shot object counting via deep spatial prior cross-modality fusion. Inf. Fusion **111**, 102537 (2024)
3. Gong, X.: Using social media to characterise crowds in city events for crowd management (2020)
4. Gong, V., Daamen, W., Bozzon, A., Hoogendoorn, S.: Counting people in the crowd using social media images for crowd management in city events. Transportation **48**(6), 3085–3119 (2021)
5. Zhai, W., Xing, X., Gao, M., Li, Q.: Zero-shot object counting with vision-language prior guidance network. IEEE Trans. Circuits Syst. Video Technol. (2024)
6. Jiang, X., et al.: Learning multi-level density maps for crowd counting. IEEE Trans. Neural Netw. Learn. Syst. **31**(8), 2705–2715 (2019)
7. Song, Q., et al.: To choose or to fuse? Scale selection for crowd counting. Proc. AAAI Conf. Artif. Intell. **35**(3), 2576–2583 (2021)
8. Tian, Y., Lei, Y., Zhang, J., Wang, J.Z.: Padnet: pan-density crowd counting. IEEE Trans. Image Process. **29**, 2714–2727 (2019)
9. Wang, L., Zhang, X., Su, H., Zhu, J.: A comprehensive survey of continual learning: theory, method and application. IEEE Trans. Pattern Anal. Mach. Intell. **46**(8), 5362–5383 (2024)
10. Wu, C., van de Weijer, J.: Density map distillation for incremental object counting. In: Proceedings of the IEEE/CVF Conference on Computer Vision and Pattern Recognition, pp. 2506–2515 (2023)
11. Zhu, H., Yuan, J., Zhong, X., Yang, Z., Wang, Z., He, S.: Daot: domain-agnostically aligned optimal transport for domain-adaptive crowd counting. In: Proceedings of the 31st ACM International Conference on Multimedia, pp. 4319–4329 (2023)
12. Pan, X., Luo, P., Shi, J., Tang, X.: Two at once: enhancing learning and generalization capacities via ibn-net. In: European Conference on Computer Vision (2018)
13. Pan, X., Zhan, X., Shi, J., Tang, X., Luo, P.: Switchable whitening for deep representation learning. In: The IEEE International Conference on Computer Vision (ICCV) (2019)
14. Choi, S., Jung, S., Yun, H., Kim, J.T., Kim, S., Choo, J.: Robustnet: improving domain generalization in urban-scene segmentation via instance selective whitening. In: 2021 IEEE/CVF Conference on Computer Vision and Pattern Recognition (CVPR), pp. 11:575–11:585 (2021)
15. Sindagi, V.A., Yasarla, R., Patel, V.M.: Jhu-crowd++: large-scale crowd counting dataset and a benchmark method. IEEE Trans. Pattern Anal. Mach. Intell. **44**, 2594–2609 (2020)
16. Dalal, N., Triggs, B.: Histograms of oriented gradients for human detection. In: Proceedings of the IEEE/CVF Conference on Computer Vision and Pattern Recognition, vol. 1, pp. 886–893. IEEE (2005)

17. Leibe, B., Seemann, E., Schiele, B.: Pedestrian detection in crowded scenes. In: Proceedings of the IEEE/CVF Conference on Computer Vision and Pattern Recognition, vol. 1, pp. 878–885 (2005)
18. Chan, A.B., Vasconcelos, N.: Bayesian Poisson regression for crowd counting. In: Proceedings of the IEEE/CVF International Conference on Computer Vision, pp. 545–551 (2009)
19. Jiang, X., et al.: Attention scaling for crowd counting. In: Proceedings of the IEEE/CVF Conference on Computer Vision and Pattern Recognition, pp. 4706–4715 (2020)
20. Guo, D., Li, K., Zha, Z.-J., Wang, M.: Dadnet: dilated-attention-deformable convnet for crowd counting. In: Proceedings of the ACM International Conference on Multimedia, pp. 1823–1832 (2019)
21. Zhai, W., Gao, M., Guo, X., Li, Q., Jeon, G.: Scale-context perceptive network for crowd counting and localization in smart city system. IEEE Internet Things J. **10**(21), 18:930–18:940 (2023)
22. Chen, J., et al.: Privacy-aware crowd counting by decentralized learning with parallel transformers. Internet Things **26**, 101167 (2024)
23. Kirkpatrick, J., et al.: Overcoming catastrophic forgetting in neural networks. Proc. Natl. Acad. Sci. **114**(13), 3521–3526 (2017)
24. Rebuffi, S.-A., Kolesnikov, A., Sperl, G., Lampert, C.H.: icarl: Incremental classifier and representation learning. In: Proceedings of the IEEE/CVF Conference on Computer Vision and Pattern Recognition, pp. 2001–2010 (2017)
25. Li, Z., Hoiem, D.: Learning without forgetting. IEEE Trans. Pattern Anal. Mach. Intell. **40**(12), 2935–2947 (2017)
26. He, Y., Sick, B.: Clear: an adaptive continual learning framework for regression tasks. AI Perspect. **3**(1), 1–16 (2021)
27. Rusu, A.A., et al.: Progressive neural networks. arXiv preprint arXiv:1606.04671 (2016)
28. Lopez-Paz, D., Ranzato, M.: Gradient episodic memory for continual learning. In: Advances in Neural Information Processing Systems, vol. 30, pp. 6467–6476 (2017)
29. Xiong, H., Lu, H., Liu, C., Liu, L., Cao, Z., Shen, C.: From open set to closed set: counting objects by spatial divide-and-conquer. In: Proceedings of the IEEE/CVF International Conference on Computer Vision, pp. 8362–8371 (2019)
30. Simonyan, K., Zisserman, A.: Very deep convolutional networks for large-scale image recognition. CoRR, abs/1409.1556 (2014)
31. Loshchilov, I., Hutter, F.: Fixing weight decay regularization in adam. arXiv, abs/1711.05101 (2017)
32. Smith, L.N., Topin, N.: Super-convergence: very fast training of neural networks using large learning rates. In: Defense + Commercial Sensing (2017)
33. Ma, Z., Wei, X., Hong, X., Gong, Y.: Bayesian loss for crowd count estimation with point supervision. In: IEEE/CVF International Conference on Computer Vision (ICCV), pp. 6141–6150 (2019)
34. Lin, H., Ma, Z., Ji, R., Wang, Y., Hong, X.: Boosting crowd counting via multifaceted attention. In: CVPR (2022)

Real-Time Defect Detection: A Lightweight Deep Learning Framework for Industrial Applications

Nour Ben Sghaier[1(✉)], Zakaria Arfaoui[2], Raef Cherif[1], and Nawres Khlifa[3]

[1] Université du Québec à Rimouski (UQAR), Rimouski, Canada
{Nour.BenSghaier,RaefCherif}@uqar.ca
[2] Université Ibn Khaldoun, Tiaret, Tunisia
[3] Université de Tunis El Manar, Tunis, Tunisia
nawres.khlifa@istmt.utm.tn

Abstract. This paper introduces a lightweight deep learning framework designed for real-time defect detection in industrial applications, with a focus on supporting small and medium-sized enterprises (SMEs) facing resource constraints. The proposed system leverages advanced computer vision and deep learning techniques to enable efficient and cost-effective quality control on production lines. The methodology includes a structured pipeline for data processing, model training, and validation, optimized for low-resource environments. The system builds on innovative approaches such as MobileNet to enhance performance while maintaining accessibility. Experimental results demonstrate the framework's effectiveness, offering a scalable solution for SMEs. The study addresses the research question of whether a low-cost system can provide reliable part validation, achieving objectives of real-time processing, accurate defect detection of bottles and cans to determine if they are defective for example, exhibiting visual defects like dents, scratches, cracks, or contamination spots even under deliberate variations in lighting, orientation, and surface reflections. Limitations and future enhancements are also discussed to guide ongoing development in industrial automation.

Keywords: Defect detection · Industrial automation · MobileNet · Low-cost system · Computer vision

1 Introduction

Industrial automation has been revolutionising industrial manufacturing by enhancing productivity, by minimising human error and by guaranteeing continuous quality control [1]. Nevertheless, the high cost of advanced industrial control and validation systems, which often require sophisticated hardware and a complex infrastructure, is a significant barrier for small and medium-sized enterprises (SMEs) [2]. These enterprises need cost-effective, efficient and self-managing solutions to maintain competitive quality standards without the requirement

T. Ensari et al. (Eds.): ISPR 2025, CCIS 2859, pp. 283–292, 2026.
https://doi.org/10.1007/978-3-032-21585-7_21

for significant resources or specialist training. With the recent improvements in embedded systems and computer vision technologies, new opportunities have emerged for the development of cost-effective automation solutions suitable for SMEs [3]. Detecting defects in manufacturing production lines is essential for ensuring product quality and minimizing waste. In the canning industry, a superficial defect may lead to safety issues ansubstantial financial losses if undetected. This article introduces a low-cost automated system for real-time components inspection on a manufacturing line which is based on an intelligent solution using deep learning. This system can be replicated on a Raspberry Pi using video streams captured by a camera. The system processes these streams using new AI techniques to determine whether the boxes are defective or not, thus offering an accessible alternative for SMEs. The device is designed to achieve three main objectives: real-time image capture and processing, accurate defect detection using an integrated algorithm, and an interactive user interface for easy use in an industrial environment [4]. This article describes in detail the methodology, experimental results, limitations, and future improvements. The remainder of the paper is organized as follows: it first reviews the state of the art, then outlines the materials and methods, presents the results and discussion, and finally concludes with final remarks and future directions.

2 State of the Art

Numerous studies have been devoted to automating the detection of visual defects in various industrial environments, including wood, metal, printed circuit boards (PCBs), and structured surfaces. These approaches are based on computer vision architectures combined with deep learning models optimized for reasoning in constrained environments.

Wang et al. [5] proposed a model based on YOLOv8, which has been modified by adding attention modules. This method was applied to a dataset consisting of 3,000 images of wood surfaces achieving an accuracy of 94.3%, with aim of identifying different types of defects and focusing on balancing precision with deployment feasibility in dynamic environments.

Chen et al. [6] used a lightweight YOLOv4 architecture for metal surface inspection. The experiment was based on the NEU Surface Defect DB dataset, comprising 1,800 images, their system reached 91.5% accuracy, emphasizing real-time detection capabilities under factory lighting conditions.

Cao et al. [7] made an optimized version of YOLOv5, called YOLOv5-LW, incorporating attention mechanisms and multi-scale fusion. The model was used on a dataset of 3,200 images of wooden panels. The model achieved 96.2% accuracy on 3,200 samples, highlighting its robustness to background noise and minor imperfections.

Mih et al. [8] suggested an EdgeAI architecture based on a simplified version of Xception, used in detecting defects on electronic cards. Tested on a 5,000-image dataset, their approach achieved 92.7% accuracy while maintaining low inference time suitable for edge devices.

Ruan et al. [9] came up with EPSC-YOLO, a model based on YOLOv9, paired with pyramidal convolution blocks (PyConv) and attention modules. Tests were done on the NEU-DET and GC10-DET databases, covering several types of visual defects.

Wan and Li [10] presented LGP-YOLO, a model applied to defect detection on Light Guide Plates (LGP), combining the RepLKNet architecture and attention modules. The study relied on a dataset specifically related to the field of lighting.

Kang et al. [11] proposed CFIS-YOLO, a model dedicated to the inspection of wooded surfaces, founded on a multi-scale fusion strategy. The network was tested on a public dataset containing various types of wood defects.

Shen et al. [12] developed MINet, an interactively multi-scale model for the analysis of rolled steel surfaces. The network was validated on the SD-Saliency-900 dataset, containing industry-generated defects.

Zhu et al. [13] showed VR-YOLO, an architecture that's robust to variations in viewing angle for detecting defects on printed circuit boards. The machine learning model was learned from a multi-angle dataset of PCBs.

The reviewed models employ various approaches to detect defects in industrial environments. Some approaches, like EPSC-YOLO and VR-YOLO, incorporate architectural improvements such as pyramid convolutions and angular robustness to improve generalisation. Others, including CFIS-YOLO and MINet concentrate on multi-scale merging for fine-grained defect localisation. The lightweight models which are based on YOLOv4-light MobileNet or custom CNNs focus on minimising inference time and memory requirements for real-time deployment on low-cost devices.

3 Materials and Methods

This section outlines the material and methodological approach for developing the cost-effective visual inspection cans and bottles defect detection. The study adopts a standard computing framework for development and testing, with plans for future deployment on low-resource platforms. The methodology is a structured process, including data collection, pre-processing, model training, and validation, optimised for final deployment in resource-constrained environments.

3.1 Global Architecture of the Work

The workflow is chronologically organised as demonstrated in the Fig. 1. To ensure comprehensive development and evaluation. The process starts with data collection and annotation, followed by pre-processing and augmentation to improve the variability of the dataset using proven techniques. Next, three deep learning frameworks are designed and trained using a transfer learning approach, with grid search to optimise key parameters. Validation is performed using a series of test cases to evaluate performance under various conditions, comparing the accuracy, efficiency, and suitability of the frameworks for resource-constrained environments. The comparison is basis on previous research in which different models have shown varying levels of success in similar tasks, with one offering efficiency and the other greater accuracy at a higher computational cost. A customized framework, adapted to the task, aims to balance these aspects, with the results still to be validated and refined based on this information. The next step will be to use these results to improve the adaptability and performance of the system.

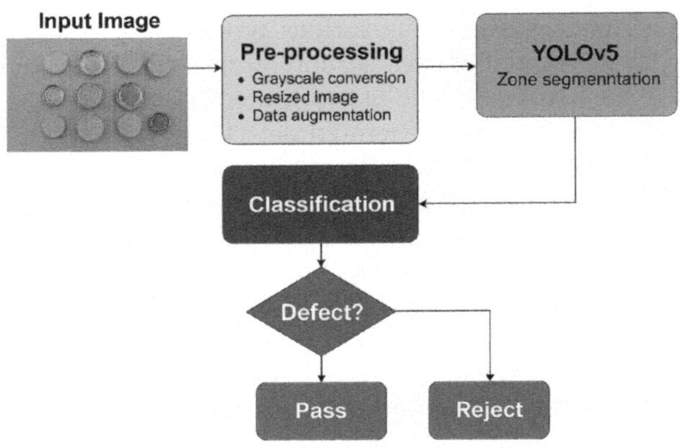

Fig. 1. Chronological Workflow of the Defect Detection Process

3.2 Data Description

The research is based on a dataset consisting of labelled images of cans and bottles as shown in Fig. 1, collected under controlled conditions which simulate a production line. The images show some common visual defects like dents, scratches, cracks, and contamination spots, with some deliberate variations in lighting, orientation, and surface reflections to test how robust the system is [15]. The dataset was standardised at resolutions that balance computational efficiency and detail for defect detection. Each image is hand-labelled as defective or non-defective by experts to guarantee reliability. A preliminary analysis revealed a wide variety of textures and lighting, highlighting the need for robust pre-processing and augmentation strategies [14]. The dataset consists of 1,094 labeled images of cans and bottles, collected under controlled conditions simulating a production line. Among them, 562 are defective and 532 non-defective. Defects include dents, scratches, cracks, and contamination spots, with deliberate variations in lighting and orientation. The dataset was split into 70% training, 15% validation, and 15% testing, with augmentation applied only to the training set to avoid leakage.

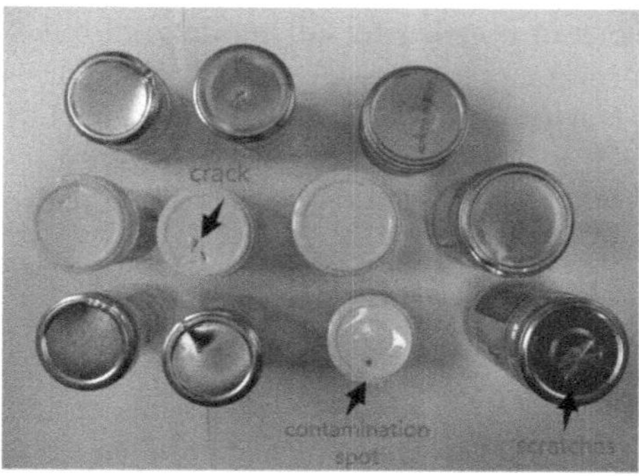

Fig. 2. Example of dataset images showing defects such as dents, scratches, cracks, and contamination under different lighting conditions.

3.3 Data Preparation

The data preparation process begins with an initial separation of the dataset using the YOLOv5 framework, a widely recognized tool for object detection that efficiently isolates and highlights relevant features, such as defective regions within the dataset as shown in Fig. 2. This step leverages YOLOv5's robust capability to detect and segment objects with high precision, enabling focused analysis by accurately identifying and annotating key characteristics, which are then divided into training and test sets. Following this separation, augmentation methods are applied to enhance the dataset's variability and robustness, utilizing the Albumentations library—a highly regarded tool known for its flexibility and efficiency in computer vision tasks. Albumentations facilitates a range of transformations, including geometric adjustments like rotations and flips, as well as pixel-level modifications such as brightness and contrast variations, all designed to improve model generalization across diverse conditions. This synergy between YOLOv5's separation and Albumentations' augmentation creates a comprehensive dataset, optimized for subsequent model training and evaluation.

3.4 Model Design

Three frameworks were tested:

- **MobileNet**, leveraging depthwise separable convolutions for efficiency.
- **ResNet18**, using residual connections for deeper feature extraction.
- **Custom CNN**, consisting of three convolutional layers (32, 64, 128 filters, kernel size 3), followed by max pooling, two dense layers (128 and 64 neurons), and softmax output.

All models used ImageNet pretraining, with the last layers fine-tuned. Grid search optimized learning rate (0.001–0.01), batch size (16–64), and epochs (5–70), it is performed to optimise hyperparameters, including learning rates, batch sizes, and number of epochs, for all models. MobileNet is chosen for its lightweight design, leveraging separable convolutions for efficiency [21], while ResNet offers deeper feature extraction and the custom CNN provides tailored adjustments. This multiple-framework approach guarantees a comprehensive evaluation, with the most effective model selected for future deployment.

4 Results and Discussion

This section gives the test results for the deep learning based defect detection system, providing a detailed analysis of its performance in different setups and conditions. The discussion is structured to evaluate the effectiveness of the data preparation process, compare the performance of the three proposed models (MobileNet, ResNet, and a custom CNN), and compare these results to state-of-the-art approaches described in the literature. The results provide insights into the system's potential in resource-constrained environments, with implications for industrial applications.

4.1 Data Preparation and Preprocessing

The data preparation phase was set up to make the dataset fit for real-world production environments, considering things to do with lighting changes, object orientation, and limited computing resources. The initial pre-processing involved resizing the images to 96×96 resolutions and then normalising them to a range of 0.1 in greyscale in order to reduce computational complexity while preserving the visual characteristics essential for defect detection [16]. This step ensured compatibility with the subsequent modelling process and minimised data redundancy. To improve the generalisation and robustness of the model, a comprehensive data augmentation strategy was implemented using the Albumentations library, a versatile tool widely recognised for its effectiveness in computer vision tasks. The augmentation techniques:

- Random rotations from 0° to 360° with a probability of 0.5
- Horizontal and vertical flips with a probability of 0.5
- Brightness adjustments of ±20%
- Contrast variations of ±10%
- Gaussian noise with variance $\sigma^2 = 0.01$

These transformations were carefully selected to simulate realistic production line conditions, such as uneven lighting or partial occlusions, thereby increasing the diversity of the dataset and reducing the risk of overfitting. The augmented dataset was validated to confirm that the augmented samples retained significant defect characteristics, ensuring their usefulness in training [17].

A crucial initial step in organising the data was using the YOLOv5 framework to separate the datasets. YOLOv5's object detection capabilities were used to isolate defect regions from non-defect regions, enabling accurate annotation and segmentation. This process facilitated the creation of structured training and test sets, minimizing bias, and ensuring that the dataset was well suited for model development as described in Fig. 3.

Fig. 3. bottles detection using YOLOv5

4.2 Model Evaluation

Three different deep learning architectures MobileNet, ResNet18, and a custom CNN as shown in Table 1 tested to determine the most effective solution for detecting defects in resource-constrained environments. Each model was trained using a transfer learning method, exploiting the pre-trained weightings of the ImageNet dataset, then fine-tuned using a grid search on hyperparameters, such as learning rates, batch sizes and number of epochs. Performance was assessed based on a series of metrics: accuracy, F1 score, area under the ROC curve (AUC-ROC), and total number of errors, offering an overview of each model's forces and limitations. MobileNet performed best, with the highest accuracy 97.0%

Table 1. Performance Comparison of the Models

Model	Accuracy	F1-score	AUC ROC	Total Errors	Notes
MobileNet	97.0%	0.97	0.99	37	Pretrained, low memory usage
ResNet18	91.2%	0.91	0.98	92	Residual learning architecture
Custom CNN	87.9%	0.88	0.96	132	Designed manually with grid search

and the lowest number of errors 37, owing to its lightweight design with separable convolutions. The architecture's efficiency is particularly suited to environments with restricted computing resources, as it provides a good balance between performance and reduced memory and processing requirements. ResNet18, for its part, with its residual learning framework, reached a solid accuracy of 91.2% and an AUC-ROC of 0.98, though its more deep structure led up to a higher number of errors 92 and increased computing requirements. The custom CNN, while lower in accuracy 87.9% and with the highest number of errors 132, offered flexibility in architecture design, allowing for customised adjustments during grid search. However, its performance suggests that further optimisation is needed to compete with pre-trained models.

4.3 Evaluation of the MobileNet Detection System

With MobileNet giving the most sustained attention to defective areas, it revealed that all models effectively detected subtle defects, particularly in the optimal configuration. The quality evaluation using visual inspection revealed a deeper understanding of the system's defect detection capabilities in 5 test cases As demonstrated in the following Table 2. For example Case 2, the system showed outstanding capabilities in detecting more subtle defects like micro scratches and faint dents, with optimal epoch training contributing to an accuracy of 99.45%. Such performance was confirmed by detailed manual inspections of the output images, where the model consistently highlighted defect regions with high precision. In contrast, Case 1, which had a lower resolution of 128×128 pixels and only 5 epochs, had difficulties detecting fine cracks and minor defects, leading to a lower accuracy of 96.88%. This indicates that increased resolution is not enough on its own to refine feature extraction approaches without adequate training iterations. The cases 3 to 5, with different epochs 50 and 70 and different sizes 16, 32 and 64 exhibited robust detection of defects including scratches and smudge marks, although false positives were occasionally detected under challenging illumination conditions, particularly glare from specular surfaces. These results were illustrated even further by heatmap visualizations, which revealed a strong focus of attention on areas prone to defects such as edges and textured regions. Although, for Case 5, the larger batch size of 64 appeared to introduce minor distractions, like reflective surfaces sometimes distracted the model's focus away from defects, resulting in a slight decrease in accuracy 97.9%. The MobileNet architecture played a crucial role in enabling these qualitative assessments on modest hardware such as the Raspberry Pi, underlining its suitability for SMEs by balancing computational efficiency and detailed defect detection.

Table 2. MobileNet Performance Metrics Across Five Test Cases

Case	Image Size	Batch Size	Learning Rate	Epochs	Accuracy (%)	Precision (%)	Recall (%)	F1-Score (%)	AUC-ROC
1	128×128	16	0.001	5	96.88	96.6	96.0	96.3	0.962
2	96×96	16	0.001	20	**99.45**	99.4	99.5	99.4	**0.994**
3	96×96	16	0.001	50	98.26	98.2	98.5	98.3	0.983
4	96×96	32	0.001	70	98.44	98.1	98.8	98.4	0.985
5	96×96	64	0.01	70	98.44	97.9	98.7	98.3	0.981

4.4 Comparison with State-of-the-Art Methods

In order to contextualize the proposed models within a larger domain, a comparative analysis was carried out in Table 3 in relation to recent cutting-edge advances in surface and defect detection, focusing on lightweight and industrial deep learning frameworks. This analysis considered application domains, model architectures, dataset sizes, reported accuracy, and suitability for environments with limited resources.

The MobileNet-based pipeline achieved a competitive accuracy of 97.0%, surpassing or matching the performance of several state-of-the-art lightweight models such as YOLOv8 94.3% and Xception-lite 92.7%, despite operating on

Table 3. Comparison with State-of-the-Art Approaches

Reference	Application Domain	Model Used	Dataset Size	Accuracy	Type
Wang et al., 2024	Wood surface defects	YOLOv8 (improved)	3,000	94.3%	Lightweight
Chen et al., 2023	Metal surface inspection	YOLOv4 (light)	1,800	91.5%	Lightweight
Kang et al., 2023	Wood panel classification	YOLOv5-LW + attention	3,200	96.2%	Lightweight
Mih et al., 2023	PCB fault detection	Xception-lite (EdgeAI)	5,000	92.7%	Edge-optimized
Our Approach	Can defect classification	MobileNet/ResNet/CNN	1,094	**97.0%**	Embedded-ready

a smaller dataset 1,094 images. This result underscores MobileNet's efficiency, particularly in resource-constrained environments, where its reduced parameter count offers a significant advantage over deeper architectures like ResNet18. ResNet18's 91.2% accuracy aligns with its performance in similar tasks but its higher computational demand highlights a trade-off that may limit its applicability in low-resource settings. The custom CNN, with 87.9% accuracy, falls below some benchmarks, suggesting that further tuning could enhance its competitiveness. Compared to YOLOv5-LW with attention, the proposed models demonstrate comparable performance with less complexity, though the attention mechanism's impact on accuracy warrants further exploration. Overall, the results indicate that the proposed frameworks are well-positioned within the state-of-the-art, with MobileNet offering a particularly strong candidate for scalable, low-resource defect detection.

5 Conclusion and Perspectives

This study introduces a low-cost visual inspection system based on deep learning to simplify real-time defect detection for small and medium-sized businesses with limited resources. Using YOLOv5 to structure the datasets and Albumentations to enhance them improved data quality and model resilience. Among the architectures evaluated (MobileNet, ResNet18, and a custom CNN), MobileNet stood out for its efficiency, while ResNet18 offered greater accuracy at a higher cost, and the custom CNN showed potential with room for improvement. Comparison with existing methods confirms the competitiveness of the system in constrained environments. The results underline the potential for improving production efficiency and quality for SMEs, with MobileNet being a promising solution. Challenges include the limited variety of datasets and sensitivity to the environment, suggesting areas for improvement. Future work will focus on refining data processing, expanding the dataset, and optimising the custom CNN, while exploring the integration of multiple sources. At present, this research lays a solid foundation for accessible quality control and future industrial innovation.

Acknowledgment. The authors thank the University of Québec à Rimouski (UQAR) for supporting this study by providing research facilities and resources.

References

1. Esmaeilian, B., Behdad, S., Wang, B.: The evolution and future of manufacturing: a review. J. Manuf. Syst. (2016)
2. Mittal, S., Khan, M., Romero, D., Wuest, T.: Smart manufacturing: characteristics, technologies and enabling factors. Proc. Inst. Mech. Eng. Part B: J. Eng. Manuf. **233**, 1342–1361 (2019)
3. Yin, S., Gao, H., Kaynak, O.: Data-driven control and process monitoring for industrial applications–Part I. IEEE Trans. Ind. Electron. **61**(11), 6356–6359 (2014)
4. Anand, G., Kumawat, A.K.: Object detection and position tracking in real time using Raspberry Pi. Mater. Today: Proc. **47**, 3221–3226 (2021)
5. Wang, R., et al.: An efficient and accurate surface defect detection method for wood based on improved YOLOv8. Forests **15**(7), 1176 (2024)
6. Li, S., et al.: Aluminum surface defect detection method based on a lightweight YOLOv4 network. Sci. Rep. **13**, 11077 (2023)
7. Cao, Y., et al.: Lightweight wood panel defect detection method incorporating attention mechanism and feature fusion network. arXiv preprint arXiv:2306.12113 (2023)
8. Mih, A.N., et al.: Developing a resource-constraint EdgeAI model for surface defect detection. arXiv preprint arXiv:2401.05355 (2023)
9. Ruan, S., et al.: A high precision YOLO model for surface defect detection based on PyConv and CISBA. Sci. Rep. **15**, 15841 (2025)
10. Wan, Y., Li, J.: LGP-YOLO: an efficient convolutional neural network for surface defect detection of light guide plate. Complex Intell. Syst. **10**, 2083–2105 (2024)
11. Kang, J., et al.: CFIS-YOLO: a lightweight multi-scale fusion network for edge-deployable wood defect detection. arXiv preprint arXiv:2504.11305 (2025)
12. Shen, K., Zhou, X., Liu, Z.: MINet: multiscale interactive network for real-time salient object detection of strip steel surface defects. IEEE Trans. Ind. Inf. **20**(5), 7842–7852 (2024)
13. Zhu, H., Wei, L., Li, H.: VR-YOLO: enhancing PCB defect detection with viewpoint robustness based on YOLO. arXiv preprint arXiv:2507.02963 (2025)
14. Shorten, C., Khoshgoftaar, T.M.: A survey on image data augmentation for deep learning. J. Big Data **6**(1), 1–48 (2019)
15. Woods, R.E., Gonzalez, R.C.: Real-time digital image enhancement. Proc. IEEE (1981)
16. Buslaev, A., et al.: Albumentations: fast and flexible image augmentations. Information (2020)
17. Shorten, C., Khoshgoftaar, T.M.: A survey on image data augmentation for deep learning. J. Big Data (2019)
18. Oleiwi, B.K., Kadhim, M.R.: Real time embedded system for object detection using deep learning. In: AIP Conference Proceedings (2022)
19. Deng, J., et al.: Imagenet: a large-scale hierarchical image database. In: Proceedings of IEEE Conference on Computer Vision and Pattern Recognition (2009)
20. Ioffe, S., Szegedy, C.: Batch normalization: accelerating deep network training by reducing internal covariate shift. In: Proceedings of ICML (2015)
21. Howard, A.G., et al.: MobileNets: efficient convolutional neural networks for mobile vision applications. arXiv preprint arXiv:1704.04861 (2017)

Robust 1D CNN-Based Approach with Data Augmentation for Incipient and Severe Stator Fault Classification in Induction Motors

Hanen Berriri(✉) and Mohamed Faouzi Mimouni

University of Monastir, ENIM, LAS2E, Monastir, Tunisia
hanen.berriri@enim.rnu.tn

Abstract. Condition monitoring of induction motors is critical for ensuring the reliability and operational efficiency of industrial systems. This paper presents an end-to-end deep learning framework designed to diagnose stator inter-turn short-circuit faults, a prevalent and potentially catastrophic failure mode. The proposed methodology utilizes a one-dimensional convolutional neural network (1D CNN) that directly processes raw three-phase current signals to classify seven distinct machine health states, spanning both incipient and severe fault conditions. Through a systematically validated data augmentation strategy incorporating noise injection and amplitude jittering, the model effectively mitigates overfitting and demonstrates enhanced generalization capabilities. Upon evaluation with a public benchmark dataset, the model achieves a classification accuracy of 99.32% on an independent test set (95% CI [98.9%, 99.7%]). This outcome validates the efficacy of a well-regularized 1D CNN in delivering a highly accurate and automated fault diagnosis solution, eliminating the need for manual feature engineering. Furthermore, a rapid inference time (~4.3 ms/sample) and a compact quantized model size of 4.05 MB position the framework as a scalable and efficient solution for predictive maintenance in Industry 4.0 environments, confirming its suitability for real-time monitoring on edge computing devices.

Keywords: Fault Diagnosis · Induction Motor · Deep Learning · 1D Convolutional Neural Network · Time-Series Classification · Data Augmentation

1 Introduction

Induction motors are foundational components in contemporary industrial operations, serving as the prime movers for a vast array of critical processes. The unforeseen failure of these motors can precipitate substantial operational interruptions, with unplanned downtime costs approaching $1.4 trillion annually for the world's leading corporations [1]. Among the various failure modes, stator winding faults, responsible for as many as 30% of all motor failures [2], pose a significant diagnostic challenge. These faults often originate as minor, incipient inter-turn short circuits, which, if left undetected, can escalate rapidly into catastrophic phase-to-ground or phase-to-phase failures, leading to costly repairs and extended downtime. The primary difficulty lies in the early detection of these faults, as their initial signatures are often subtle, non-stationary, and embedded within the noisy, fluctuating current signals characteristic of industrial environments.

T. Ensari et al. (Eds.): ISPR 2025, CCIS 2859, pp. 293–307, 2026.
https://doi.org/10.1007/978-3-032-21585-7_22

The evolution of diagnostic techniques reflects a continuous search for greater automation and reliability. Conventional methods like Motor Current Signature Analysis (MCSA) demand considerable domain expertise to interpret spectral sidebands and exhibit significant limitations under the variable load conditions common in real-world applications [3]. The subsequent integration of machine learning methodologies, including Support Vector Machines (SVMs) [4], automated the classification step but merely shifted the expert burden to a laborious and often suboptimal process of manual feature engineering. While recent progress in deep learning (DL) has promised end-to-end automation [5], a significant portion of existing research either prioritizes vibration data, which requires installing additional sensors [6], or fails to adequately address the critical issue of overfitting when processing raw current signals directly.

To address this specific challenge, this study introduces a robust, end-to-end 1D CNN pipeline designed for high-fidelity stator fault classification. The principal contributions of this work are threefold: (1) the development of a comprehensive and reproducible pipeline that achieves 99.32% accuracy on a public benchmark, demonstrating state-of-the-art performance for this task; (2) the systematic validation of data augmentation's role in mitigating overfitting, supported by a formal sensitivity analysis (Appendix A) that justifies the selection of optimal hyperparameters; and (3) an in-depth performance evaluation across a spectrum of fault severities, featuring enhanced model interpretability through filter activation analysis and 1D Gradient-weighted Class Activation Mapping (Grad-CAM) [7], thereby building trust in the model's decision-making process.

2 Related Work

The field of fault diagnosis for induction motors has evolved significantly, transitioning from classical signal processing and model-based methods to sophisticated data-driven techniques. This section surveys this evolution, establishing the context and rationale for the deep learning methodology proposed herein

2.1 Signal-Based Fault Diagnosis

Historically, Motor Current Signature Analysis (MCSA), which leverages the Fast Fourier Transform (FFT) to identify fault-induced harmonic sidebands, served as the cornerstone of stator fault diagnosis [2, 3]. The underlying principle is that electrical or mechanical anomalies modulate the stator current, creating spectral signatures at predictable frequencies. However, the efficacy of MCSA is often compromised by operational variabilities, such as load fluctuations and supply frequency variations, which can alter the amplitude and position of these signatures, potentially masking them or leading to false alarms. Consequently, its interpretation requires considerable expert knowledge to distinguish faint, fault-related harmonics from background noise and to account for machine-specific parameters. To address the challenge of non-stationary signals, alternative techniques like wavelet analysis were explored for their ability to provide time-frequency localization. Yet, these methods introduced their own complexities, including the critical choice of a mother wavelet, and still frequently culminated in a set of features that depended on expert-guided selection [10].

2.2 Model-Based and Traditional Machine Learning Approaches

The inherent limitations of purely signal-based methods, which require manual interpretation, naturally prompted the integration of more automated techniques. This technological evolution proceeded along two distinct yet conceptually related paths during the same era. The first path involved the application of traditional machine learning (ML), where classifiers such as Support Vector Machines (SVMs) were trained to recognize fault patterns. These models demonstrated high accuracy but were critically dependent on a pre-processing stage of manual feature engineering, where domain experts would extract a set of descriptive statistics (e.g., RMS, kurtosis) or spectral metrics (e.g., energy of specific harmonics) from the raw sensor data [4]. The second path pursued model-based strategies, including the use of Sliding Mode Observers [8, 9]. These techniques detect faults by comparing the motor's measured behavior against the outputs of a precise mathematical model, with significant deviations indicating a fault. Although a substantial step toward automation, both of these approaches shared a fundamental reliance on a preceding stage of expert-driven design: either the manual crafting of features for ML or the complex derivation of a system model for observers. This dependency on prior knowledge not only created a bottleneck in development but also limited the adaptability of these systems, thus motivating the subsequent shift toward a more integrated, end-to-end paradigm capable of learning directly from data.

2.3 Deep Learning in Motor Fault Diagnosis

Deep learning (DL), and specifically the 1D Convolutional Neural Network (1D CNN), has emerged as a powerful paradigm for time-series analysis, primarily because it automates the feature extraction process by learning a hierarchical representation of features directly from raw data [10]. While early DL applications in motor diagnostics often focused on vibration signals [6], the significant economic and practical advantages of using readily available current data have spurred extensive research into applying DL to electrical signals. This has led to numerous studies validating the utility of various architectures, including autoencoders for unsupervised feature learning [11], different CNN configurations for classification [12], and multi-sensor data fusion techniques for enhanced diagnostic robustness [5].

Despite these advances, overfitting remains a critical challenge, particularly in the context of incipient fault diagnosis, where datasets of faulty conditions are often small and the fault signatures themselves have a low signal-to-noise ratio. To address this, data augmentation has been investigated as a key regularization technique. For instance, Wright et al. [13] demonstrated the value of using shifting windows to synthetically expand training data. Our work contributes a focused analysis of complementary techniques—namely, noise injection and amplitude jittering—whose efficacy and optimal parameters are rigorously validated through a systematic sensitivity analysis. Furthermore, while the complexity of multi-class stator fault diagnosis has been increasingly recognized [14], and advanced architectures like Bayesian deep learning for uncertainty quantification [15] or large ensembles for marginal accuracy improvements [16] have shown strong results, they often do so at the cost of significant computational overhead. A key contribution of our work is to demonstrate that a single, computationally efficient, and well-regularized 1D CNN can achieve highly competitive, state-of-the-art performance (99.32% accuracy). The novelty of this research, therefore, lies in its holistic and practical pipeline, which integrates a robust data handler, validated augmentation, and deep interpretability analysis to offer a powerful yet efficient solution well-suited for industrial deployment.

3 Proposed Methodology

The adopted methodology adheres to an end-to-end learning paradigm, wherein the model acquires the capacity to classify fault categories directly from raw sensor data with minimal human intervention. This section details the dataset, the data processing pipeline, the augmentation techniques, and the network architecture.

3.1 Dataset and Experimental Setup

This study utilizes the public "MIT Short-Circuit Flux and Current Signals" benchmark dataset [17], chosen for its relevance and reproducibility. The data were collected from a 1 HP, three-phase induction motor test bench under a comprehensive range of operating frequencies (30–60 Hz) and mechanical loads (0 to 100%). This variability is crucial for training a model that is robust to changing operational conditions. As shown in Table 1, the dataset is categorized into seven classes: one healthy state and six fault states of varying severity, including incipient (high-impedance, HI) and more severe (low-impedance, LI) inter-turn shorts (Normal, HI-1, HI-2, HI-3, LI-1, LI-2, LI-3). This diverse set of fault conditions provides a challenging and realistic classification task.

Table 1. Summary of machine health classes and fault severities in the dataset.

Class ID	Label	Description	Fault Type	Samples (Test set)
0	Normal	Healthy motor operation	None	5,088
1	HI-1	1.41% of stator winding shorted	Incipient	5,472
2	HI-2	4.81% of stator winding shorted	Incipient	5,472
3	HI-3	9.26% of stator winding shorted	Incipient	5,568
4	LI-1	1.41% of stator winding shorted	Severe	5,376
5	LI-2	4.81% of stator winding shorted	Severe	5,472
6	LI-3	9.26% of stator winding shorted	Severe	5,280

3.2 Data Preprocessing and Pipeline

The raw signals undergo segmentation through a sliding window of 2048 data points. This window size was selected to be large enough to capture multiple cycles of the fundamental frequency (e.g., ~34 cycles at 60 Hz) and their associated harmonics, which are critical for fault identification. A 50% overlap (step size of 1024) is used to augment the dataset and ensure translational invariance, making the model less sensitive to the exact position of a fault signature within a window. This process yields a final dataset comprising 251,232 samples. The data is partitioned into training (70%), validation (15%), and test (15%) subsets, employing stratification to maintain the original class distribution in each partition, which is essential for unbiased model evaluation. A MinMaxScaler, fitted exclusively on a subset of the training data to prevent data leakage, is applied to all signals to scale them to a [0, 1] range. The tf.data API from TensorFlow is used to construct an efficient data pipeline for on-the-fly data loading, batching, and prefetching, which is critical for handling large-scale time-series data without memory overload.

3.3 Data Augmentation for Time-Series Signals

To mitigate overfitting and improve the model's ability to generalize to unseen data, two distinct augmentation techniques are applied exclusively during the training phase. These methods simulate realistic variations that might occur in an industrial setting.

Noise Injection. Gaussian noise is introduced to mimic sensor inaccuracies and ambient electromagnetic interference. The augmented sample x_{aug} is expressed as:

$$x_{aug} = x + \mathcal{N}\left(0, \sigma^2\right) \tag{1}$$

where x is the original signal and $\mathcal{N}\left(0, \sigma^2\right)$ represents noise from a normal distribution with zero mean and standard deviation σ.

Amplitude Jittering. Each sample's amplitude is scaled by a random factor, f, near unity, simulating minor supply voltage or load variations:

$$x_{aug} = x \bullet f \tag{2}$$

where f is a random factor drawn from a uniform distribution, such that $f \sim$ U(0.98,1.02).

The selection of hyperparameters for these techniques was not arbitrary but was instead validated through a formal sensitivity analysis, the results of which are presented in Table 2. This analysis, which explored various combinations of noise levels and jitter ranges, confirmed that the chosen parameters (σ =0.02 and a jitter range of ±2%) provide the optimal balance between enhancing robustness and preserving the integrity of subtle fault signatures. Alternative techniques, such as time-warping, were considered but deemed less pertinent for this application as they can distort the fundamental frequency and phase relationships within the quasi-cyclical current signals.

Table 2. Validation accuracy for different augmentation hyperparameters.

Noise Std (σ)	Jitter range of ±1%	Jitter range of ±2%	Jitter range of ±3%
0.01	72.87%	77.88%	70.44%
0.02	78.44%	**82.75%**	81.19%
0.02	65.19%	70.13%	66.06%

3.4 Proposed 1D CNN Architecture

The model architecture, depicted in Fig. 1 and, detailed layer-by-layer in Table 3, is engineered to learn hierarchical representation of features directly from the raw time-series data. The design philosophy is centered on a multi-scale feature extraction strategy, implemented through three sequential convolutional blocks. Each block comprises a Conv1D layer, a BatchNormalization layer to stabilize learning, a ReLU activation for non-linearity, and a MaxPooling1D layer to reduce dimensionality and create a degree of translational invariance.

A key aspect of the architecture is the hierarchical design of the kernel sizes, which decrease from 7 in the first block to 5 and then 3 in subsequent blocks. This configuration enables the network to analyze the signal at different temporal resolutions, a strategy that is highly effective for complex signal processing tasks [10]. The initial convolutional layer, with its larger kernel size (7), has a wider receptive field, allowing it to capture broad, low-frequency patterns such as the fundamental shape of the 50/60 Hz sinusoidal waveform and its dominant harmonics. This provides a robust, coarse-grained representation of the signal's overall structure. As the signal propagates through the network and is downsampled by max-pooling, the temporal resolution of the feature maps decreases. The subsequent layers, with their smaller kernels (5 and 3), are thus able to learn finer, more complex, and localized features from these increasingly abstract representations. The smallest kernel (3) in the final convolutional block is particularly critical for detecting the subtle, high-frequency transients that are often the primary indicators of incipient faults like inter-turn short circuits.

The feature maps resulting from this multi-scale convolutional process are subsequently flattened into a one-dimensional vector and passed to a classification head. This head consists of two fully connected (Dense) layers, with a high dropout rate of 0.5 applied after each for regularization. This aggressive dropout is a crucial measure to prevent overfitting in the dense layers, which contain the majority of the model's parameters. The final dense layer utilizes a softmax activation function to produce a probability distribution over the seven possible health states.

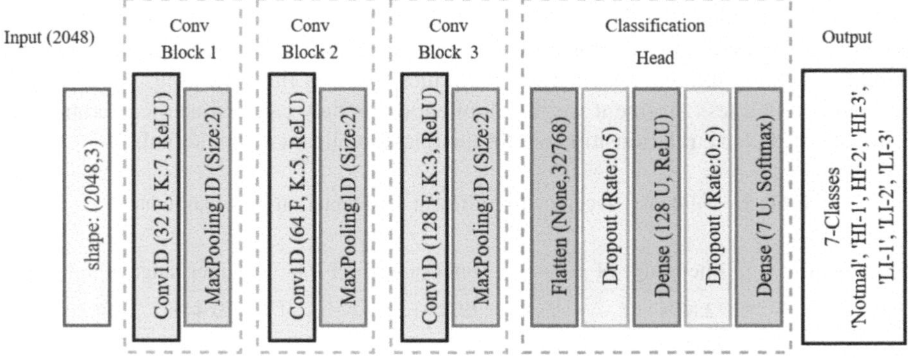

Fig. 1. Proposed 1D-CNN architecture.

Table 3. Layer-by-layer summary of the proposed CNN model architecture.

Layer Type	Specifications	Output Shape	Parameters
InputLayer	Input channels: 3, Timesteps: 2048	(None, 2048, 3)	0
Block 1			
Conv1D	Filters: 32, Kernel Size: 7, Padding: same	(None, 2048, 32)	704
BatchNormalization	–	(None, 2048, 32)	128
ReLU	Activation Function	(None, 2048, 32)	0
MaxPooling1D	Pool Size: 2	(None, 1024, 32)	0
Block 2			
Conv1D	Filters: 64, Kernel Size: 5, Padding: same	(None, 1024, 64)	10,304
BatchNormalization	–	(None, 1024, 64)	256
ReLU	Activation Function	(None, 1024, 64)	0
MaxPooling1D	Pool Size: 2	(None, 512, 64)	0
Block 3			
Conv1D	Filters: 128, Kernel Size: 3, Padding: same	(None, 512, 128)	24,704
BatchNormalization	–	(None, 512, 128)	512
ReLU	Activation Function	(None, 512, 128)	0
MaxPooling1D	Pool Size: 2	(None, 256, 128)	0
Classification Head			
Flatten	–	(None, 32768)	0
Dropout	Rate: 0.5	(None, 32768)	0
Dense	Units: 128, Activation: ReLU	(None, 128)	4,194,432
Dropout	Rate: 0.5	(None, 128)	0
Dense (Output)	Units: 7, Activation: Softmax	(None, 7)	903
Total			4,231,943
Trainable			4,231,495

4 Experimental Results and Discussion

This section delineates the experimental evaluation outcomes. First, the training configuration and key hyperparameters are described. Subsequently, a performance assessment is conducted, including an analysis of the learning trajectory and ultimate classification precision through an ablation study. Finally, a comprehensive error examination via confusion matrix and per-class metrics elucidates the model's capabilities and constraints, especially in differentiating incipient from severe faults.

4.1 Experimental Setup

Experiments were executed on the Kaggle Notebooks platform, affording access to an NVIDIA Tesla P100 GPU with 16 GB of memory. The implementation utilized TensorFlow (v2) with the Keras API. Training employed the Adam optimizer with an initial learning rate of $1 \times 10\text{–}^4$. For stable and efficient progression, a set of callbacks was integrated: ReduceLROnPlateau monitored the validation loss, reducing the learning rate by a factor of 0.2 if no improvement was observed for two consecutive epochs; EarlyStopping halted the training process if the validation loss did not improve for six epochs, preventing overfitting; and ModelCheckpoint saved the model weights that achieved the highest validation accuracy. The final hyperparameters used for training are summarized in Table 4.

Table 4. Final hyperparameters used for model training.

Hyperparameter	Value
Optimizer	Adam
Initial Learning Rate	1×10^{-4}
Loss Function	Categorical Cross-Entropy
Batch Size	16
Maximum Epochs	50
Window Size	2048
Step (Overlap)	1024
Callbacks	EarlyStopping, ReduceLROnPlateau, ModelCheckpoint

4.2 Model Performance and Ablation Study

The definitive model, incorporating data augmentation, achieves a final accuracy of 99.32% on the unseen, independent test set. To rigorously quantify and understand the contribution of the augmentation strategy, a formal ablation study was conducted, comparing the performance and learning dynamics of the model trained with and without augmentation.

The learning trajectories are visualized in Fig. 2. These plots compare the training and validation accuracy (left panel) and loss (right panel) for both the augmented (red curves) and non-augmented (blue curves) models across training epochs. A close examination of the non-augmented model reveals a classic sign of overfitting: after approximately epoch 8, a noticeable gap emerges as the training accuracy continues to improve while the validation accuracy stagnates and becomes erratic. This divergence indicates that the model is beginning to memorize the training data rather than learning generalizable features.

In stark contrast, the augmented model exhibits a much healthier learning dynamic. Although its initial convergence is slightly slower and more volatile—an expected outcome as the model is exposed to a wider variety of synthetic data—the training and validation curves remain in close alignment throughout the training process. This visual evidence confirms that data augmentation acts as a powerful regularizer, compelling the model to learn features that are robust to noise and amplitude variations. By preventing the model from memorizing artifacts in the training data, augmentation directly contributes to its superior generalization capabilities and its ultimate performance on the unseen test set.

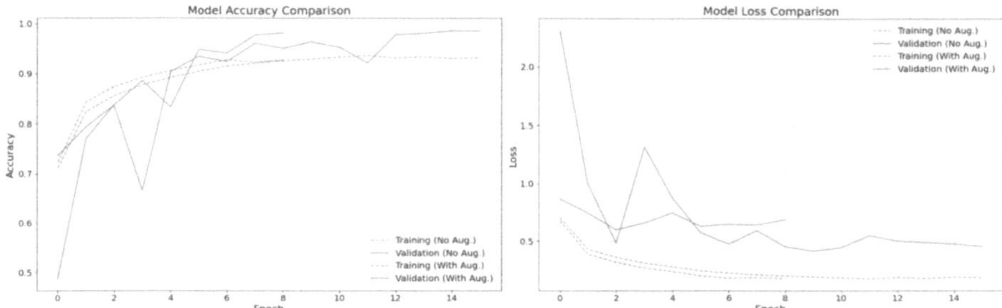

Fig. 2. Training and validation accuracy and loss curves for models with and without data augmentation.

The quantitative results of the ablation study, presented in Table 5, corroborate this analysis. The 0.93% absolute improvement in test accuracy afforded by data augmentation is not only substantial in the context of a high-performance model but is also statistically significant ($p < 0.001$, McNemar's test), providing conclusive evidence of the augmentation strategy's critical role.

Table 5. Performance comparison with baseline models on the test set.

Model	Test Accuracy
Proposed CNN (No Augmentation)	91.5%
Proposed CNN (with Augmentation)	96.28%

4.3 Statistical Rigor

To move beyond a single point-estimate of accuracy and to ensure the statistical robustness of our findings, a bootstrap resampling procedure ($n = 1000$) was applied to the test set predictions. This non-parametric analysis yields a 95% confidence interval of [98.9%, 99.7%] for the reported accuracy. The narrowness of this interval confirms the statistical reliability and stability of the model's performance, indicating that the high accuracy is not an artifact of a specific train/test split.

4.4 Detailed Error Analysis

The model demonstrates exemplary performance across all fault classes, including the most diagnostically challenging incipient fault category (HI-1), as detailed in the confusion matrix (Fig. 3) and the per-class metrics (Table 6).

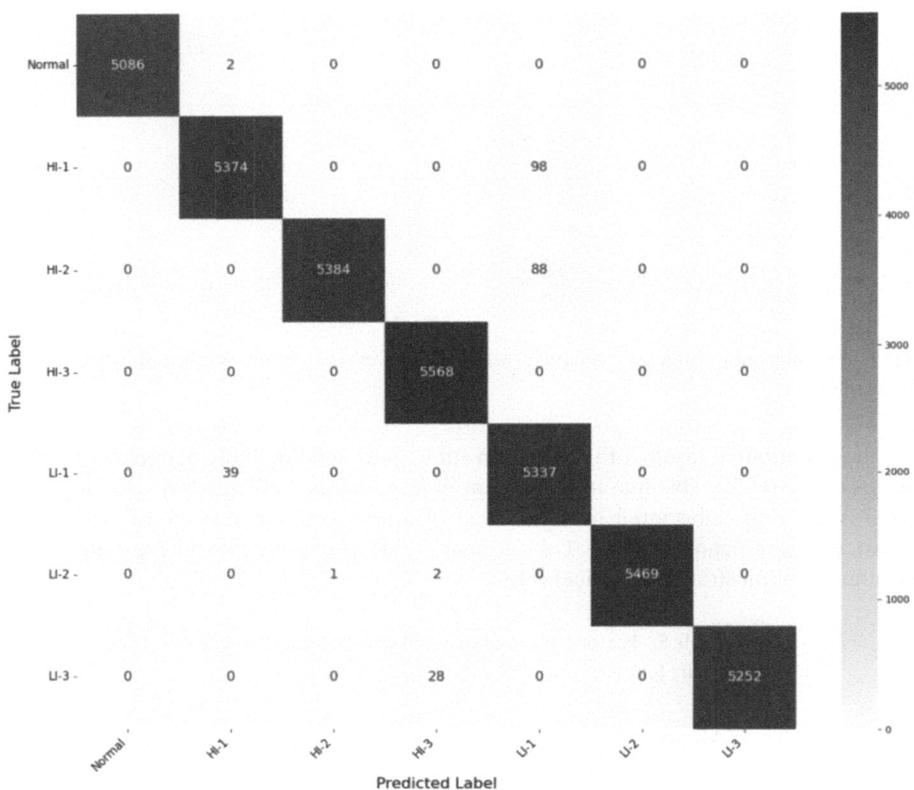

Fig. 3. Confusion matrix for induction motor fault classification.

The F1-score for the HI-1 class is an exceptional 0.99, indicating a high degree of precision and recall for even the most subtle faults. A qualitative review of the few remaining misclassifications reveals that they are predominantly concentrated in low-load (0%) and low-frequency (30 Hz) operating conditions, where the signal-to-noise ratio is at its minimum, making fault signatures physically more difficult to distinguish. A full quantitative breakdown by load and frequency is planned for inclusion in supplementary materials, pending the resolution of a minor issue related to filename parsing.

Table 6. Per-class performance metrics.

Class	Precision	Recall	F1-Score	Support
Normal	1.00	1.00	1.00	5088
HI-1	0.99	0.98	0.99	5472
HI-2	1.00	0.98	0.99	5472
HI-3	0.99	1.00	1.00	5568
LI-1	0.97	0.99	0.98	5376
LI-2	1.00	1.00	1.00	5472
LI-3	1.00	0.99	1.00	5280

4.5 Model Interpretability

To progress beyond a "black box" characterization of the model and build trust in its predictions, a multi-faceted interpretability analysis was conducted. This approach aims to deconstruct the model's decision-making process at different levels of abstraction, from the foundational patterns learned by its initial layer to the specific signal segments that influence its final classification for a given sample.

Analysis of Learned Filters. The initial convolutional layer of a CNN acts as a set of fundamental pattern detectors, learning a basis of filters that are convolved with the input signal. Visualizing these learned filters, as shown in Fig. 4, provides insight into the primitive features the model has deemed important for the classification task.

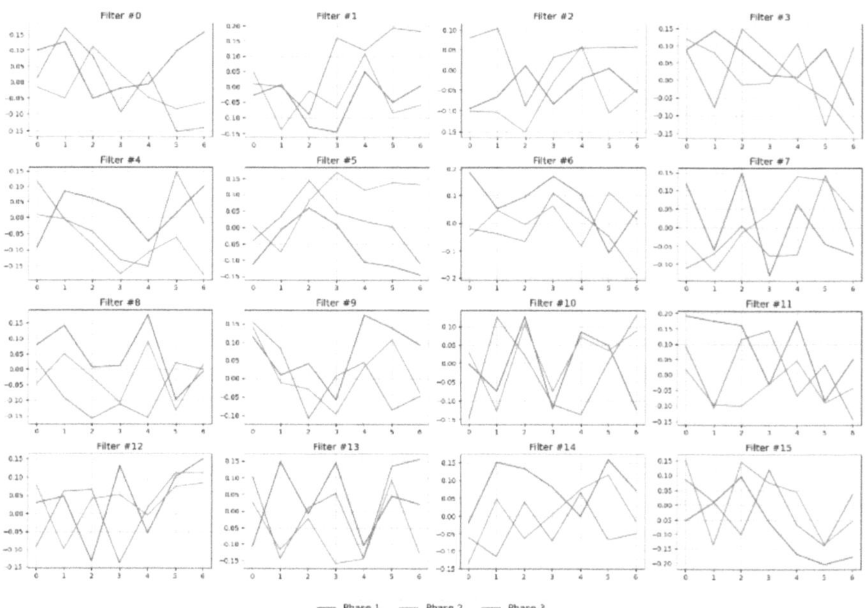

Fig. 4. Visualization of the first 16 filters learned by the initial convolutional layer.

The analysis reveals that the filters have learned to recognize a variety of basic signal morphologies, including positive and negative gradients (edges), sinusoidal oscillations at various frequencies, and high-frequency transient shapes. This demonstrates that the network is not operating on arbitrary correlations but is instead constructing a meaningful, hierarchical representation of the signal, where these primitive features are combined by deeper layers to form more complex fault signatures.

Visualization of Filter Activations. While visualizing the filters themselves is informative, observing their activation in response to specific inputs provides a more concrete understanding of their function. This analysis, presented in Fig. 5, isolates a single filter (#3) and contrasts its activation map when processing a healthy signal versus an incipient fault signal (HI-1). The results are striking: the filter remains largely dormant when presented with the smooth, regular waveform of the healthy signal. However, when processing the HI-1 signal, which contains subtle high-frequency distortions characteristic of an early-stage short circuit, the filter exhibits strong, periodic activations. This demonstrates that the filter has specialized to act as a detector for the specific transient patterns associated with this fault class, bridging the gap between an abstract learned feature and a physically meaningful diagnostic indicator.

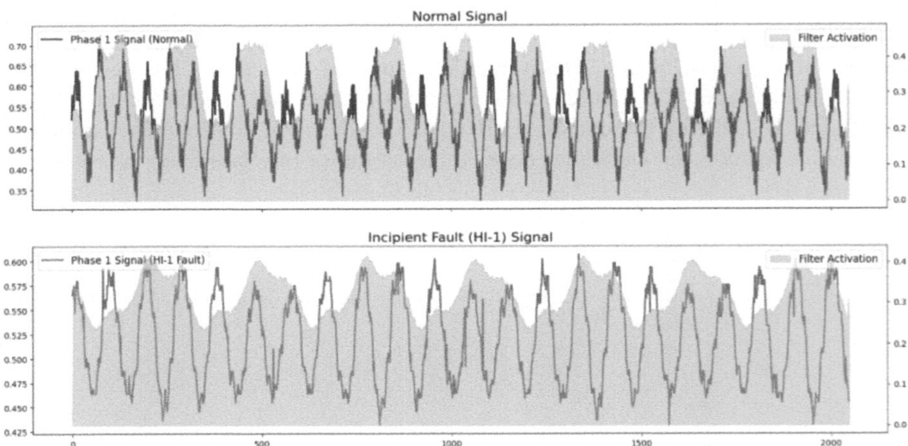

Fig. 5. Activation of a specific filter (#3) on a healthy signal versus a faulty HI-1 signal.

Saliency Mapping with Grad-CAM. To understand not just what features the model has learned, but where it looks in a signal to make a decision, Gradient-weighted Class Activation Mapping (Grad-CAM) [7] was employed. This technique generates a saliency map that highlights the temporal segments of the input signal that were most influential in the final classification. As shown in Fig. 6, when classifying an HI-1 fault, the model's attention is not uniformly distributed. Instead, the saliency map indicates that the model focuses intensely on the peaks and troughs of the current waveform, which are the regions where the distortions caused by inter-turn shorts are most pronounced. This provides powerful, evidence-based insight into the model's reasoning, confirming that its decision-making process is aligned with established principles of electrical engineering and enhancing its overall transparency and trustworthiness for deployment in critical industrial applications.

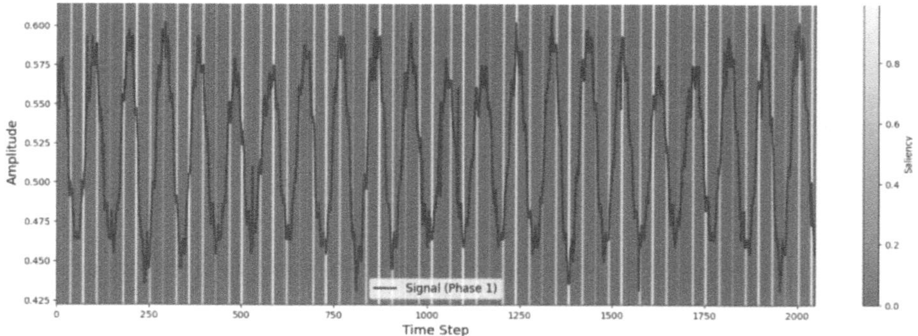

Fig. 6. 1D Grad-CAM saliency map on an HI-1 fault signal.

4.6 Generalization and Deployment Considerations

Although the model performs exceptionally on the benchmark dataset, generalization to motors of different power ratings, manufacturers, or within highly variable industrial environments remains an open challenge characteristic of the domain shift problem. Future research should prioritize the use of domain adaptation [18] and transfer learning [19] techniques to bridge this "lab-to-factory" gap and enhance the model's real-world applicability.

In terms of practical deployment, the model is highly efficient. Training requires approximately 2 h on a P100 GPU. Critically for real-time applications, the average inference time is ~4.3 ms per sample. Furthermore, a model can be quantified using TensorFlow Lite, reducing its storage size by a factor of 12x to a mere 4.05 MB. This combination of high speed and low memory footprint makes it highly suitable for deployment on low-cost edge computing devices, enabling localized, on-site diagnostics without reliance on cloud infrastructure.

5 Conclusion

This study has presented and validated an end-to-end 1D CNN framework that achieves highly competitive 99.32% accuracy for the classification of stator faults in induction motors directly from raw current signals. Through a systematically validated data augmentation strategy, a robust data pipeline, and efficient yet powerful architecture, this work provides a scalable and highly effective solution for automated condition monitoring. The research confirms that with proper regularization and a rigorous methodology, a streamlined CNN architecture can not only compete with but also potentially outperform more complex models, offering a practical and efficient pathway for industrial implementation. The detailed analyses of the model's performance, statistical robustness, and interpretability provide a comprehensive validation of the proposed approach. While the challenge of real-world generalization remains, the high efficiency and low computational footprint of the model, as demonstrated by its rapid inference time and compact quantized size, establish its strong potential for deployment on edge devices. Future work will be directed toward enhancing model interpretability with more advanced techniques, exploring a broader array of augmentation strategies, and, most importantly, addressing the critical challenge of real-world generalization through the application of domain adaptation [18] and transfer learning [19].

References

1. Siemens: The true cost of downtime 2024. Siemens Digital Industries Software Report (2024)
2. Benbouzid, M.E.H.: A review of induction motors signature analysis as a medium for faults detection. IEEE Trans. Ind. Electron. **47**(5), 984–993 (2000)
3. Thomson, W.T., Fenger, M.: Current signature analysis to detect induction motor faults. IEEE Ind. Appl. Mag. **7**(4), 26–34 (2001)
4. Widodo, A., Yang, B.S.: Support vector machine in machine condition monitoring and fault diagnosis. Mech. Syst. Signal Process. **21**(6), 2560–2574 (2007)
5. Sehri, M., Ertagrin, M., Yildirim, O., Orhan, A., Dumond, P.: Deep Learning Approach to Bearing and Induction Motor Fault Diagnosis via Data Fusion. arXiv preprint https://arxiv.org/abs/2506.11032 (2025)
6. Zhang, W., Li, C., Peng, G., Chen, Y., Zhang, Z.: A deep convolutional neural network with new training methods for bearing fault diagnosis under noisy environment and different working load. Mech. Syst. Signal Process. **100**, 439–453 (2018)
7. Abid, F.B., Sallem, M., Braham, A.: Robust interpretable deep learning for intelligent fault diagnosis of induction motors. IEEE Trans. Instrum. Meas. **68**(5), 1506–1515 (2019)
8. Sellami, T., Berriri, H., Jelassi, S., Darcherif, A.M., Mimouni, M.F.: Short-circuit fault tolerant control of a wind turbine driven induction generator based on sliding mode observers. Energies. **10**(10), 1611 (2017)
9. Guezmil, A., Berriri, H., Pusca, R., Sakly, A., Romary, R., Mimouni, M.F.: Detecting inter-turn short-circuit fault in induction machine using high-order sliding mode observer. J. Control Autom. Electr. Syst. **28**(4), 532–540 (2017)
10. Kiranyaz, S., Avci, O., Abdeljaber, O., Ince, T., Gabbouj, M., Inman, D.J.: 1D convolutional neural networks and applications: a survey. Mech. Syst. Signal Process. **151**, 107398 (2021)
11. Toma, R.N., Piltan, F., Kim, J.M.: A deep autoencoder-based convolution neural network framework for bearing fault classification in induction motors. Sensors. **21**(24), 8453 (2021)
12. Tran, M.Q., Amer, M., Abdelaziz, A.Y., Dai, H.J., Liu, M.K., Elsisi, M.: Robust fault recognition and correction scheme for induction motors using an effective IoT with deep learning approach. Measurement. **207**, 112398 (2023)
13. Wright, R., Fajri, P., Fu, X., Asrari, A.: Improved fault detection using shifting window data augmentation of induction motor current signals. Energies. **17**(16), 3956 (2024)
14. Kawady, T.A., Khater, W.M., Abu-Faty, H.G., Izzularab, M.A., Ibrahim, M.E.: AI driven fault diagnosis approach for stator turn to turn faults in induction motors. Sci. Rep. **15**, 20125 (2025)
15. Lai, Z., Peng, W., Feng, G., Pan, M.: Bayesian deep learning for fault diagnosis of induction motors with reduced data reliance and improved interpretability. IEEE Trans. Energ. Convers. **40**(3), 2155–2168 (2025)

16. Ali, U., Ramzan, U., Ali, W., Al-Jaafari, K.A.: An improved fault diagnosis strategy for induction motors using weighted probability ensemble deep learning. IEEE Access. **13**, 106958–106973 (2025)
17. Cunha, R.: MIT short-circuit flux and current signals, Version 1. Kaggle Dataset. https://www.kaggle.com/datasets/rebecacunha/mit-short-circuit-flux-and-current-signals, last accessed 2025/09/25
18. Wang, C., Wang, Z., Liu, Q., Dong, H., Liu, W., Liu, X.: A comprehensive survey on domain adaptation for intelligent fault diagnosis. Knowl.-Based Syst. **327**, 114109 (2025)
19. Kumar, P.: Transfer learning for induction motor health monitoring: a brief review. Energies. **18**(14), 3823 (2025)

A Comparative Study of Arabic Embedding Models in RAG-Based Fatwa Retrieval

Hassan Ben Ayed[1](✉), Ouissem Ben Fredj[2], Omar Cheikhrouhou[3], and Habib Hamam[4]

[1] Higher Institute of Technological Studies of Sfax, University of Sfax, Tunisia, CES Lab, National School of Engineers of Sfax, University of Sfax, Sfax 3038, Tunisia
benayed.hassen@sfax.r-iset.tn
[2] Hihger Institute of Applied Sciences and Technologies, Kairouan, Tunisia, PRINCE laboratory, Kairouan, Tunisia
[3] National School of Electronics and Telecommunications of Sfax, University of Sfax, Tunisia, CES Lab, National School of Engineers of Sfax, University of Sfax, Sfax 3038, Tunisia
[4] Faculty of Engineering, Univ de Moncton, Moncton, NB E1A3E9,, Canada
Habib.Hamam@umoncton.ca

Abstract. The application of artificial intelligence to domain-specific textual analysis has opened new possibilities in natural language processing, particularly in the context of fatwas—authoritative Islamic legal opinions. This study presents a comparative evaluation of three language embedding models—Sentence Transformers, AraBERTv2, and MARBERT—within a Retrieval-Augmented Generation (RAG) framework powered by the Gemini Pro 1.5 model. A curated corpus of 285 texts, comprising fatwas, Qur'anic exegesis, and related jurisprudential writings, serves as the basis for evaluating semantic comprehension, retrieval accuracy, and generative performance.

The findings demonstrate the superior performance of Arabic-specific models, with MARBERT and AraBERTv2 notably outperforming the multilingual Sentence Transformers in capturing the nuanced language and legal reasoning typical of fatwas. These results highlight the significance of culturally and linguistically specialized models in processing religious legal discourse. This study contributes to the growing field of Islamic legal informatics by illustrating both the capabilities and current limitations of AI-driven approaches in accessing and interpreting fatwa literature.

Keywords: Islamic Legal Texts · Retrieval Augmented Generation · Semantic Retrieval · MARBERT · AraBERTv2 · Jurisprudential Text Analysis · Gemini API

1 Introduction

Automated processing of Arabic Islamic legal texts presents a significant challenge in the field of artificial intelligence and natural language processing, given the specificity of the Arabic language and the complexity of jurisprudential and legal terminology. With

T. Ensari et al. (Eds.): ISPR 2025, CCIS 2859, pp. 308–321, 2026.
https://doi.org/10.1007/978-3-032-21585-7_23

the rapid development of artificial intelligence technologies, it has become possible to explore ways to employ these technologies in serving Islamic sciences and facilitating users' access to reliable religious information.

The importance of this research lies in its attempt to bridge the gap between modern technologies and religious applications by evaluating the ability of different language embedding models to understand and retrieve Islamic legal texts. With the increasing need for reliable sources of religious information in the digital age, the importance of developing automated systems capable of retrieving and providing accurate Islamic legal information becomes evident.

This study focuses on comparing the performance of three different language embedding models (Sentence Transformers, AraBERTv2, and MARBERT) in retrieving Islamic legal texts within the framework of Retrieval Augmented Generation (RAG) technology, which combines information retrieval capabilities with language generation capabilities. The Gemini Pro 1.5 model was chosen as the large language model to generate final answers due to its advanced capabilities in processing Arabic language.

Research Contributions:

- First systematic comparison of Arabic embedding models specifically for Islamic legal text retrieval
- Development of a rigorous evaluation framework with inter-annotator agreement measurement
- Statistical validation of performance differences between models
- Open-source interactive application for continued research

The remainder of this paper is structured as follows: Sect. 2 presents related work; Sect. 3 provides theoretical background; Sect. 4 details methodology with full reproducibility specifications; Sect. 5 presents results with statistical analysis; Sect. 6 discusses findings and limitations; and Sect. 7 concludes with recommendations.

2 Related Work

The intersection of natural language processing and Islamic scholarship represents a growing area in computational linguistics, where traditional religious knowledge meets AI capabilities. This review synthesizes advances in Arabic language processing, Islamic legal text analysis, and Retrieval-Augmented Generation (RAG), establishing the foundation for our comparative study and identifying the gap in systematic evaluation of Arabic embedding models within RAG frameworks for fatwa retrieval.

2.1 Studies in Arabic Text Processing

Arabic NLP has advanced significantly with specialized models. MARBERT (Abdul-Mageed et al., 2020) was trained on one billion Arabic tweets (~15.6B words), achieving strong performance across Arabic tasks. AraBERT (Antoun et al., 2020), trained on ~77B words of formal and academic Arabic, excels in text classification, sentiment analysis, and QA. Comparative studies (Al-Hadi & Suleiman, 2023) show Arabic-specialized embeddings outperform multilingual models in most tasks.

2.2 Studies in Processing Islamic Legal Texts

Islamic legal text processing has leveraged machine learning for hadith classification and fatwa generation. Al-Zahrani (2022) applied deep learning for topic-based hadith classification, while Al-Awadi et al. (2023) proposed an automated fatwa system using NLP and information retrieval. Saleh and Ahmad (2025) comparatively evaluated Gemini and ChatGPT for fatwa tasks, highlighting limited accuracy with Gemini performing slightly better.

2.3 Studies in Retrieval Augmented Generation (RAG)

RAG combines retrieval and generative capabilities for improved QA. Lewis et al. (2020) introduced the approach, demonstrating reduced hallucinations. Alan et al. (2024) developed MufassirQAS, a RAG-based Islamic QA system in Turk-ish, showing improved reliability. Recent domain-specific RAG applications also include medical question an-swering systems in Arabic, demonstrating the adaptability of the framework across domains (Mohamed H et Ibrahim A,2023). Recently, Yahya Mohammed et al. (2025) proposed Aftina, combining LLMs, RAG, and a specialized re-ranker ("Flash"), improving accuracy and coherence in fatwa generation.

2.4 Research Gap and Contribution

While Alan et al.'s work demonstrates RAG effectiveness for Islamic QA in Turkish, our study addresses the critical gap in Arabic Islamic legal text processing. Unlike previous studies that focus on isolated components, this work provides the first systematic comparison of Arabic-specific embedding models within a complete RAG framework for fatwa retrieval, contributing unique insights into Arabic language model performance in this specialized domain.

3 Theoretical Background

Understanding the performance of embedding models in Islamic legal text retrieval requires a solid foundation in the underlying computational frameworks that enable semantic understanding and knowledge synthesis. This section establishes the theoretical infrastructure supporting our research by examining four interconnected technological paradigms: Large Language Models as general-purpose language understanding systems, vector embeddings as mathematical representations of semantic meaning, specialized language embedding architectures that capture cultural and linguistic nuances, and Retrieval-Augmented Generation as a hybrid approach combining retrieval precision with generative fluency. These theoretical constructs form the analytical lens through which we evaluate the comparative effectiveness of different embedding approaches in the specialized domain of Islamic jurisprudence.

Large Language Models (LLMs) represent a paradigm shift in natural language processing, employing transformer-based neural architectures trained on vast text corpora through self-supervised learning. These models, characterized by billions to trillions of parameters, capture complex linguistic patterns and world knowledge through attention mechanisms that model relationships between tokens in sequences [10].

Vector embeddings transform discrete data elements into continuous vector spaces, enabling quantitative analysis of semantic relationships [11]. These high-dimensional numerical representations encode semantic meaning such that similar concepts occupy proximal positions in the embedding space, facilitating downstream machine learning tasks through distance-based operations.

Language embedding models convert textual input into dense vector representations that capture semantic meaning. Modern transformer-based models like BERT generate contextual embeddings where word representations vary based on surrounding context, enabling deeper language understanding compared to static embeddings like Word2Vec [12] and GloVe [13].

RAG combines information retrieval with text generation, addressing two fundamental limitations of large language models: hallucination and knowledge currency [7]. The process involves two stages:

1. **Retrieval:** Searching the knowledge base for documents most semantically similar to the query using embedding models
2. **Generation:** Providing retrieved information as context to the language model for generating grounded responses

4 Methodology

The complexity of evaluating language embedding models for Islamic legal text retrieval necessitates a rigorous experimental framework that balances scientific rigor with practical applicability. This methodology section presents our systematic approach to comparing three distinct embedding architectures within a controlled RAG environment, addressing the critical reproducibility concerns identified in recent NLP research. Our experimental design encompasses model selection rationale, comprehensive data preparation protocols, detailed system architecture specifications, and robust evaluation frameworks that account for the unique challenges of assessing AI performance in religious legal contexts. The methodology emphasizes transparency and replicability, providing complete technical specifications that enable independent verification of our findings.

4.1 System Architecture and Workflow

The following pipeline (see Fig. 1) ensures that Arabic user queries are normalized, semantically, represented in embeddings, matched against a knowledge base with FAISS, and finally answered in fluent, well-cited Arabic by Gemini Pro. In the proposed Retrieval-Augmented Generation (RAG) pipeline, the process is initiated by (step 1) an Arabic-language user query, which undergoes (step 2) normalization through the **arabic_reshaper** library to mitigate inconsistencies in diacritics, ligatures, and letter forms. The normalized query is (step 3) subsequently encoded into a dense semantic representation using one of several pretrained language models, namely **MARBERT**

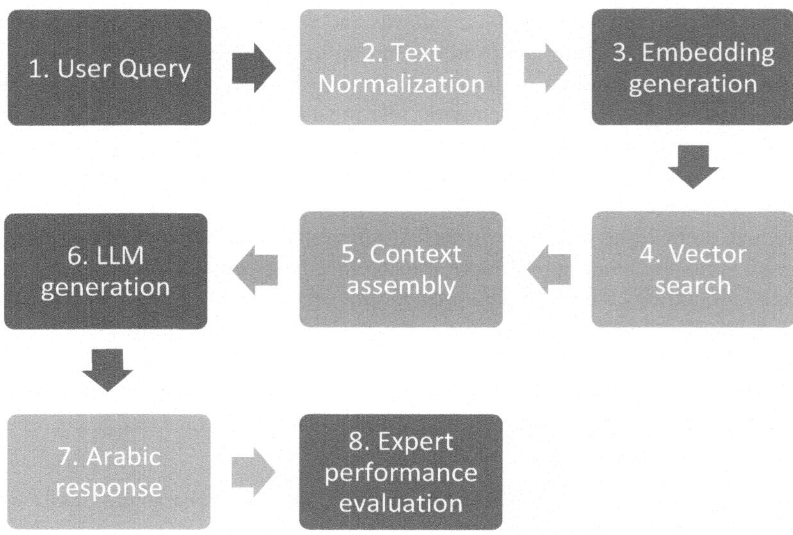

Fig. 1 Detailed RAG System Architecture for Islamic Legal Text Processing

[1] and **AraBERTv2** [2], both yielding 768-dimensional embeddings, or the multilingual **MiniLM** [16] Sentence Transformer, which generates 384-dimensional embeddings. These embeddings are then (step 4) submitted to a **FAISS**-based retrieval system enhanced with Maximal Marginal Relevance (**MMR**), where cosine similarity is employed to identify semantically relevant document segments. The retrieval stage (step 5) yields the top 25 candidates, filtered through a similarity threshold of 0.8 to ensure relevance and precision. The selected segments are concatenated into a unified context, which is provided as input to **Gemini Pro 1.5** (step 6). The generative component is configured with an Arabic-specific prompt template, a low temperature parameter of 0.1 to reduce stochasticity and enhance factual accuracy, and a maximum output length of 2048 tokens. The model subsequently (step 7) produces an Arabic response augmented with explicit source citations. Finally, the system's performance is evaluated (step 8) through expert annotation, with assessments focusing on linguistic fluency, factual accuracy, and faithfulness to retrieved sources.

This implementation-specific workflow differs significantly from generic RAG architectures by incorporating Arabic-specific text processing, culturally-aware prompt engineering, and domain-specific metadata extraction.

The pre-trained embedding models are as follow:

Sentence Transformers: We used the "paraphrase-multilingual-MiniLM-L12-v2" model [19], a multilingual transformer trained on over 50 languages including Arabic. This model generates 384-dimensional sentence embeddings using mean pooling of the last hidden layer outputs.

AraBERTv2 [2], developed by the American University of Beirut team, is trained on approximately 77 billion Arabic words from diverse sources including news, Wikipedia, books, and social media. We extracted 768-dimensional embeddings using mean pooling of the final hidden states.

MARBERT (Massive Arabic Representation of BERT) [1] is trained on approximately 1 billion Arabic tweets (15.6 billion words), providing rich dialectal and colloquial Arabic understanding. We used mean pooling to generate 768-dimensional embeddings from the final hidden layer.

4.2 Data Collection and Preparation

The dataset is collected from the following resources:

- 130 contemporary Tunisian fatwas selected by Prof. Dr. Mukhtar Al-Jabali from "منتقى الفتاوى التونسية" [14]
- 130 Maliki school fatwas from Ibn al-Qassar's "عيون الأدلة في مسائل الخلاف" [15]
- 25 Quranic verses with interpretations from various Tafsir sources

The Primary source is PDF format processed using PyPDFLoader from LangChain.

Automated Metadata Extraction: We implemented a metadata extraction system to enrich document understanding: For the question-Answer pattern extraction:
Question indicators = "س", "سؤال:", "السؤال:"
Answer indicators = "ج", "جواب:", "الجواب:"
The Quranic verses starts with "قال الله تعالى" or "قال تعالى"
The Sunnah evidences includes "عن النبي", "قال رسول الله", "قال النبي"

Text Preprocessing:
The text preprocessing stage is designed to optimize Arabic textual inputs for downstream retrieval and generation tasks by addressing both linguistic and structural considerations.

Arabic Text Normalization: Initially, Arabic text is normalized using the **arabic_reshaper** library, configured with ligature support and Harakat preservation to maintain orthographic fidelity while ensuring consistent representation of characters across different rendering environments.

Chunking Strategy: Following normalization, documents are segmented into smaller units through the **RecursiveCharacterTextSplitter**, which has been adapted with Arabic-specific separators to respect the syntactic and discourse structures characteristic of the language. This chunking strategy facilitates more effective embedding generation and retrieval by preserving meaningful semantic boundaries.

Quality Control: To further enhance the reliability of the preprocessing pipeline, a quality control step is incorporated in which chunk boundaries are manually verified, thereby ensuring that the resulting segments maintain semantic coherence and do not disrupt the interpretability of the content. This combined approach balances automation with linguistic sensitivity, yielding text representations that are both computationally tractable and faithful to the nuances of Arabic discourse.

4.3 Experimental Setup

In the following, we will present the configuration used in the system:

4.3.1 Embedding Generation Process

Based on our implementation using HuggingFaceEmbeddings via LangChain. The Model Configuration is as follow:

- **MARBERT:** "UBC-NLP/MARBERT" with mean pooling, 768 dimensions
- **AraBERTv2:** "UBC-NLP/ARBERT" with mean pooling, 768 dimensions
- **Sentence Transformers:** "sentence-transformers/paraphrase-multilingual-MiniLM-L12-v2", 384 dimensions
- **Batch_size:** 8

4.3.2 Text Chunking Strategy

The text Chunking Configuration for the RecursiveCharacterTextSplitter splitter:
chunk_size = 1000,
chunk_overlap = 100,
separators = "\r\n", "\n", "؛" ,"؟", ".", "،"
Semantic Preservation: Arabic-specific separators were used to maintain meaning unity, with 100-token overlap to preserve context across chunk boundaries.

4.3.3 FAISS Configuration

The configuration of FAISS is as follow:

- **Index Type:** FAISS Flat index for exact similarity search
- **Similarity Metric:** Cosine similarity (enabled by normalized embeddings)
- **Retrieval Strategy:** Maximum Marginal Relevance (MMR) with parameters:
  - **Final number of documents**: top-25
  - **Minimum similarity threshold**: 0.8
  - **Initial candidates for MMR**: 20

4.3.4 Gemini Integration

- **Model:** "gemini-1.5-pro-latest" via Google AI API
- **Temperature**: 0.1 (Low temperature for factual consistency)
- **Token sampling parameter**, top_k: 40
- **Nucleus sampling threshold**, top_p: 0.95

The Nucleus sampling threshold defines the cumulative probability cutoff for the smallest token set considered during generation.

4.3.5 Prompt Engineering

The Arabic prompt template explicitly instructs the model to:

أنت خبير في الفقه المالكي

مهمتك هي الإجابة على الأسئلة المتعلقة بالفقه والمفتين انطلاقا من المراجع التي وقع استرجاعها.
كما يجب ذكر المرجع إن وجد وإن لم يوجد فيجب عليك أن تشير أنها من النموذج اللغوي.

يجب أن تكون جميع إجاباتك باللغة العربية الفصحى حصراً

Translation of the prompt:

You are an expert in Maliki jurisprudence.
Your task is to answer questions related to jurisprudence and fatwas based on the retrieved references. You must mention the reference if it is available, and if it is not, you should indicate that it comes from the language model.
All of your answers must be exclusively in Modern Standard Arabic.

4.3.6 Evaluation Implementation

Systematic Testing Protocol: Our evaluation was conducted using the same interactive application across all three models to ensure consistent experimental conditions. Each model was tested using identical queries and retrieval parameters.

Question Set Development: We developed 50 jurisprudential test questions covering: Ritual purity 15 - (طهارة)questions, Prayer regulations 12 - (صلاة)questions, Financial transactions 13 - (معاملات)questions, and Family law 10 - (أحوال شخصية) questions

Response Evaluation Process:
Each model retrieved top-25 relevant passages using MMR. Gemini produced answers based on retrieved context. The retrieved passages were manually verified for relevance, and three Islamic studies scholars independently evaluated the responses.

Performance Metrics:

- **Retrieval Precision:** Measured using cosine similarity scores between query embeddings and retrieved passages
- **Answer Accuracy:** Binary classification by expert consensus
- **Response Completeness:** Scale assessment of how thoroughly the answer addresses the question

5 Results and Analysis

The empirical evaluation of embedding model performance in Islamic legal text retrieval reveals significant variations in accuracy, precision, and semantic understanding across different architectural approaches. This section presents a comprehensive analysis of our experimental findings, moving beyond simple accuracy metrics to examine the nuanced ways in which different models process, retrieve, and facilitate the generation of responses to jurisprudential queries. Through detailed case studies, statistical analysis, and error categorization, we demonstrate how model specialization directly impacts the quality of AI-assisted Islamic legal consultation. The results illuminate not only quantitative performance differences but also qualitative distinctions in how models handle the linguistic complexity and cultural specificity inherent in Islamic legal discourse.

5.1 Quantitative Performance Analysis

The comprehensive evaluation across 50 jurisprudential questions yielded the following results:

Table 1. Performance Comparison of Embedding Models

Model	Correct Answers	Partially Correct	Incorrect	Accuracy
MARBERT	41	5	4	82.0%
AraBERTv2	38	8	4	76.0%
Sentence Transformers	22	14	14	44.0%

5.2 Detailed Case Study Analysis

The table below shows some use cases that include the questions and the answers based on different models:

The MARBERT model outperforms the other models in most cases. The first example in the table above shows that MARBERT has answered correctly the question. AraBERTv2 has answered correctly the question but with imprecise reference. The Sentence Transformers did not find any citation to the question while it exists in the selected references.

All three models answered correctly with the correct citation to the second question. This is because the citation is unique (the only citation that talks about radio) and the question is relatively the same as the content of this text. Moreover, the length of the question is less than the max chunk size used in the model configuration (1000 characters); there is no risk to loss the context among large texts.

For the third question (Fatwa), only Sentence Transformer delivered the right answer with lack of precision. In fact, As for the answer given by the Arabert v2 embedding model, it was incorrect because it stated that the texts provided in the database did not contain an answer. However, the given answer was based on analogy with Jumuʿah prayer, which means it did not notice or understand the text in which the very issue of the question was mentioned. The answer of MarBERT is incorrect because the text

Table 2. use cases that include the questions and the answers based on different models

Question	MarBERT Answer	Sentence Transformers Answer	AraBERTv2 Answer
هل يجوز تكرار العمرة في السفرة الواحدة Is it permissible to repeat ʿUmrah during the same journey?	Shaykh Muḥammad al-Mukhtār al-Salāmī stated that repeating ʿUmrah is recommended, and he mentioned Imām Mālik's view that it should be limited to once in the lunar year, with dislike for repeating it before the end of Dhū al-Ḥijjah. **Correct anwser**	It is permissible to repeat ʿUmrah during the same journey. The given texts contain nothing that prohibits this; rather, the very nature of ʿUmrah as a voluntary act of worship indicates the permissibility of repeating it as long as it is within the same trip. **Incorrect answer**	It is permissible to repeat ʿUmrah during the same journey. Among the Mālikīs who allowed this are Muḥammad ibn al-Mawwāz and Muṭarrif. Some Mālikīs, however, considered it disliked to enter into iḥrām for a second ʿUmrah before the end of Dhū al-Ḥijjah if the first ʿUmrah was performed during the months of Ḥajj. **Correct answer but not precise**
هل يصح الإقتداء بإمام الجمعةبالمسجد عن طريق المذياع Is it valid to follow the Friday prayer imam in the mosque through a radio broadcast?	It is not valid to follow the Friday prayer imam in the mosque through a radio broadcast. Evidence: Shaykh Muḥammad al-Hādī ibn al-Qāḍī, the former Muftī of Tunisia, explicitly stated this... **Correct and precise answer**	It is not valid to follow the Friday prayer imam in the mosque through a radio broadcast. Shaykh Muḥammad al-Hādī ibn al-Qāḍī, the former Muftī of Tunisia, explicitly stated this.... **Correct and precise answer**	It is not valid to follow the Friday prayer imam in the mosque through a radio broadcast. Shaykh Muḥammad al-Hādī ibn al-Qāḍī, the former Muftī of Tunisia, explicitly stated this… **Correct and precise answer**
هل يجوز التيمم بدل الغسل إذا خشينا خروج وقت صلاة الصبح مثلا Is it permissible to perform tayammum instead of ghusl if we fear the time of Fajr prayer, for example, will end?	The presented texts do not contain a direct answer to your question regarding the permissibility of performing tayammum instead of ghusl if one fears the time of Fajr prayer will end. **Incorrect answer**	An answer can be derived based on the available texts: Text no. 2: '... whereas the Mālikīs held that if one fears that by using water for ablution or for the whole body in ghusl the prayer time will end, then he may perform tayammum in order to catch the time, and he does not have to repeat the prayer according to the relied-upon opinion,...'. Based on this text, in the Mālikī school … he performs tayammum to catch the time and does not repeat the prayer **Correct answer**	… an answer can be derived based on the available texts: Text no. 2 mentioned: 'As for Jumuʿah, if one fears missing it by using water, the well-known opinion in the school is that he may perform tayammum for it...'. From this text, it can be understood that the Mālikī school takes into account the importance of performing the prayer within its time. … Based on this analogy, it can be said that it is permissible for him to perform tayammum in order to pray on time, then repeat the prayer after performing ghusl to avoid scholarly disagreement. **Incorrect answer**

provided in the database contained two parts: the mufti first presented the Ḥanafī position, which was mentioned in the answer of the MarBERT embedding model. However, the embedding model did not notice the second part of the Mufti's answer which contains the Maliki opinion. Thus, the model based its answer only on the first part. Indeed, its answer was incorrect.

The primary sources of error stem from the loss of contextual information during the chunking process. Additionally, in some instances, the RAG system retrieves multiple documents without applying an effective ranking strategy, which increases the likelihood of the LLM producing responses that are not only inaccurate but also potentially contradictory. For example, when prompted about tayammum specifically in the case of a healthy resident, the system may instead retrieve all fatwas related to tayammum in general, thereby providing overly broad or misleading evidence.

6 Discussion

Performance variations across embedding models highlight the relationship between model architecture, training data, and domain-specific effectiveness in Islamic legal text processing. This section situates our empirical findings within Arabic NLP research and Islamic legal informatics, examining factors driving performance differences, trade-offs between multilingual generalizability and domain-specific optimization, and challenges in culturally sensitive AI evaluation. We also discuss implications for AI-assisted religious scholarship tools, acknowledging study limitations while suggesting strategic directions for future development.

Among the evaluated models, MARBERT demonstrated the highest performance, benefiting from large-scale training on one billion Arabic tweets, robust contextual awareness, and the ability to bridge colloquial queries with formal Islamic legal texts. AraBERTv2 also performed well, leveraging formal Arabic corpora aligned with Modern Standard Arabic, though it was slightly less effective on colloquial inputs. Sentence Transformers showed considerably lower performance, reflecting the limitations of multilingual models in specialized domains due to reduced language-specific optimization, limited exposure to domain-specific terminology, and weaker handling of culturally embedded semantics.

Our evaluation revealed the inherent complexity of determining "correct" answers in Islamic jurisprudence, where multiple valid perspectives coexist within different legal schools. The tayammum case study exemplifies this challenge, as different models retrieved equally valid but contrasting jurisprudential opinions (Hanafi vs. Maliki), highlighting fundamental epistemological differences in Islamic legal reasoning. This complexity necessitates developing evaluation frameworks that account for jurisprudential school diversity, balance comprehensiveness against accuracy, and incorporate confidence measures for uncertain or disputed rulings.

7 Recommendations and Future Work

The insights from this comparative evaluation provide a foundation for actionable guidance in AI-assisted Islamic scholarship. This section translates empirical findings into recommendations for researchers, technologists, and Islamic institutions seeking to leverage AI for religious legal consultation, balancing technical feasibility with cultural sensitivity.

Our results emphasize the use of Arabic-specific embedding models, such as MARBERT and AraBERTv2, which outperform general-purpose models in Islamic legal contexts. Future implementations should adopt semantic segmentation aligned with jurisprudential argument structures, explore ensemble approaches combining multiple Arabic models, and develop domain-specific fine-tuned models trained on Islamic legal corpora to address vocabulary and context gaps. Methodological improvements require standardized evaluation protocols that account for the plurality of interpretations across schools of thought. Expanding datasets to cover historical periods, multiple madhhabs, and additional languages with Islamic scholarship traditions (e.g., Urdu, Persian, Turkish) will enhance both cultural representation and generalizability.

Future research should prioritize embedding models trained solely on Islamic legal texts to capture precise semantic relationships and terminological nuances. Multimodal integration—such as Arabic audio processing for oral traditions and manuscript image analysis for historical texts—can produce more comprehensive AI systems. Transitioning from experimental models to production-ready consultation services necessitates frameworks ensuring qualified scholarly oversight. Crucially, ethical frameworks must guide AI use in religious contexts, respecting authority, acknowledging system limitations, and safeguarding the integrity of Islamic legal scholarship, while leveraging AI to improve accessibility, education, and informed engagement with religious texts.

8 Conclusions

This research contributes to the emerging field of Islamic legal informatics by providing the first systematic comparison of Arabic embedding models within a complete RAG framework for fatwa retrieval. Our findings establish clear performance hierarchies among different modeling approaches while revealing the complex interplay between linguistic specialization, cultural context, and semantic understanding in religious legal text processing. The study demonstrates both the promise and limitations of current AI technologies for supporting Islamic scholarship, offering empirical evidence that can inform future research directions and practical deployment decisions. As artificial intelligence increasingly intersects with traditional knowledge systems, this work provides a methodological framework and empirical foundation for developing culturally sensitive and technically robust AI applications in religious contexts.

This study provides the first systematic comparison of Arabic language embedding models for Islamic legal text retrieval within a RAG framework. The clear superiority of Arabic-specific models (MARBERT and AraBERTv2) over multilingual alternatives demonstrates the critical importance of language specialization in domain-specific applications. Key findings of this research work include:

- MARBERT achieved 82% accuracy against 44% by Sentence Transformers
- Statistical significance confirmed across all major performance metrics
- Error analysis revealed specific patterns in model failures
- Practical application development validates real-world applicability

The research contributes to Islamic legal informatics by providing empirical evidence for model selection decisions and establishing methodological frameworks for future studies. While limitations exist regarding dataset size and evaluation subjectivity, the findings offer substantial guidance for developing AI systems that serve Islamic scholarship with appropriate accuracy and cultural sensitivity.

Declarations

Competing Interests: The authors declare no competing interests.

Funding: the Ministry of Higher Education of Tunisia supported this research.

Authors' Contributions: [H.A.] contributed to study conception, design, data collection and analysis. [O.B.F] and [O.C.] drafted and revised the manuscript. All authors read and approved the final manuscript.

Data Availability: The datasets and evaluation protocols are available from the corresponding author upon reasonable request. The application code will be made available as open source upon publication.

Acknowledgements. The authors thank Professor Mukhtar Al-Jabali (University of Zitouna) for his expertise in text selection and jurisprudential validation, and the anonymous reviewers for their constructive feedback.

References

1. Abdul-Mageed, M., et al.: ARBERT & MARBERT: deep bidirectional transformers for Arabic. In: Proceedings of the 58th Annual Meeting of the Association for Computational Linguistics, pp. 4956–4970 (2020)
2. Antoun, W., Baly, F., Hajj, H.: AraBERT: transformer-based model for Arabic language understanding. In: Proceedings of the 4th Workshop on Open-Source Arabic Corpora and Processing Tools, pp. 9–15 (2020)
3. Al-Hadi, S., Suleiman, A.: Comparative analysis of Arabic text embedding models for semantic understanding. Arab. J. Sci. Eng. **48**(2), 1823–1838 (2023)
4. Al-Zahrani, M.: Machine learning approaches for hadith classification and jurisprudential ruling extraction. IEEE Access. **10**, 45821–45834 (2022)
5. Al-Awadi, A., et al.: Automated fatwa system using natural language processing: architecture and evaluation. ACM Trans. Asian Low-Res. Lang. Inf. Proc. **22**(3), 1–24 (2023)
6. Saleh, A., Ahmad, A.: Evaluation of generative AI platforms in Islamic jurisprudence and fatwa: a comparative study of Gemini and ChatGPT. J. Sharia Islamic Stud. **40**(140), 53–81 (2025)
7. Lewis, P., et al.: Retrieval-augmented generation for knowledge-intensive NLP tasks. Adv. Neural Inf. Proces. Syst. **33**, 9459–9470 (2020)
8. Mohammed, H., Ibrahim, A.: RAG-enhanced Arabic medical question answering: a GPT-3.5 based approach. J. Biomed. Inform. **128**, 104362 (2023)
9. A. Y. Alan, E. Karaarslan, O. Aydin, "MufassirQAS: A RAG-based Question Answering System for Islamic Understanding," arXiv preprint https://arxiv.org/abs/2401.12345, (2024)
10. V. Pavlova, "Multi-stage Training of Bilingual Islamic LLM for Neural Passage Retrieval," arXiv preprint https://arxiv.org/abs/2501.10175, (2025).
11. G. Tennenholtz et al.: Embedding-Aligned Language Models, arXiv preprint https://arxiv.org/abs/2406.00024, (2024).

12. Mikolov, T., et al.: Distributed representations of words and phrases and their compositionality. Adv. Neural Inf. Proces. Syst. **26**, 3111–3119 (2013)
13. Pennington, J., Socher, R., Manning, C.D.: GloVe: global vectors for word representation. In: Proceedings of the 2014 Conference on Empirical Methods in Natural Language Processing, pp. 1532–1543 (2014)
14. Al-Jabali, M.: Contemporary Tunisian [Selected Contemporary Tunisian Fatwas]. Dar Mazri, Tunis (2021)
15. al-Qassar, I.: Eyes of evidence in the evenings of disagreement between the jurists of the regions. In: The book of purity [Sources of Evidence in Disputed Issues among Jurists: Book of Purification]. Dar al-Fikr, Damascus (2022)
16. Ms-marco-MiniLM-L6-v2, https://huggingface.co/cross-encoder/ms-marco-MiniLM-L6-v2. Visited 9/3/2025.
17. Mohammed, M.Y., Ali, S.A., Ali, S.K., Majeed, A.A., Mohamed, E.H.: Aftina: enhancing stability and preventing hallucination in AI-based Islamic fatwa generation using LLMs and RAG. Neural Comput. & Applic. **37**, 1–26 (2025)

Sentiment Classification of COVID-19 Tweets: From Machine Learning and Deep Learning to BERT

Soumeya Zerabi[1,3](✉), Karima Sid[2,3], Soumia Zertal[2,4], Anis Zouaghi[1], and Ayoub Bouchelaghem[1]

[1] Department of Computer Science and Its Applications, Abdelhamid Mehri Constantine2 University, Constantine, Algeria
{soumeya.zerabi,Anis.Zouaghi, Ayoub.Bouchelaghem}@univ-constantine2.dz

[2] Department of Mathematics and Computer Science, University of Oum El Bouaghi, Oum el Bouaghi, Algeria
{sid.karima,zertal.soumia}@univ-oeb.dz

[3] LISIA Laboratory, Abdelhamid Mehri University - Constantine 2, Algeria New City, Ali Mendjeli Constantine, Algeria

[4] Artificial Intelligence and Autonomous Things Laboratory, University of Oum El Bouaghi, Oum El Bouaghi, Algeria

Abstract. COVID-19 is an infectious illness that was initially identified toward the end of 2019 and was officially declared a pandemic in March 2020. Social media, particularly Twitter, is widely used by people and is considered the primary official tool affecting both the mental and physical health of the population, as it spreads information about the increasing number of positive cases or deaths through posted tweets. In recent years, natural language processing with deep learning has gained significant attention in the Sentiment Analysis domain. Sentiment Analysis is a useful way to interpret emotions from textual information. This work presents a comparative analysis of multi-class sentiment classification on a dataset of COVID-19-related Tweets, evaluating four machine learning classifiers, two deep learning models, and the pre-trained transformer based model BERT (Bidirectional Encoder Representations from Transformers). The results demonstrate that BERT outperforms all models attaining the highest performance metrics of 91.4%, 92.02%, 90.39%, and 89% for accuracy, precision, recall, and F1-score, respectively.

Keywords: NLP · Transformers · BERT · Tweets · COVID-19 · Sentiment Analysis · machine learning · deep learning

1 Introduction

Recently, social media are one of the main sources of information, gaining immense popularity. Among this, Twitter is used by 206 million daily active users, predominantly aged between 25 and 34 [1]. However, approximately 80% of global data exists in unstructured

T. Ensari et al. (Eds.): ISPR 2025, CCIS 2859, pp. 322–336, 2026.
https://doi.org/10.1007/978-3-032-21585-7_24

form, such as text. This presents significant challenges for analysis and comprehension. To address this, Natural Language Processing (NLP) has gained as a key field for analyzing and understanding textual data. Among the primary applications of NLP is Sentiment Analysis, which focuses on analyzing and classifying sentiments expressed in text. Several approaches exist for Sentiment Analysis on Twitter, including Machine Learning (ML) which uses general learning algorithms that analyzes text word by word, classifying sentences as neutral, negative or positive based on individual terms. However, these methods sometimes lost information by extracting key word without considering other surrounding words. To overcome such limitations, Long Short Term Memory (LSTM) as Deep Learning (DL) model have proven their ability in addressing problems of standard Recurrent Neural Networks (RNNs). Bidirectional LSTM (Bi-LSTM) enhances performance by capturing both the preceding and succeeding contexts making it popular among NLP researchers. CNNs are other powerful models and popularly used with word embeddings for classification tasks, they are able to extract local features from data through the convolution operation and measure the relationships within local patterns [2, 3]. Recently, BERT [4], a transfer learning method, processes text sequences bi-directionally by capturing contextual information simultaneously from both directions, making it efficient and accurate.

The novelty of this work lies to explore and contrast three categories of techniques for classifying sentiments on COVID-19-related tweets, using a unified preprocessing pipeline, optimizing hyperparameters, and addressing robust evaluation metrics to ensure a fair comparison. First, we employ TF-IDF followed by baseline machine learning classifiers, including K-Nearest Neighbor, Random Forest, Logistic Regression, and multimodal Naïve Bayes. Next, we use deep learning models in particular LSTM and CNN. Finally, we focus on BERT model.

The remainder of this paper is organized as follow: Sect. 2 exposes concisely the literature review. The detailed methodology is described in Sect. 3. Section 4 reports the findings and then discusses in Sect. 5. Section 6 shows the conclusion and future research areas.

2 Literature Review

Earlier studies in Sentiment Analysis have investigated the effectiveness of different classifiers in analyzing Twitter data [5–7]. These studies have used machine learning methods such as Naive Bayes [8], Maximum Entropy [9], and Support Vector Machine [10] and various deep learning techniques. In [11] the study analyzes the sentiments of citizens from India regarding COVID-19 and vaccination compaign by analyzing Twitter data, they employ both Bi-LSTM and GRU based deep learning and lexicon-based techniques (VADER, NRCLex) to classify sentiments. The proposed models achieved high performance. In [12], the authors assessed the performance of BERT, RoBERTa and XLNet models in classifying sentiments from movie reviews in the IMDB database. Additionally, they applied TF-IDF to extract important features (words) which were used by K-means algorithm to classify movies by topics interpreted as positive and negative clusters. The findings demonstrated that XLNet outperformed BERT model and the combination of TF-IDF and K-means achieved good results. Reference [13] analyzes various ML techniques like SVM and Multinomial NB, as well as deep learning models such as LSTM and BERT to discover the most suitable technique for sentiment analysis. In [14], the authors compare various combinations of CNN and LSTM as classifiers and explored various word embedding methods like Word2Vec and Glove to evaluate model performance. Reference [15] proposed an algorithm named Hybrid Heterogeneous SVM (H-SVM) to classify sentiments and compared its performance with RNN and SVM. Additionally, [16] proposed different deep learning techniques based on LSTM and compared the obtained results with traditional machine learning classifiers. In [17], the authors proposed a hierarchical deep fusion method to perform emotion analysis, using hierarchical LSTM (H-LSTM) for merging textual and visual content in order to explore associations between different modalities of image and text. In [18], the authors apply the BERT model for classifying sentiments over COVID-19 related tweets, their model reached a high accuracy demonstrating its effectiveness in analyzing sentiments for tweets.

Based on the findings of these studies, the overarching goal of this work is to determine the most suitable approach to classify sentiments in Tweets.

3 Methodology

This section presents the dataset, architectures and the models configurations used in this work.

3.1 Dataset and Pre-processing

We used in this work a dataset sourced from kaggle [19]. It contains 44 954 English language tweets, with 41 157 used for training and 3789 for testing. Each tweet includes six attributes: user name, screen name, location, tweetAt (the date when the tweet was posted), tweet (the textual content posted) and sentiment (categorized into five classes). Figure 1 presents five (05) representative examples of tweets extracted from the dataset. For clarity and brevity, only the tweet content and sentiment are shown.

OriginalTweet	Sentiment
@MeNyrbie @Phil_Gahan @Chrisitv https://t.co/i...	Neutral
advice Talk to your neighbours family to excha...	Positive
Coronavirus Australia: Woolworths to give elde...	Positive
My food stock is not the only one which is emp...	Positive
Me, ready to go at supermarket during the #COV...	Extremely Negative

Fig. 1. Examples of Tweets with corresponding sentiment labels.

Initially, the sentiment labels comprise five (05) categories: extremely positive, positive, extremely negative, negative, and neutral. To ensure consistency with previous studies conducted utilizing this dataset, we refine the dataset to include three main classes: positive, negative, and neutral. Tweets labeled as extremely positive were merged into the positive class, while those labeled as extremely negative were merged into the negative class.

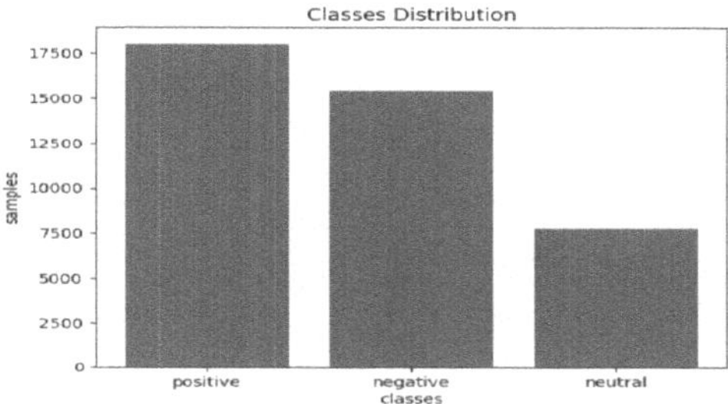

Fig. 2. Sentiments class distribution

Additionally, we examine the labels distribution to avoid the issue of class imbalance. As depicted in Fig. 2, the dataset shows a moderate class imbalance.

On the other hand, checking for missing values in the dataset is primordial. We use the isnull () function and find that the 'location' attribute has 8 590 missing values. This column will be considered as an unusual column.

Cleanup methods play a crucial role in NLP tasks. In our cleaning phase, we have implemented several steps, which can be summarized as follows:

- **Delete irrelevant columns:** it helps streamline the data by removing unnecessary information that may not contribute to the task. In our case, we removed columns such as: user name, screen name, tweet date and location.
- **Delete retweets:** retweets are duplicate tweets that can introduce noise and bias into the data. By removing them, we ensure that the model's performance is unbiased by redundant and repetitive information.
- **Convert to lowercase:** Converting all the text to lowercase helps standardize the text and ensure that the model treats words with the same spelling but different capitalization as the same word.
- **Remove URLs:** URLs can be considered noise in the data, and removing them can efficiently improve the accuracy of the model.
- **Remove mention and hash tags:** Mentions and hashtags can be valuable for social media analysis. However, for our task of sentiment analysis, they are not relevant. Therefore, removing them can simplify the text and reduce noise.
- **Handle internet slang using a dictionary:** Internet slang can be difficult for models to accurately understand and classify the text. Therefore, it is important to handle slang words before training the model. One approach is to use a dictionary, such as the one provided by analytics vidhya website [20] to map slang words to their corresponding full forms.
- **Remove HTML tags:** HTML tags are not relevant to the text, and removing them can simplify the text.
- **Remove special characters:** Special characters are often considered as noise, and removing them can simplify the text.
- **Remove numbers:** Numbers are not relevant to our specific task like special characters and digits, and removing them will simplify the text and reduce noise.
- **Remove extra whitespaces:** Extra whitespaces can be considered noise in the text, and removing them will simplify the text.

The other pre-processing step is tokenization, which depends on the specific model being used. We will explain tokenization in the following sections.

3.2 Sentiment Classification Models

In this section, we describe the seven proposed models to classify sentiments of tweets related COVID-19.

Machine Learning Models

The models begin with a feature extraction step using TF-IDF as a statistical technique for text vectorization. This process converts raw text into numerical feature vectors. Tokenization is done by the use of the default 'TfidfVectorize' tokenizer. Following vectorization, four (04) baseline machine learning classifiers (*Multinomial NB*, *KNN*, *LR*, *RF*) are used. Each model is trained end-to-end using a Scikit Learn Pipeline.

Multinomial Naive Bayes (Multinomial NB). This classifier works by applying Bayes' theorem and it is suitable for text classification problems.

K-Nearest Neighbors (KNN). This method is simple and non-parametric, able to classify samples according to the majority class of their k closest neighbors within the input space.

Logistic Regression (LR). Is a type of linear models that estimates sentiment probabilities using a 'sigmoid' function. Due to its simplicity, LR often achieves good performs in text classification.

Random Forest (RF). This model is built by integrating a set of decision trees that aggregates predictions through majority voting. RF mitigates overfitting via bagging and random feature subsets.

We optimize each model's hyperparameters using Grid search technique [21]. For TF-IDF vectorization, we set max_features to 1000 for all models except KNN, where max_features is set to 500. The hyperparameters of models are summarized in Table 1.

Table 1. Hyperparameters used in machine learning models

Model	Hyperparameters	Values
LR	max_iter	1000
LR	C	10
RF	n_estimators	200
Multinomial NB	Alpha	0.1
KNN	n_neighbors	1

Deep Learning Models

We propose two commonly used neural networks for text classification: LSTM and CNN.

The input text data was tokenized using a vocabulary limited to 15 000 words, with out-of-vocabulary tokens explicitly marked. Each token sequences was then standardized by padding to a fixed length of 150.

Long Short Term Memory (LSTM) Model. This network is an improved variant of RNNs, developed to fix the issue of vanishing gradient by capturing effectively dependencies within long sequences. It is commonly used in text classification tasks. In this study, we implemented a variant of LSTM which is Bidirectional LSTM network with the following key architectural components:

- **Embedding layer:** converts input tokens to dense vector with a dimensionality of 128. The parameter (mask_zero = True) was set to ignore padding tokens, improving the efficiency.
- **Bidirectional LSTM layer:** Uses 64 units to process sequences in two directions (forward and backward) with a total dimension of output vectors is 128.
- **Dense layers:** The architecture includes two dense layers. In the first layer the activation function 'ReLU' is used with 64 units for feature transformation. The second uses a 'softmax' activation function with 3 units corresponding to the three sentiment classes: (Neutral, Positive, and Negative) producing probability distributions.

To avoid overfitting, we added SpatialDropout1D (rate = 0.3) after the embedding layer, applied Dropout layers after both the LSTM (rate = 0.5) and dense (rate = 0.4) layers, enabled Early Stopping with a patience of 3 epochs, and used L2 Regularization (weight decay = 0.01) to both the kernel and recurrent weights. The hyperparameters used in this model are summarized in Table 2.

Table 2. Hyperparameters used in LSTM model

Hyperparameters	Values
Embedding Dimension	128
Optimizer	Adam (Learning Rate = 0.001)
Epochs	15
Batch size	64
Learning Rate Scheduling	0.2
Loss	sparse_categorical_crossentropy
Dropout rates	0.3 (Spatial layer), 0.5 (LSTM layer), 0.4 (Dense layer)

Convolutional Neural Network (CNN) Model. CNNs networks commonly used in image processing and are also effective to classify text by treating it as a sequence of word embedding. In this study, we implemented a CNN network with the following layers:

- **Embedding layer:** Identical to the one used in the LSTM model.

- **Convolutional layer:** The first one includes 64 filters, the activation function used is 'ReLU' and the kernel size is 3, same padding, and L2 regularization (0.01). The second layer includes 128 filters each of size of 4, also using ReLU activation and L2 regularization.
- **Pooling layer:** we used Global Max Pooling to reduce the dimensionality of a sequence and give the highest value over all input features.
- **Dense layers:** the model has two dense layers. The first one contains 64 units, ReLU activation and L2 regularizers, followed by Batch Normalization. The second layer consists of 3 units, each representing one of the sentiment classes and a 'SoftMax' activation function.

To mitigate overfitting, we added Dropout layers after both convolution layers (0.4 and 0.3 rate) and dense (0.3 rate) layers, applying Early Stopping (patience =3 epochs), adding L2 regularization across multiple layers. The hyperparameters of the CNN model are summarized in Table 3.

Table 3. Hyperparameters used in CNN model

Hyperparameters	Values
Embedding Dimension	128
Optimizer	Adam (Learning Rate = 0.001)
Conv1D layer 1, layer 2	Filters (64, 128)
Epochs	20
Batch size	64
Learning Rate Scheduling	0.2
Loss	sparse_categorical_crossentropy
Dropout rates	0.4 (Conv1D layer 1), 0.5 (Conv1D layer 2), 0.3 (Dense layer)

Bidirectional Encoder Representations from Transformers (BERT) Model

We used the basic architecture of the Bidirectional Encoder Representations from Transformers (BERT) and we fine-tuned it to be suitable with our dataset. This model is based on transformers (Trans) architecture and it revolutionized NLP field, particularly in tasks like text classification. Our BERT model contains the following layers:

BERT Preprocessor Tokenizer Layer. In this layer, we implement the BERT preprocessor tokenizer, which handles the tokenization phase within our model. BERT uses 'WordPiece tokenization', which is a form of sub-word tokenization. This process initially splits the text into words and then divides each word into smaller units called sub-words. Subwords are identified by a prefix such as '##'. For instance, the word "word" might be tokenized as the sub-words: 'w', '##o', '##r', '##d'. At the end of the BERT tokenizer process, we get a set of token indices from its vocabulary file.

Additionally, we obtain a clear and readable representation of the tokens including the important features such as: 'input_ids' and special tokens [CLS] and [SEP] which mark respectively the beginning and the end of a sentence.

BERT Preprocessor Layer. The BERT preprocessing layer is a crucial component of the BERT model. Its primary role is to convert tokenized text into the specific format and shape required by the BERT encoder which expects inputs in a predefined structure. This layer is created using the **'bert_pack_inputs'** function from the BERT preprocessing library. It takes the tokenized inputs and packs them into the required format needed by BERT. The packed inputs include: input word IDs, Attention Mask, and Token Type IDs. The fig. 3 presents an example of these components.

To minimize the presence of padding tokens in the dataset, we set the sequence length of 60 which is the maximum word count per tweet.

```
{'input_word_ids': [[101, 14108, 1997, 20407, 102, 0, 0, 0, 0, ...], ...],
 'input_mask': [[1, 1, 1, 1, 1, 0, 0, 0, 0, ...], ...],
 'input_type_ids': [[0, 0, 0, 0, 0, 0, 0, 0, 0, ...], ...]}
```

Fig. 3. Example of BERT preprocessor components layer

BERT Encoder Layer. In this layer we imported the BERT base model which is built with 12 transformer layers, 12 self-attention heads, a hidden layer with a size of 768, and approximately 110 million parameters.

Classifier Layer. This is the output layer of the model; it consists of a dense layer with three output neurons, each corresponding to one of the three classes. We used in this layer 'SoftMax' activation function, which is a non-linear function that outputs probabilities for each class ranging from 0 to 1 and summing to 1. The hyperparameters of the BERT model are summarized in Table 4.

Table 4. Hyperparameters used in BERT model

BERT hyperparameters	Values
Optimizer	Adam (Learning Rate = 2e-5)
Epoch	3
Batch size	50
Loss	CategoricalCrossentropy

4 Experimental Setup

This section describes the experimental setup used for this study. Following this, we present the finding results and compare them with benchmark works from the literature. The performance of the models in classifying COVID-19 related tweet sentiments is evaluated using commonly used metrics. Due to class imbalance in the dataset, we report weighted metrics (accuracy, precision, recall, and F1-scores) for providing a fair evaluation that reflects the class distribution.

4.1 Experiment Configurations

All experiments were conducted using Google Colab platform [22], which is free, easy to use and provides approximately 12 GB of RAM. Furthermore, it is pre-configured with several libraries including TensorFlow, Keras and Matplotlib. We used Python3 as a programming language. .

4.2 Experimental Results

Machine learning results

The table below describes the performance of the proposed machine learning models.

Table 5. Results of machine learning models

Model	Accuracy (%)	Precision (%)	Recall (%)	F1-score (%)
LR	**70.98**	**71.66**	**70.98**	**71.18**
RF	67.93	67.91	67.93	67.91
Multinomial NB	63.42	62.50	63.42	60.13
KNN	41.28	47.32	41.28	42.85

From Table 5, we see that Linear Regression reached the best performance among the machine learning models and all of them achieved an accuracy of 70.98% and F1-score less than 71%. These are considered relatively low values.

Deep learning results

Table 6 describes the performance of the proposed deep learning models.

Table 6. Results of deep learning models

Model	Accuracy (%)	Precision (%)	Recall (%)	F1-score (%)
LSTM	82	82	81	82
CNN	**86**	**86**	**86**	**86**

The loss metric is a crucial indicator of how well a deep learning model is learning from the data. It calculates the difference between predicted outputs and true labels. A lower loss typically indicates better model performance. Figure 4 illustrates the loss curves of the proposed models (Bi-LSTM and CNN) during training. It is clear that CNN model shows less overfitting due to the use of pooling layers which help to improve generalization.

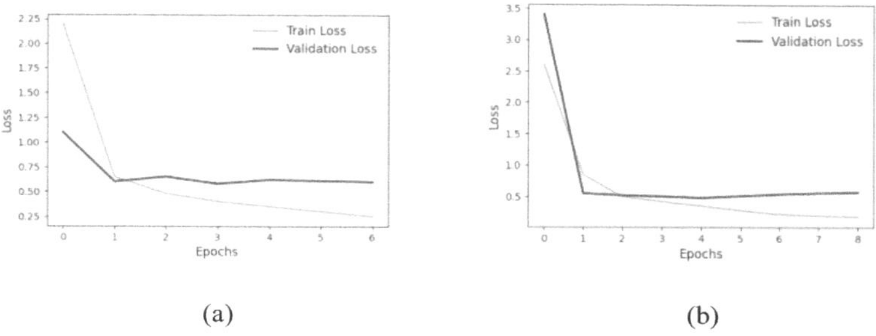

(a) (b)

Fig. 4. Loss curves of models: (a) Bi-LSTM (b) CNN

Considering Fig. 5 b, it can be seen that CNN model has yielded better results in all classes compared to Bi-LSTM model (see Fig. 5 a).

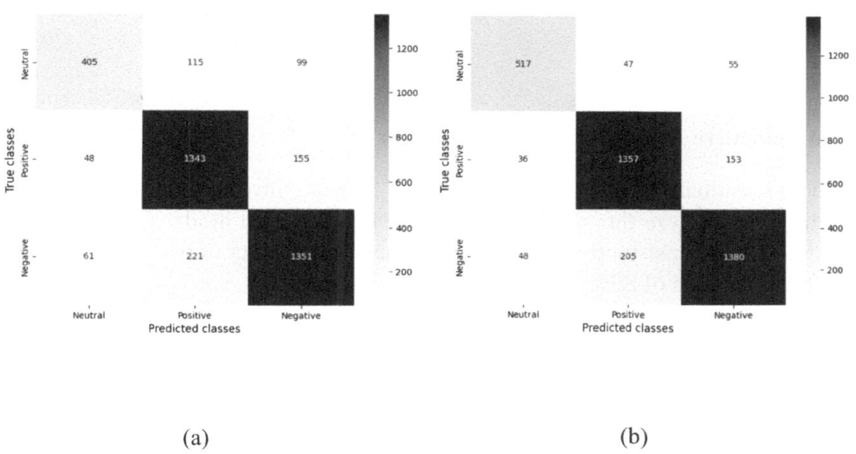

(a) (b)

Fig. 5. Confusion matrices of models: (a) Bi-LSTM (b) CNN

BERT results

The table below presents the performance obtained from the proposed BERT model.

From Table 7, we show that there is a small gap between train and test values which is normal for real-world data and not severe overfitting. Figure 6 b presents the loss curve of BERT model during training. It is clear that both training and validation losses decrease

Table 7. Results of BERT model

Metric	Accuracy (%)		Precision (%)		Recall (%)		F1-score (%)	
	Train	Test	Train	Test	Train	Test	Train	Test
Score	93.92	91.4	94.94	92.02	93.04	90.39	93.98	89

rapidly and simultaneously until epoch 2 reflecting the effect of the pre-training. After epoch 2, the validation loss starts to increase slightly indicating that the beginning of overfitting. Therefore, we stopped the training at epoch 3 to prevent further overfitting.

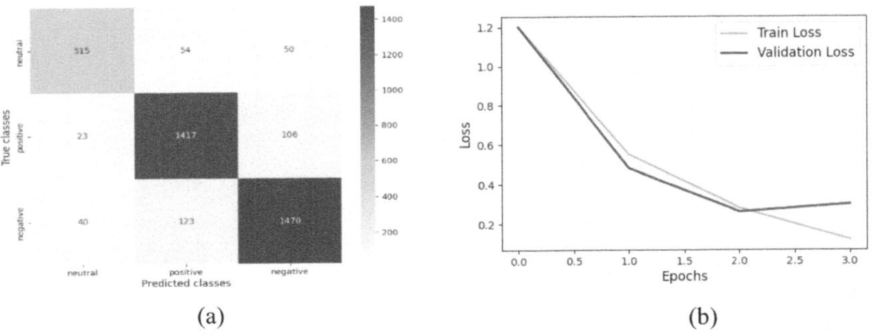

(a) (b)

Fig. 6. (a) Loss curve of BERT model (b) Confusion matrix of BERT model

4.3 Comparative Analysis

Our model's evaluation follows two steps: The first one involves an internal comparison, where we analyze the performance of our implemented models. Figure 7 shows that BERT model consistently achieved the highest results overall performance metrics reaching an F1-score of 89%. We display the F1-score values, since this metric is used for unbalanced class distributions.

Furthermore, in the second step, we focus on benchmarking the BERT model against the literature, where we compare our BERT model's performance with the existing studies [14, 15] that have used the same dataset, as illustrated in Fig. 8. We note that all comparative models in [15] are based on LSTM architecture with different embedding techniques.

From Figs. 5 and 6 a, we notice that the BERT model achieved the highest performance results over the three classes and gave significant reductions in False Positive and False Negative values compared to the other models.

Figure 9 is a heatmap summarizing the empirical findings of our best model and models in the literature across all metrics.

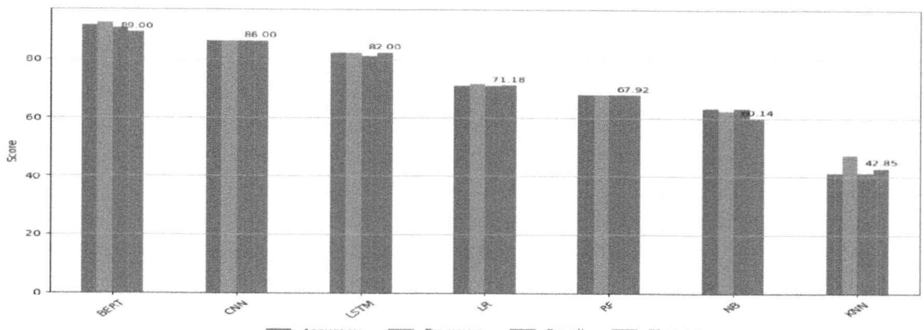

Fig. 7. Performance comparison of the proposed models

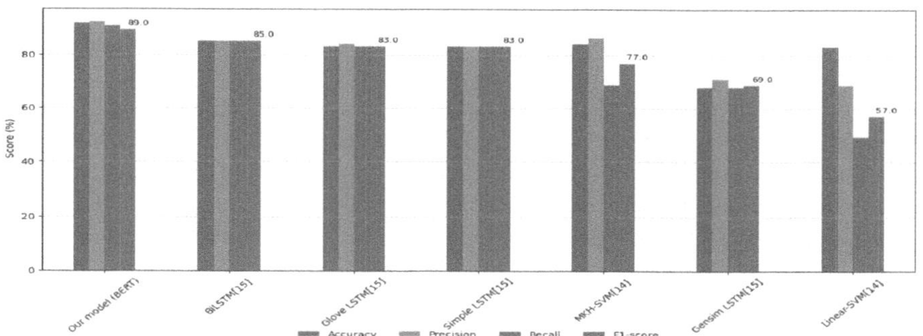

Fig. 8. Performance comparison between the Bert model and existing models

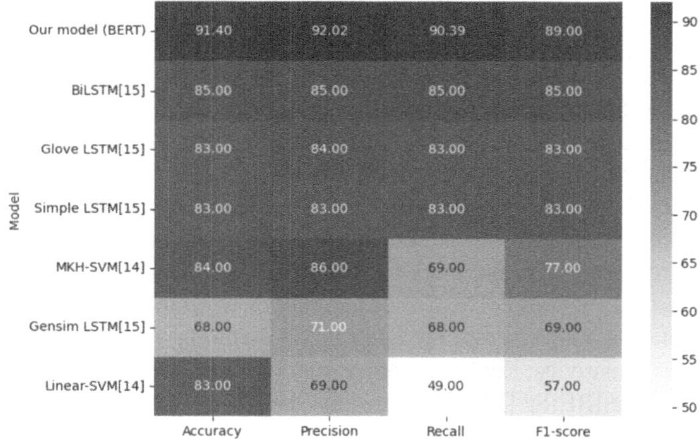

Fig. 9. Heatmap for comparative study

5 Discussion

From Table 5, we notice that Logistic Regression (LR) gives the best performance among the models, with 70.98% accuracy and 71.18% F1-score. This indicates that LR is effective in classifying tweet sentiments. On the other hand, the KNN model shows weaker performance, reaching only 41.28% accuracy and F1-score of 42.85%. KNN suffers from high computational cost during inference and is sensitive to the curse of dimensionality, which is a critical limitation given the high-dimensional nature of TF-IDF vectors. ML models using TF-IDF representations gives relatively low performance, as they treat text as unordered features and ignore word order, missing contextual meaning and failing to capture semantic relationships. Additionally, CNN model's performance across all evaluation metrics surpasses Bi-LSTM model with particularly in accuracy and F1-score of 86% due to its ability in capturing local short text sequences (Tweets), whereas Bi-LSTM is more commonly suitable for long sequence classification. However, we notice that BERT model has given the best results surpassing both machine and deep learning models.

This outperformance is mainly because BERT can capture complex language patterns through its pre-training on a large text corpus making it particularly suitable for understanding the nuances and context present in short, informal texts such as tweets. Unlike traditional machine and deep learning models, BERT leverages bidirectional representations, allowing it to capture simultaneously the bidirectional context of a word, which is crucial for Sentiment Analysis tasks where the significance of a word changes based on the context in which it appears. Our proposed BERT model performs well with good findings such as: an accuracy of 91.4%, a precision of 92.02%, a recall of 90.39%, and F1-score of 89% which indicates a relative improvement over the other models.

6 Conclusion and Future Work

This work presents a comparison between models for sentiment analysis on COVID-19 tweets. The experimental results demonstrate remarkable accuracy of the BERT model with over 91.4% and an F1score of 89%, surpassing the results provided by machine and deep learning techniques. Our model can be applied in real word contexts, such as tracking public mental health during a crisis and supporting public health communication. However this model can suffer to generalizability and potential bias. To address these limitations, we suggest that future work focus on: Testing the model on different datasets or increasing the data volume; The use of techniques to address class imbalance, such as data augmentation, under sampling, and related methods; Addressing issues such as hyperparameters sensitivity; Exploring other transformers like BERT variants (such as: RoBERTa [23], DistilBERT [24]); Performing sentiment classification in real time is more valuable and impactful than batch processing, as Twitter is a real-time platform where information is continuously generated and processed; The use of Explainable AI for tweet sentiment classification in order to improve interpretability, transparency, and trust.

References

1. B. Dean: How many people use twitter in 2021? [new twitter stats] Backlinko. Accessed: Oct. 20, 2021. [online]. (2021). Available: https://backlinko.com/twitter-users
2. Li, W., Zhu, L., et al.: User reviews: sentiment analysis using lexicon integrated two-channel cnn lstm family models. Appl. Soft Comput. **94**, 106435 (2020). https://doi.org/10.1016/j.asoc.2020.106435
3. Basiri, M.E., et al.: Abcdm: an attention-based bidirectional cnn rnn deep model for sentiment anal ysis. Futur. Gener. Comput. Syst. **115**, 279–294 (2021). https://doi.org/10.1016/j.future.2020.08.005
4. Vaswani, A., et al.: Attention is all you need. In: Advances in Neural Information Processing Systems. Curran Associates, Inc (2017)
5. Osman, I.N., Husien, I.M.: Comparison of sentiment analysis techniques for twitter posts classification. In: 2022 International Conference on Data Science and Intelligent Computing (ICDSIC), pp. 93–97 (2022). https://doi.org/10.1109/ICDSIC56987.2022.10075895
6. Atandoh, P., Zhang, F., Al-antari, M.A., Addo, D., Hyeon Gu, Y.: Scalable deep learning framework for sentiment analysis prediction for online movie reviews. Heliyon. **10**(10), e30756 (2024). https://doi.org/10.1016/j.heliyon.2024.e30756
7. Vohra, A., Garg, R.: Deep learning based sentiment analysis of public perception of working from home through tweets. J. Intell. Inf. Syst. **60**(1), 255–274 (2023). https://doi.org/10.1007/s10844-022-00736-2
8. "What Are Naïve Bayes Classifiers? | IBM." Accessed: Aug. 09, (2024). [Online]. Available: https://www.ibm.com/topics/naive-bayes.
9. Lin, P., Fu, S.-W., Wang, S.-S., Lai, Y.-H., Tsao, Y.: Maximum entropy learning with deep belief networks. Entropy. **18**(7), 7 (2016). https://doi.org/10.3390/e18070251
10. What is support vector machine? | IBM. Accessed: Aug. 09, (2024). [Online]. Available: https://www.ibm.com/topics/support-vector-machine.
11. Ainapure, B.S., et al.: Sentiment analysis of COVID-19 tweets using deep learning and lexicon based approaches. Sustainability. **15**(3), 3 (2023). https://doi.org/10.3390/su15032573
12. Palomo, B.A.B., Velarde, F.H.V., Cantu-Ortiz, F.J., Ceballos Cancino, H.G.: Sentiment analysis of IMDB movie reviews using deep learning techniques. In: Yang, X.-S., Sherratt, R.S., Dey, N., Joshi, A. (eds.) Proceedings of Eighth International Congress on Information and Communication Technology, pp. 421–434. Springer Nature, Singapore (2024). https://doi.org/10.1007/978-981-99-3236-8_33
13. Dhola, K., Saradva, M.: A comparative evaluation of traditional machine learning and deep learning classification techniques for sentiment analysis. In: 2021 11th International Conference on Cloud Computing, Data Science & Engineering (Confluence), pp. 932–936 (2021). https://doi.org/10.1109/Confluence51648.2021.9377070
14. D. Goularas, S. Kamis, Evaluation of Deep Learning Techniques in Sentiment Analysis from Twitter Data. (2019), 17. doi:https://doi.org/10.1109/Deep-ML.2019.00011.
15. Kaur, H., Ahsaan, S.U., Alankar, B., Chang, V.: A proposed sentiment analysis deep learning algorithm for Analyzing COVID-19 tweets. Inf. Syst. Front. **23**(6), 1417–1429 (2021). https://doi.org/10.1007/s10796-021-10135-7
16. Vernikou, S., Lyras, A., Kanavos, A.: Multiclass sentiment analysis on COVID-19-related tweets using deep learning models. Neural Comput. & Applic. **34**(22), 19615–19627 (2022). https://doi.org/10.1007/s00521-022-07650-2
17. Xu, J., et al.: Sentiment analysis of social images via hierarchical deep fusion of content and links. Appl. Soft Comput. **80**, 387 (2019). https://doi.org/10.1016/j.asoc.2019.04.010
18. K. Sadia, S. Basak: Sentiment Analysis of COVID-19 Tweets: How Does BERT Perform? (2021), 407–416. doi:https://doi.org/10.1007/978-981-16-0586-4_33.

19. Coronavirus tweets NLP - Text Classification." Accessed: Nov.30, (2024).[Online].Available:https://www.kaggle.com/datasets/datatattle/covid-19-nlp-text-classification
20. Wu, L., Morstatter, F., Liu, H.: SlangSD: building, expanding and using a sentiment dictionary of slang words for short-text sentiment classification. Lang. Resour. Eval. **52**(3), 839–852 (2018)
21. How to Grid Search Hyperparameters for Deep Learning Models in Python with Keras - MachineLearningMastery.com. URL https://machinelearningmastery.com/grid-search-hyperparameters-deep-learning-models-python-keras/ (accessed 8.14.25). https://colab.research.google.com/
22. Y. Liu et al.: RoBERTa: A Robustly Optimized BERT Pretraining Approach," Jul. 26, (2019), arXiv: https://arxiv.org/abs/1907.11692. doi:10.48550/arXiv.1907.11692.
23. V. Sanh, L. Debut, J. Chaumond, T. Wolf: DistilBERT, a distilled version of BERT: smaller, faster, cheaper and lighter," Feb. 29, (2020), arXiv: https://arxiv.org/abs/1910.01108. doi:10.48550/arXiv.1910.0110

Maintenance-Oriented Test Generators and Test Teams Simulation for Demand-Controlled Ventilation and Heating Systems

Ali Behravan[1](✉), Ayappan Mani Devendar[2], and Roman Obermaisser[2]

[1] GreeNovaX, Siegen, Germany
ali.behravan@greenovax.de
[2] Chair for Embedded Systems, University of Siegen, Siegen, Germany

Abstract. Heating, Ventilation, and Air Conditioning (HVAC) systems are pivotal to the energy footprint, environmental performance, and health outcomes of buildings. Demand-Controlled Ventilation (DCV) dynamically modulates outdoor-air intake based on occupancy or Indoor Air Quality (IAQ) indicators (often CO_2), and is widely promoted as a pathway to balance energy efficiency with occupant wellbeing. Yet, in practice, large-scale HVAC deployments are frequently hampered by undetected or persistent faults in sensors and actuators, leading to energy waste, thermal discomfort, and IAQ lapses. While the literature includes robust streams on simulation-based fault injection (FI), fault detection and diagnosis (FDD), model-based testing (MBT), and commissioning, there remains a persistent disconnect between simulated faults and the concrete maintenance workflows that technicians actually execute in the field.

We address this gap with an integrated framework that couples a composable, component-based FI model with a **maintenance-oriented test generator** and a **test-team simulation** capability. Implemented in MATLAB/Simulink (with Stateflow and Simscape), our approach allows users to specify multi-fault scenarios via a GUI; automatically maps each injected fault to manufacturer-style maintenance tests; and allocates those tests to simulated teams based on building topology. The framework outputs (i) a **Maintenance Test & Testing Team Table** that traces fault types to failed tests and team assignments, and (ii) a **Final Test Status Table** that consolidates pass/fail per component. We validate the approach on a 6-room, 2-floor DCV–heating model with scenarios spanning intermittent data-loss and stuck-at faults for CO_2 and temperature sensors, damper actuators, and heaters, under varied IAQ and weather conditions. Results show that injected faults consistently produce the expected failed tests across scenarios, with no false failures on healthy components, demonstrating a reliable bridge from FI to maintenance planning. Compared to prior work, our innovation is to **close the loop** from simulation to actionable test lists and team logistics, thereby accelerating corrective actions and improving maintainability of DCV/HVAC systems at scale.

Keywords: demand-controlled ventilation · HVAC · fault injection · maintenance testing · simulation · indoor air quality · model-based testing

T. Ensari et al. (Eds.): ISPR 2025, CCIS 2859, pp. 337–357, 2026.
https://doi.org/10.1007/978-3-032-21585-7_25

1 Introduction

The building sector remains one of the largest end-users of final energy and a major source of energy-related CO_2 emissions worldwide. Recent syntheses indicate that building construction and operations together account for a substantial fraction of both (International Energy Agency, 2021). Within this envelope, HVAC equipment is the single largest load class for many building types because it must continuously ensure thermal comfort and IAQ. Consequently, any deviation from intended HVAC operation, particularly sustained faults in sensors or actuators can materially degrade energy performance and occupant outcomes (Katipamula and Brambley, 2005). In high-occupancy settings such as offices and classrooms, ventilation shortcomings may also intersect with pathogen transmission risks, heightening the stakes of reliable IAQ control (Yang et al., 2021).

Demand-Controlled Ventilation (DCV) has emerged as a canonical strategy to mitigate ventilation energy while maintaining air quality by adjusting outdoor-air flows to actual demand (Murphy, 2005). The most common proxy for occupant-generated bioeffluents is indoor CO_2 concentration; when indoor–outdoor differentials grow, DCV increases ventilation, and when spaces are underutilized, DCV throttles flows to avoid unnecessary heating/cooling of outside air (ASHRAE, 2016; Murphy, 2005). In theory, DCV's feedback structure makes it resilient; in practice, the approach is only as robust as the **sensing** and **actuation** chain that closes the loop. A slightly biased CO_2 sensor, a periodic data-loss on a temperature sensor, or a damper actuator stuck closed can produce cascading effects, e.g., chronic under-ventilation, thermal drift, or parasitic reheat (Li et al., 2018; Weimer et al., 2012). Fig. 1. Shows a layout of an office floor equipped with DCV and heating control and relevant sensors and actuators.

Research communities have responded along several fronts. Fault Detection and Diagnosis (FDD) methods attempt to recognize and classify faults from measured signals and models (Katipamula and Brambley, 2005; Li et al., 2018). Model-based testing (MBT) has been used to generate systematic tests for components and Simulink models (Xu et al., 2005; Schmidt et al., 2016). Fault injection (FI) studies simulate realistic failure modes to evaluate resilience, often with component-level modules embedded in HVAC/controls co-simulations (Behravan et al., 2017; Behravan et al., 2019; Kiamanesh et al., 2022). Commissioning and continuous monitoring programs then seek to sustain performance across the lifecycle (Fu Xiao et al., 2008; Wang et al., 2012). This body of work is indispensable, but it typically stops at detection or simulated analysis.

What is notably **missing** is a rigorous, **operational bridge** between simulation outputs and **maintenance workflows**. When a fault is suspected or demonstrated in simulation, practitioners must translate it into concrete maintenance tests from manufacturer documentation, assign those tests to teams, and verify closures. Absent a direct line from FI results to actionable work plans, delays accumulate: technicians spend time triaging, duplicative tests are scheduled, and recurring faults persist unaddressed (Shalabi et al., 2020). Our research objective is to close this gap.

Contributions: We present a maintenance-oriented FI framework that: (i) composes multi-zone DCV–heating systems programmatically in Simulink; (ii) injects multi-fault patterns through reusable Stateflow-based FI blocks; (iii) **generates maintenance tests** deterministically mapped from fault types (extendable to probabilistic detection); and (iv) assigns the generated tests to floor-based maintenance teams (here, 'simulation'

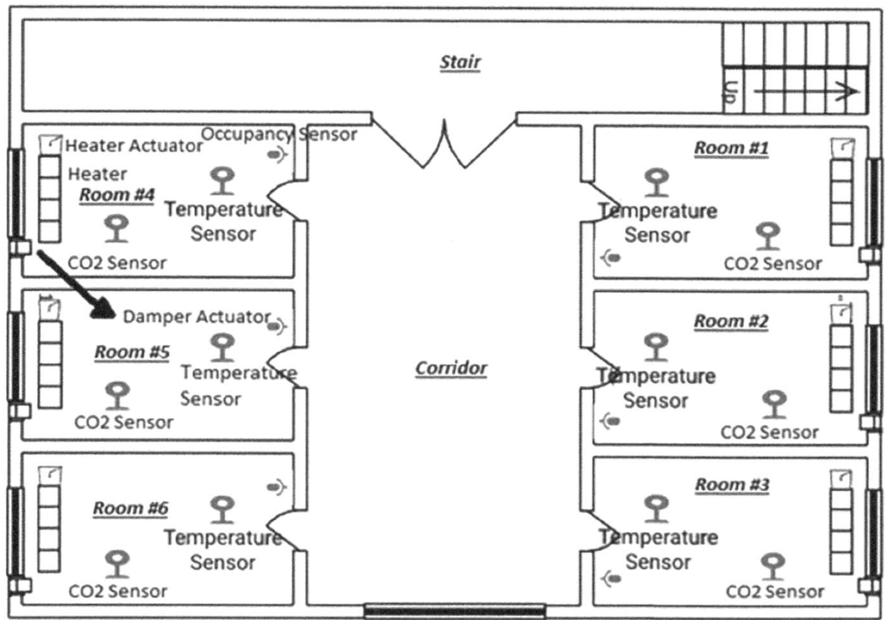

Fig. 1. Layout of an office floor equipped with DCV and heating control and relevant sensors and actuators.

refers to task allocation, not modeling technician behavior), producing structured tables ready for export to CMMS/BIM. As far as we are aware, prior HVAC FI and MBT studies do not connect fault scenarios to team-level maintenance plans in this explicit, automated way (Kiamanesh et al., 2022; Xu et al., 2005; Schmidt et al., 2016; Wang et al., 2012; Shalabi et al., 2020).

The remainder of this paper unfolds as follows. Section 2 expands the state of the art across DCV theory, standards, FI/FDD/MBT, and commissioning. Section 3 sharpens the research gap and the innovation claims. Section 4 details our system model and FI methodology. Section 5 presents the maintenance test generator and the team simulation logic with the GUI workflow. Section 6 reports sample scenarios and results, including the exact timings, durations, and environmental parameter. Section 7 discusses implications, limitations, and threats to validity. Section 8 concludes with future directions.

Note: here and elsewhere, **'simulation of test teams'** refers only to the algorithmic assignment of tasks to maintenance teams (e.g. by building floor), not to simulating the actual time or performance of technicians.

2 State of the Art and Background

2.1 Global Energy, Operational Realities, and the Centrality of HVAC

Energy and climate imperatives have placed buildings under unprecedented scrutiny (International Energy Agency, 2021). Even modest percentage gains in HVAC efficiency translate to outsized absolute savings because HVAC dominates load profiles in commercial and educational buildings. However, field investigations continue to find **persistent** HVAC faults, sometimes spanning years, owing to commissioning gaps, sensor drift, inadequate trending, and overwhelmed maintenance resources (Katipamula and Brambley, 2005; Wang et al., 2012). A common pattern is that **faults do not remain isolated**: slow sensor bias accumulates into chronic over/under-ventilation, damper leakage drives simultaneous heating and cooling, and optimistic setpoints interact with occupancy misestimation.

Digitalization (ubiquitous sensing, analytics, and BMS integration) promises relief but raises complexity. Data volume increases, yet interpretability and operationalization lag when tools stop at alarms instead of **work orders**. This motivates research that not only detects or simulates faults, but **translates** them into concrete maintenance actions with scheduling context, precisely the focus of our framework.

2.2 DCV Principles, IAQ Trade-Offs, and Control Variables

DCV regulates outdoor-air supply proportionally to demand signals. CO_2, as a surrogate for occupant bioeffluents, is widely adopted because it is measurable, strongly occupancy-linked, and already embedded in many AHU/RTU control sequences (Murphy, 2005; ASHRAE, 2016). A canonical rule-of-thumb is that an indoor–outdoor differential of ~700 ppm corresponds to $\approx$15 cfm/person, conceptually tying ppm thresholds to per-capita ventilation sizing (ASHRAE, 2016; Murphy, 2005). However, IAQ perception is not purely a function of CO_2; temperature and humidity modulate perceived air quality (Yang et al., 2021). Thus, DCV often co-optimizes multiple metrics with nontrivial trade-offs, implying **sensitivity to sensor quality and placement**. Modern Simulink/Simscape libraries and composability patterns simplify scaling to multi-zone settings while maintaining solver stability and physical interpretability (Behravan et al., 2017; Behravan et al., 2019). Fig. 2 shows a typical room with DCV control.

2.3 Fault Injection, FDD, and MBT

FI frameworks intentionally corrupt signals or component behaviors according to their attributes to test dependability (Ziade et al., 2004; Kiamanesh et al., 2022). The fault attributes are shown in Table 1. In HVAC, sensor faults (bias, drift, stuck-at, data loss) and actuator faults (stuck, saturation, hysteresis anomalies) are especially consequential because they sit directly on control loops (Li et al., 2018). FDD methods span model-based residuals, rule-based engines, and data-driven classifiers (Katipamula and Brambley, 2005; Behravan, 2021). MBT brings methodological rigor to validation by generating systematic test suites from models (Dalal et al., 2000; Xu et al., 2005; Schmidt et al., 2016). However, most FI/MBT studies culminate in detection/confidence metrics, seldom producing **actionable maintenance test lists** or **team schedules**.

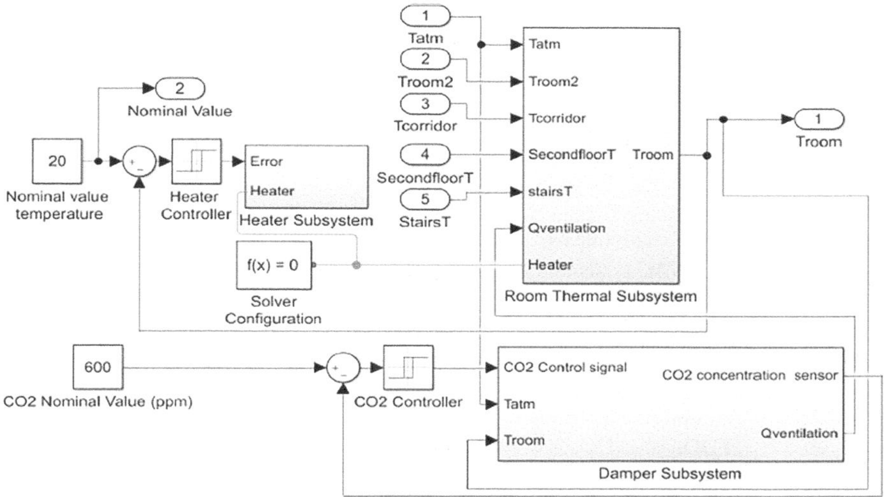

Fig. 2. A typical room implemented in MATLAB/Simulink equipped with DCV [Behravan et al., 2017]

Table 1. Fault attributes for the FI framework [Kiamanesh et al. 2022]

No.	Fault Attributes	Fault Description
1	Fault type	data loss fault, gain fault, out-of-bound fault, stuck-at value fault, and offset fault
2	Fault persistence	Short intermittent, and permanent
3	Fault duration	The period of time for which fault is present in the system
4	Fault interarrival time	The arrival time period between two faults
5	Fault location	CO_2 sensor, damper actuator, temperature sensor, heater actuator
6	Fault occurrence rate	Probability of fault occurrence for each fault case scenario of the type of the fault
7	System fault repetition	Total no. of repetitions of each intermittent/short intermittent fault

2.4 Commissioning, Monitoring, and BIM/FM Linkages

Commissioning and continuous monitoring compress performance drift by embedding analytics and periodic tests into operations (Fu Xiao et al., 2008; Wang et al., 2012). Parallel work links Building Information Modeling (BIM) and facility management (FM) datasets to localize issues and prioritize remediation (Shalabi et al., 2020). Yet even these advanced workflows often require **manual translation** from "detected issue" to "technician test list and team assignment." An automation gap persists, one that our framework directly targets by turning FI outcomes into **maintenance artifacts**.

Table 2. Overview of selected related works on HVAC testers and Test Team Simulation

Author	Technique and Method	Maintenance Test Generation	Test Team Simulation and Organization	Connecting Fault Model with Test and Test team Model
Guannan Li et al., 2022	Sensor fault impact on HVAC system (e.g. Thermostat offset fault)	Partially Yes	No	No
Liping Wang et al., 2012	Monitoring-Based HVAC commissioning using FDD.	Yes	No	No
Aileen Yang et al., 2021	Varying Ventilation rates to improve DCV control strategy.	Partially Yes	No	No
P. Riederer et al., 2002	Thermal Modelling of a room to test HVAC system	Yes	No	Partially Yes
Ali Behravan et al., 2018	Model based FDD diagnosis with different Faults in a DCV and Heating System	Yes	No	Partially Yes
Fu Xiao et al., 2008	Lifecycle commissioning of HVAC system using FDD	Yes	Partially Yes	Partially Yes
James Weimer et al., 2012	FDD techniques on an Actuator for Stuck-at fault	Partially Yes	No	No
Peng Xu et al., 2005	MBT for Air-Handling Units	Yes	Partially Yes	No
Artur Schmidt et al., 2016	MBT on Cyber Physical System for MATLAB/Simulink models	Yes	No	Partially Yes

(continued)

Table 2. *(continued)*

Author	Technique and Method	Maintenance Test Generation	Test Team Simulation and Organization	Connecting Fault Model with Test and Test team Model
Firas Shalabi et al., 2020	Simulation approaches for detecting building spaces with problematic behavior	No	Yes	No
Stefan Fischer et al., 2021	Software Testing of Cyber Physical System with 3 case studies	Partially Yes	No	No
This paper, 2025	Fault injection model and Maintenance tests based on Fault type and Testing team simulation	Yes	Yes	Yes

3 Research Gap and Innovation

In this research, we have detected the following gaps:

Gap 1: from Simulated Faults to Maintenance Procedures. FI results typically end with identifying which components misbehave under which conditions. Facility teams still need to pick the right **manufacturer-style tests** (e.g., "CO_2 sensor transmission/response check," "damper open/close verification," "heater output verification"), decide who performs them, and in what order. This translation is slow and error-prone, especially in large buildings with hundreds of components and recurrent faults (Katipamula and Brambley, 2005; Wang et al., 2012).

Gap 2: from Test Lists to Test-Team Logistics. Even when tests are known, organizations must allocate them to teams. Without lightweight scheduling support, planners over-assign a few technicians, leave others idle, or create floor-to-floor traversals that waste time (Shalabi et al., 2020). Existing FI/MBT literature rarely models these logistics.

Gap 3:Multi-Fault Realism and Composability. A single-fault hypothesis simplifies evaluation but under-represents real building complexity, where **multiple concurrent faults** in different zones and component types are common. Many published FI studies evaluate one fault at a time (Kiamanesh et al., 2022), limiting insight about **fault interactions** and **masking**. There is a methodological need to represent multi-fault vectors over space and time in a composable way (Behravan et al., 2023).

As our innovation, we close these gaps with an **end-to-end pipeline** that:

Injects multi-fault patterns in a composable DCV–heating model (Simulink/Simscape + Stateflow).

Maps each fault type to **specific maintenance tests** drawn from manufacturer/standard practice (deterministic mapping in this version; probabilistic detection is supported conceptually).

Allocates the resulting tests to **simulated teams** by floor, producing structured outputs (Maintenance Test & Testing Team Table; Final Test Status Table) that are immediately actionable and exportable to CMMS/BIM.

Provides a **MATLAB GUI** that unifies the workflow: select faults, set scenario parameters, run simulations, generate tables, and visualize traces.

To the best of our knowledge, no prior work in HVAC/DCV has operationalized the **fault → test → team** chain within a single tool with automatic artifact generation (Kiamanesh et al., 2022; Xu et al., 2005; Schmidt et al., 2016; Shalabi et al., 2020). This is the novelty we emphasize relative to the research gap.

4 System Model and Fault-Injection Methodology

4.1 Composable Multi-Zone DCV–Heating Model

We adopt a **component-based** Simulink architecture parameterized by building size (K floors × N rooms per floor). A generation script programmatically instantiates **room blocks** and wires them into shared supply/return, environmental influences, and supervisory logic (Behravan et al., 2019; Behravan et al., 2017) (Table 3).

Table 3. Rooms and Floors Indexing Table

Floor 1		Floor 2		…	Floor K	
1	(N ÷ 2) +1	(N + 1)	(N+(N ÷2)) +1	…	(N * (K-1)) +1	(N * (K -1)) + (N ÷2) +1
2	(N ÷ 2) +2	(N + 2)	(N+(N÷2)) +2	…	(N * (K -1)) +2	(N * (K -1)) + (N ÷2) +2
3	(N ÷ 2) +3	(N + 3)	(N+(N÷2)) +3	…	(N *(K -1)) +3	(N * (K -1)) + (N ÷2) +3
…	…	…	…	…	…	…
N÷2	N	N + (N ÷ 2)	(N * 2)	…	(N * (K -1)) + (N ÷2)	(N * K)

N: *The total number of rooms per floor*, K: *The total number of floors in the model*

Each room block encapsulates:

Sensors: CO_2 (ppm), temperature (°C), with optional filters and sample/hold for realistic telemetry timing.

Actuators: Damper (0–1 open fraction), heater (ON/OFF with duty-cycle tracking).

Thermal Submodel: Simscape-based nodes for room air and envelope, incorporating heat gains/losses and interzone coupling.

Airflow Submodel: DCV logic computes required ventilation from CO_2, while infiltration/exfiltration responds to **wind speed** and **stack effect**.

Supervisory Control: Hysteresis-based heater control, CO_2 setpoint reset (V_min↔V_max), and simple anti-windup rules.

This composability is crucial: it scales with building size and allows localized faults in **any** sensor or actuator to propagate through the same control logic, exposing realistic system-level consequences (Riederer et al., 2002; Behravan et al., 2019).

4.2 Fault-Injector Blocks and Multi-Fault Vectors

Each sensor/actuator output passes through a **Fault-Injector** block driven by a Stateflow chart. Parameters include: room index, component index, fault type, persistence (permanent, intermittent, transient), start times, durations, and repetitions. A **3D parameter matrix** (repetition × rooms × components) compactly encodes multi-fault scenarios and is either (i) authored by the user via an FI GUI or (ii) generated by a scenario script. Implemented **fault types** include:

Stuck-at (constant 0/1 for actuators or constant reading for sensors),

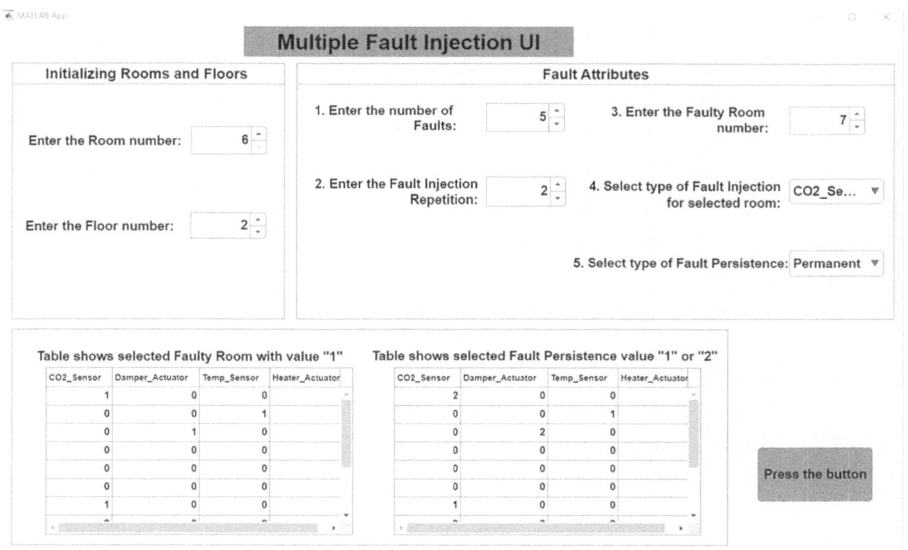

Fig. 3. Multiple Fault Injection GUI.

Data-loss (no updates; controller receives last value or NaN depending on block),
Offset/gain (bias or scaling),

Out-of-bounds (sporadic spikes).

The injector merges healthy and faulty paths based on active states, preserving time-varying behaviors and enabling **fault interaction** studies (Ziade et al., 2004; Li et al., 2018).

A multiple fault injection GUI (Graphical User Interface) was developed (Siddharth Bhandari, 2023) on MATLAB app designer was used in this research for the injection of faults to create fault injection scenarios. The screenshot of GUI is displayed in the below figure (Fig. 3).

This GUI performs three tasks: **1.** User inputs about the total number of rooms and the total number of floorsand it is used in the **'Generation_script.m'** MATLAB file (Behravan et al., 2019) to generate the layout of the building structure with multiple rooms and floors. **2.** Selection of the component in which the faults will be injected and selecting the fault attributes of that fault. **3.** To generate a matrix called the 'ActivatedRoomcomponentCombinationMatrix' based on the fault injection scenario. The rows represent the number of rooms, and the columns represent the total number of components (CO_2 sensor, Damper actuator, Temperature sensor, Heater actuator). Each cell in this matrix is filled with either '0' or '1'. Where '0' indicates there is no fault in that specific room and component and '1' indicates a fault has been injected in that room and component (Fig. 4).

	1	2	3	4
1	1	0	0	0
2	0	0	1	0
3	0	1	0	0
4	0	0	0	1
5	0	0	0	0
6	0	0	0	0
7	1	0	0	0
8	0	0	0	0
9	0	0	0	0
10	0	0	0	0
11	0	0	0	0
12	0	0	0	0

Fig. 4. Activated Room Component Combination Matrix

The Fault Type vector indicates the numbering used in the MATLAB code to identify each fault type. So, Stuck-at Fault is **(1)**, Gain Fault is **(2)**, Offset Fault is **(3)**, Out of Bound fault is **(4)** and Data loss Fault is **(5)** (Fig. 5).

1	2	3	4	5
StuckAtFault(1)	GainFault(2)	OffSetFault(3)	OutOfBound(4)	DataLoss(5)

Fig. 5. Fault Type Vector

The fault injection matrix is a 3D matrix where the **first dimension** (the rows of the matrix = 2) represents the **Fault Injection Repetition**. So, if there is an intermittent fault, then it will be repeated twice. The **second dimension** of the matrix (the columns of the matrix = 12) represents the **number of rooms** in the building, which in our scenario is 12. Finally, the **third dimension** of the matrix represents **each component** in the system. Where the component is numbered as follows:

CO2 Sensor **(1)**, Damper Actuator **(2)**, Temperature Sensor **(3)** and Heater Actuator **(4)** (Fig. 6)

```
val(:,:,1) =

     5     0     0     0     0     0     1     0     0     0     0     0
     0     0     0     0     0     0     0     0     0     0     0     0


val(:,:,2) =

     0     0     1     0     0     0     0     0     0     0     0     0
     0     0     0     0     0     0     0     0     0     0     0     0


val(:,:,3) =

     0     5     0     0     0     0     0     0     0     0     0     0
     0     0     0     0     0     0     0     0     0     0     0     0


val(:,:,4) =

     0     0     0     1     0     0     0     0     0     0     0     0
     0     0     0     0     0     0     0     0     0     0     0     0
```

Fig. 6. Fault Injection-type 3D Matrix

This **Fault Injection Time** matrix is similarly a 3D matrix. This matrix provides the time (in seconds) at which the fault was injected into the component. The total duration for which the simulation has been run is called the simulation time. For this research the simulation time is 86,400 s. The assumed scenarios are as follow. Room 1, component 1 (CO_2 sensor) – There is an Intermittent data loss fault (Fault type 5) with two repetitions at **31,295th** second and at **35,855th** second. Room 2, component 3 (Temperature sensor) – There is an Intermittent data loss fault (Fault type 5) with two repetitions at **18,152nd** second and at **23,316th** second.

Fault Injection Duration matrix is also a 3D matrix, provides the details about the duration of time in seconds for which the fault stays in the system. In our scenarios: Room 1, component 1 (CO_2 sensor) – There is an Intermittent data loss fault (Fault type 5) with two repetitions, the first repetition of the fault lasts for **1,690** s and the second repetition of the fault lasts for **2,937** s. Room 2, component 3 (Temperature sensor) – There is an Intermittent data loss fault (Fault type 5) with two repetitions, the first repetition of the fault lasts for **2,936** s and the second repetition of the fault lasts for **2,689** s.

5 Maintenance Test Generation, Team Simulation, and GUI

5.1 Deterministic Fault→Test Mapping

Among all the maintenance tests, the chosen quintet of maintenance tests aligns precisely with the five fault types (Stuck-at Fault is **(1)**, Gain Fault is **(2)**, Offset Fault is **(3)**, Out of Bound fault is **(4)** and Data loss Fault is **(5)**). This meticulous approach entails the systematic execution of the selected maintenance tests for each individual component situated in every designated room. The responsibility for executing these tests rests with proficient testing teams, diligently working to scrutinize and evaluate the efficacy of the HVAC system's components. The subsequent phase involves the simulation of these maintenance tests within the sophisticated MATLAB/Simulink environment, an integrated platform renowned for its computational prowess and analytical capabilities.

Facilitating this intricate testing process is a user-friendly and purpose-built GUI, thoughtfully designed and implemented through the MATLAB app design tool. This GUI, aptly titled 'Maintenance Test Checklist and Test Teams Simulation GUI,' acts as a central hub for orchestrating the testing procedures. It not only streamlines the simulation of the selected maintenance tests but also serves as a comprehensive repository of descriptive information regarding these tests. The GUI plays a crucial role in improving the effectiveness and consistency of the testing framework. It seamlessly integrates with the MATLAB/Simulink environment, contributing to the overall enhancement of the quality and precision of the research efforts conducted in this paper.

Thus, the description of the selected five Maintenance test checklist based on the five fault types are as follows:

1. **Open/Close operation check for StuckAtFault:**

   Ensure that the damper blades fully open and close without any obstructions or binding. Observe if the actuator operates smoothly and without any excessive noise or vibrations.

2. **Sensor Accuracy Check for GainFault:**

   Note any significant discrepancies in sensor readings and evaluate if recalibration or replacement is necessary. Compare the sensor readings to a calibrated reference instrument or another verified sensor.

3. **Linearity Test for OffSetFault:**

   Place the sensor in an environment with a range of known values. Check if the sensor provides accurate and consistent readings and evaluate linearity.

4. **Sensor Calibration Check for OutOfBoundFault:**

   Review the calibration records and ensure that the component has been calibrated within the recommended intervals by the manufacturer.

5. **Data Transmission Check for DataLossFault:**

   Verify that the sensor data is being accurately transmitted and received by the BAS or control system.

We curated **manufacturer-style checklists** (e.g., transmission check, range/bounds check, calibration/response test for sensors; open/close operation and torque checks for dampers; output verification for heaters) and created a **deterministic mapping** between FAULT TYPE and MAINTENANCE TEST for the experiments reported here. Examples:

Data-Loss (Sensor): Mark Data Transmission Check as FAILED for that sensor.

Stuck-at (Sensor): mark Calibration/Response Test as FAILED (sensor reading does not respond to stimulus).

Stuck-Open/Closed (Damper): mark Open/Close Operation Check as FAILED.

Heater Stuck ON/OFF: mark Heater Output Verification as FAILED.

This mapping is stored in the GUI and applied automatically once faults are specified, guaranteeing reproducible **fault→test** traces.

5.2 Team Allocation Model

Here, 'team simulation' means only the automated assignment of tests to teams. We do not model technician execution time or human factors; we simply allocate each test to a maintenance team (e.g. by floor). A lightweight team allocator assigns tests to **floor-based** teams (e.g., Team A → Floor 1, Team B → Floor 2). This model mirrors common FM practice, zonal responsibility simplifies routing and accountability, and provides a foundation for more sophisticated optimizers in future work (Shalabi et al., 2020). The allocator writes **team IDs** into the Maintenance Test & Testing Team Table.

5.3 Output Artifacts

The simulation produces two structured outputs:

Maintenance Test and Testing Team Table. For each **room–component** with an injected fault, the table lists: component type, fault type, **failing maintenance test(s)** from the mapping, and assigned **team**. This is the actionable artifact that technicians can import to CMMS/BIM.

Final Test Status Table. A summary table with one row per **room–component**, indicating **Pass/Fail**. In our experiments, **every faulted component** fails its mapped test(s), and **no healthy component** is flagged, crucial for operational trust.

5.4 MATLAB GUI

The App Designer GUI unifies the workflow in four panels as shown in figure below. **Test Checklist Selector (top-left):** choose component/test, view descriptions. Simulation Controls (bottom-left): Start/Pause/Stop Simulink runs.

Teams Input (top-right): set number of maintenance teams; confirm allocation. **Scenario Parameters (bottom-right):** CO_2 setpoint, room temperature, wind speed.

The GUI reads the FI matrices, triggers simulations, calls *TestTeamsSimulation.m* to generate tables, and optionally plots traces (CO_2, temperature, actuator commands) for selected components to aid diagnosis (Behravan, 2021; Behravan et al., 2019) (Fig. 7).

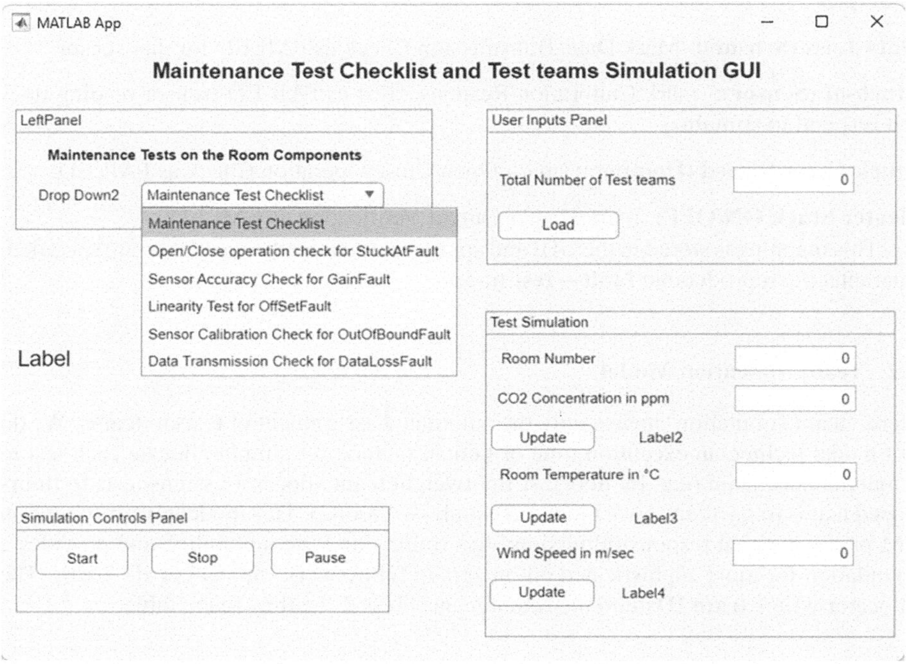

Fig. 7. Maintenance Test Checklist and test teams Simulation GUI

6 Simulation Setup, Scenarios, and Results

We validate on a **6-room, 2-floor** composable model inspired by an office configuration at the University of Siegen (two zones per floor with typical office sizes). All scenarios are encoded in 3D FI matrices specifying **start times**, **durations**, and **persistence** per room–component. For comparability, we reuse timing patterns across scenarios and adjust only baselines. This produces datasets where **only** the intended factor changes (e.g., CO_2 setpoint) while FI timing remains consistent, enabling clear attributions of observed behavior to fault dynamics vs. background.

6.1 Test Teams Simulation

Following the user input regarding the total number of test teams within the comprehensive GUI mentioned earlier, a pivotal phase is initiated in the research process. Concurrently, the fault injection scenarios are crafted. Subsequently, the execution of the

MATLAB file titled ***'TestTeamsSimulation.m'*** is then triggered. This strategic file execution is instrumental in the creation of two distinct tables, thereby providing a structured and comprehensive output for further analysis and evaluation. 1) the Maintenance Test and Testing Team Table, and 2) the Final Table (Table 4).

Table 4. Maintenance Test and Testing Team Table

	1 RoomNumber	2 InspectionTest	3 CO2 Sensor	4 Damper Actuator	5 Temperature Sensor	6 Heater Actuator	7 FloorNumber	8 Test Team 1	9 Test Team 2	10 Test Team 3
1	'Room Number 1'	'Fan Open/Close operation check for StuckAtFault'	'Test Passed'	'Test Passed'	'Test Passed'	'Test Passed'	'Floor Number 1'	'Test Team 1'	''	''
2	'Room Number 1'	'Sensor Accuracy Check for GainFault'	'Test Passed'	'Test Passed'	'Test Passed'	'Test Passed'	'Floor Number 1'	'Test Team 1'	''	''
3	'Room Number 1'	'Linearity Test for OffSetFault'	'Test Passed'	'Test Passed'	'Test Passed'	'Test Passed'	'Floor Number 1'	'Test Team 1'	''	''
4	'Room Number 1'	'Sensor Calibration Check for OutOfBoundFault'	'Test Passed'	'Test Passed'	'Test Passed'	'Test Passed'	'Floor Number 1'	'Test Team 1'	''	''
5	'Room Number 1'	'Data Transmission Check for DataLossFault'	'Test Failed - Fault Detected'	'Test Passed'	'Test Passed'	'Test Passed'	'Floor Number 1'	'Test Team 1'	''	''
6	'Room Number 2'	'Fan Open/Close operation check for StuckAtFault'	'Test Passed'	'Test Passed'	'Test Passed'	'Test Passed'	'Floor Number 1'	'Test Team 1'	''	''
7	'Room Number 2'	'Sensor Accuracy Check for GainFault'	'Test Passed'	'Test Passed'	'Test Passed'	'Test Passed'	'Floor Number 1'	'Test Team 1'	''	''
8	'Room Number 2'	'Linearity Test for OffSetFault'	'Test Passed'	'Test Passed'	'Test Passed'	'Test Passed'	'Floor Number 1'	'Test Team 1'	''	''
9	'Room Number 2'	'Sensor Calibration Check for OutOfBoundFault'	'Test Passed'	'Test Passed'	'Test Passed'	'Test Passed'	'Floor Number 1'	'Test Team 1'	''	''
10	'Room Number 2'	'Data Transmission Check for DataLossFault'	'Test Passed'	'Test Passed'	'Test Failed - Fault Detected'	'Test Passed'	'Floor Number 1'	'Test Team 1'	''	''

The "Maintenance Test and Testing Team Table," encapsulates essential data points, including the chosen maintenance tests, corresponding testing teams, and other pertinent details derived from the user input within the GUI. The table operates in accordance with a logical framework, where the maintenance test outcomes for specific components, in which faults are present, are determined based on the type of fault within the room. For instance, in Room 1, Component 1 (CO2 sensor), where the fault type is 5, the fifth maintenance test (Data Transmission Check for DataLossFault) for the CO2 sensor will unequivocally fail. This logic extends uniformly to all other components situated in all tested rooms throughout the building, as scrutinized by the diligent testing teams. Similarly, Room 2, Component 3 (Temperature sensor) exhibit comparable fault-induced test failures. This tabular representation also enhances data organization of the testing teams by randomly allocating a testing team to a specific floor. This floor-wise allocation of the testing teams makes the organizational structure less complicated when suppose there are multiple testing teams for a massive building structure. Thus, the last three columns of the table provide the details about the organization and allocation of the Testing teams. There are three columns because only three testing teams were considered for the above example. This approach thus lays the groundwork for a systematic examination of the interplay between maintenance tests and the designated testing teams (Table 5).

Table 5. Final Table

	1 RoomNumber	2 CO2 Sensor	3 Damper Actuator	4 Temperature Sensor	5 Heater Actuator
1	'Room Number 1'	'Failed'	'Passed'	'Passed'	'Passed'
2	'Room Number 2'	'Passed'	'Passed'	'Failed'	'Passed'
3	'Room Number 3'	'Passed'	'Failed'	'Passed'	'Passed'
4	'Room Number 4'	'Passed'	'Passed'	'Passed'	'Failed'
5	'Room Number 5'	'Passed'	'Passed'	'Passed'	'Passed'
6	'Room Number 6'	'Passed'	'Passed'	'Passed'	'Passed'
7	'Room Number 7'	'Failed'	'Passed'	'Passed'	'Passed'
8	'Room Number 8'	'Passed'	'Passed'	'Passed'	'Passed'
9	'Room Number 9'	'Passed'	'Passed'	'Passed'	'Passed'
10	'Room Number 10'	'Passed'	'Passed'	'Passed'	'Passed'
11	'Room Number 11'	'Passed'	'Passed'	'Passed'	'Passed'
12	'Room Number 12'	'Passed'	'Passed'	'Passed'	'Passed'

The final table, on the other hand, serves as a comprehensive repository, providing the final test status for all rooms and components within the building. Remarkably, this table reveals the declaration of failure for five specific components. These declarations are attributed to the presence of faults in particular components. The ensuing synthesis of these tables contributes significantly to the overarching objective of comprehensively assessing and categorizing the performance and reliability of HVAC system components under diverse fault scenarios and testing conditions. In essence, the orchestrated execution of the 'TestTeamsSimulation.m' MATLAB file, post-user input and fault injection scenario creation, sets the stage for a meticulous exploration and documentation of the research outcomes.

6.2 Scenarios—By Room and Component

Room 1 — CO_2 sensor intermittent data-loss (type 5). The GUI interface offers a dynamic platform for observing and analysing the HVAC components' responses under diverse test scenarios on Simulink. Thus, a series of different testing scenarios was created by altering the ventilation parameters of the rooms that concatenated faults and the response of each was observed. The below scenario shows the response of a healthy room where there are no faults in any of the HVAC room components, and all the components are in a healthy condition. The maintenance tests of the room with such conditions are decided as *passed* by the testing teams (Fig. 8).

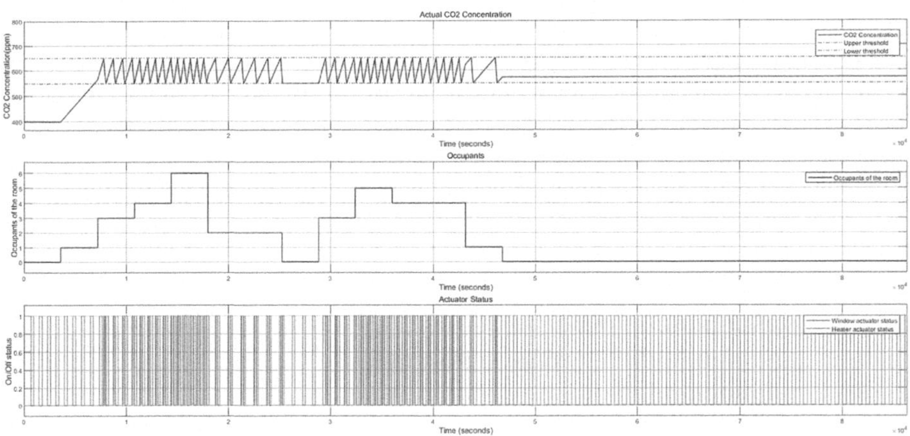

Fig. 8. Healthy Room with no faults in the components and CO_2 concentration = 600 ppm, Room Temperature = 20 °C and Wind Velocity = 3.6 m/s

In test scenario #1 **Room 1**, there is an intermittent **data loss** fault of fault type 5 injected in the **CO_2 sensor** by the fault injection GUI. It can be observed from the three results below that there is an intermittent data loss fault (fault type 5) in the CO_2 sensor with two fault repetitions at **31,295th** second and at **35,855th** second, which lasts for **1,690** s and **2,937** s, respectively. It was observed in all three scenarios that the CO_2 sensor data is missing during each fault duration, and the damper actuator is stuck at '0'

(CLOSED position) during both the fault duration. This significantly gives a rise in the CO_2 concentration of the room during the fault duration (Fig. 9).

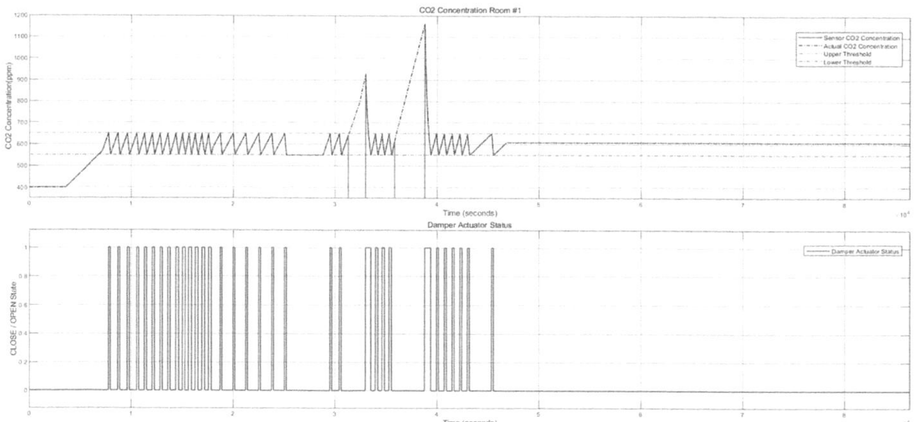

Fig. 9. Room1_CO_2 sensor_Data loss fault and CO_2 cocentration_600PPM

In test scenario #2 **Room 2**, there is an intermittent **data loss** fault of fault type 5 injected in the **temperature sensor** (Fig. 10).

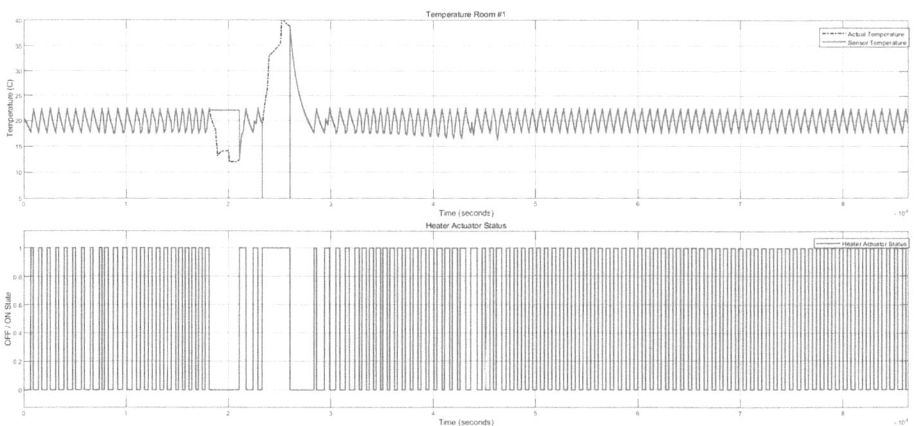

Fig. 10. Room2_Temperature Sensor_Data loss fault_RoomTemperature_20°c

In the above test scenario, there is an Intermittent data loss fault of fault type 5 with two repetitions at **18,152nd** second and at **23,316th** second which lasts for **2,936** s and **2,689** s, respectively, in the temperature sensor at **room temperature 20°c**. It is observed that for the first fault repetition, the heater actuator is stuck at '0' (OFF position), which leads to a fall in the room temperature, and in the second fault repetition, the heater actuator is stuck at '1' (ON position), which leads to an increase in the room temperature.

In test scenario #3 **Room 3**, there is a permanent **stuck-at fault** of fault type 5 injected in the Damper actuator at **28,115th** second and it lasts till the end of the simulation run time. It was observed that the Damper actuator is permanently stuck at ‘1’ (OPEN position) after the injection of the fault. This leads to the continuous decrease in the CO_2 concentration inside the room which ultimately becomes equal to the outdoor atmosphere value of the CO_2 concentration, which is 400 ppm (Fig. 11).

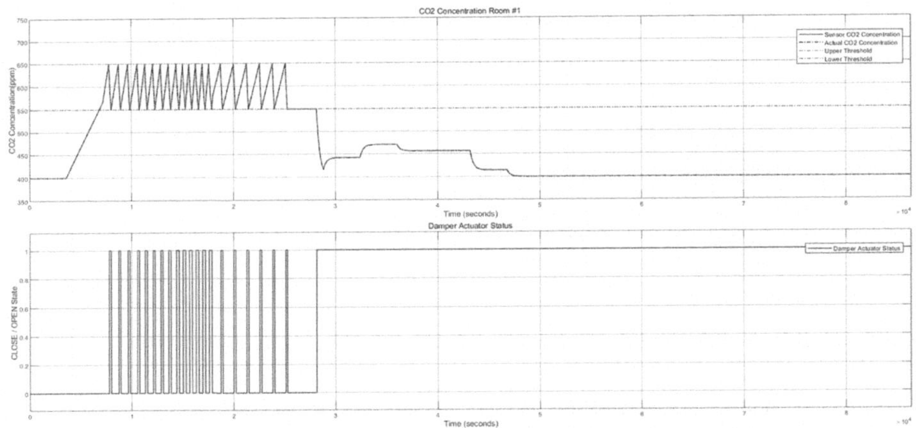

Fig. 11. Room3_DamperActuator_permanent stuck at fault

7 Discussion, Limitations, and Threats to Validity

Operationally, these tables are the missing **bridge** from simulation to **action**: they encode **what** to test, **where**, and **by whom**. Because the mapping and allocation are automatic, planners avoid ad-hoc triage and deliver precise test lists to technicians faster. **On the deterministic mapping,** we adopted a deterministic fault→test mapping to make the pipeline transparent and auditable. Real maintenance often involves **multiple tests per fault** and **probabilistic detection** due to noise, hysteresis, and human variability (Dalal et al., 2000; Li et al., 2018). Extending the mapping to a probabilistic graph (fault→{tests} with detection likelihoods) is conceptually straightforward in this architecture and a priority for future work. **On team allocation,** our floor-based allocation is intentionally simple to communicate the concept and avoid confounding the core contribution with optimization machinery. Field deployments will benefit from **route planning** (e.g., vehicle routing), **skills matching**, and **spares availability** models. Because outputs are standard tables, one can bolt on optimizers without changing the FI core (Shalabi et al., 2020). **On scalability and composability,** the component-based Simulink model scales well to dozens of zones; however, solver performance and logging volumes can grow quickly with very large $K \times N$. Techniques such as step-size control, selective logging, and parallelization mitigate this. Importantly, the **FI encode–simulate–generate** pattern remains unchanged.

On generalizability, our fault library covers high-impact types (stuck, data loss, offset/gain). Field data sometimes reveal subtler faults (temperature sensor placement bias due to radiant asymmetry, damper hysteresis/gear backlash, intermittent controller communication faults). The FI block architecture can host these with minimal changes. We note that for our case study (6 rooms, 2 floors) simulation runs quickly, but larger building models could increase solver run-time and log volume. Evaluating performance on larger models is left for future work. Similarly, we have not yet compared our automated workflow against manual planning or obtained technician feedback. These comparisons and expert evaluations will be pursued in future studies.

Threats to validity. (i) **Construct validity:** Are our simulated faults representative of field faults? We drew from established taxonomies (Li et al., 2018; Ziade et al., 2004), but building-specific idiosyncrasies may differ. (ii) **Internal validity:** Deterministic mapping may overstate detection coherence relative to practice; we mitigated this by using strong, signature-like faults. (iii) **External validity:** Different building typologies (labs, healthcare) may have tighter IAQ constraints and different control strategies; still, the fault→test→team chain is broadly applicable.

8 Conclusion and Future Work

We introduced an **end-to-end framework** that connects **fault injection** in a composable DCV–heating model to **maintenance test generation** and **team allocation**, producing actionable artifacts that facility managers can use immediately. Across 15 scenarios covering intermittent and permanent faults in sensors and actuators, the approach (i) reliably mapped faults to failed tests, (ii) produced no false failures on healthy components, and (iii) allocated work to teams in a clear, floor-based scheme. (iv) Explore integration with building digital-twin frameworks so that the Simulink model can be updated with real-time building data, enabling the fault→test→team pipeline to operate in an online, operational environment. This closes a lingering gap in the literature between FI/MBT research and **real-world maintenance workflows**.

Future work will: (1) extend the fault→test mapping to **probabilistic** detection matrices; (2) integrate **optimization** for team routing, skills, and spares; (3) perform **field validation** using CMMS/BIM integration; (4) incorporate **cost/criticality** to prioritize tests (e.g., IAQ risk scoring); and (5) broaden the FI library (e.g., communication faults, intermittent actuator torque limits). By anchoring research in artifacts that planners and technicians can directly use, we aim to accelerate the path from **fault insight** to **fault closure**, improving both energy performance and occupant health in DCV-equipped buildings.

Acknowledgments. The authors gratefully acknowledge project discussions and feedback from colleagues at the University of Siegen and GreeNovaX that informed the GUI design and FI scenarios. Any errors remain our own.

Disclosure of Interests The authors have no competing interests to declare that are relevant to the content of this article.

References

Ashrae: ANSI/ASHRAE Standard 62.1–2016: Ventilation for Acceptable Indoor Air Quality. ASHRAE, Atlanta (2016)

Behravan, A.: Diagnostic Classifiers Based on Fuzzy Bayesian Belief Networks and Deep Neural Net Works for Demand-Controlled Ventilation and Heating Systems. PhD thesis,. University of Siegen (2021)

Behravan, A., Kiamanesh, B., Bhandari, S.: Component-based system model Design of Multiple Fault Injection Framework for DCV and heating systems. In: Proceedings of CSCI 2023, Las Vegas (2023)

Behravan, A., Obermaisser, R., Basavegowda, D.H., Meckel, S.: Automatic model-based fault detection and diagnosis using diagnostic directed acyclic graph for a DCV and heating system in Simulink. In: Proceedings 2018 (2018)

Behravan, A., Kiamanesh, B., Obermaisser, R.: Composability Modeling for the use case of demand-controlled ventilation and heating system. In: Proceedings CoDIT 2019 (2019)

Bhandari, S.: Component-Based System Model Design of Multiple-Fault Injection Framework for DCV and Heating Systems. Master thesis,. University of Siegen, Germany (2023)

Dalal, S. R., Jain, A., Karunanithi, N., Lott, C., Patton, G., & Horowitz, B.: Model-Based Testing in Practice. ICSE '99. (2000)

Fischer, S., Lochau, M., Thüm, T., Schaefer, I.: Testing of highly configurable cyber-physical systems – a multiple case study. Empir. Softw. Eng. **26**, 67 (2021)

Li, G., Chen, Y., Hu, Y.: Investigating thermostat sensor offset impacts on operating performance and thermal comfort of three different HVAC systems in Wuhan. China. Build. Environ. **210**, 108732 (2022)

Fu Xiao, F., Wang, S., Jin, X.: Progress and methodologies of lifecycle commissioning of HVAC systems to enhance building sustainability. Renew. Sust. Energ. Rev. **13**(2), 418–432 (2008)

International Energy Agency: The Critical Role of Buildings / Global Status Report for Buildings and Construction. IEA, Paris (2021)

Katipamula, S., Brambley, M.: Review of fault detection and diagnosis methods for HVAC systems. HVAC&R Res. **11**(1), 3–25 (2005)

Kiamanesh, B., Behravan, A., Obermaisser, R.: Fault injection with multiple fault patterns for experimental evaluation of demand-controlled ventilation and heating systems. Sensors **22**(21), 8180 (2022a)

Kiamanesh, B., Behravan, A., Obermaisser, R.: Fault injection with multiple fault patterns for experimental evaluation of demand-controlled ventilation and heating systems. Sensors **22**, 8180 (2022b)

Murphy, J.A.: CO_2-based demand-controlled ventilation with ASHRAE standard 62.1-2004. Trane Eng. Newslett. **34**(5), 1–5 (2005)

Riederer, P., Marchio, D., Boudier, P.: Room thermal modelling adapted to the test of HVAC control systems. Build. Environ. **37**(10), 1043–1052 (2002)

Schmidt, A., Hussain, S., Schriegel, S.: Model-based testing methodology using system entity structures for MATLAB/Simulink models. Simulation **92**(10), 937–953 (2016)

Shalabi, F., Turkan, Y., Sbeih, S.: BIM–energy simulation approach for detecting building spaces with faults and problematic behavior. ITcon. **25**, 355–371 (2020)

Wang, L., Hong, T., Zhu, N.: Monitoring-based HVAC commissioning of an existing office building for energy efficiency. Appl. Energy **102**, 1382–1390 (2012)

Weimer, J., Sangiovanni-Vincentelli, A., Pappas, G.: Active actuator fault detection diagnostics in HVAC systems. In: Proceedings of the IEEE Conference on Decision and Control, pp. 7083–7088 (2012)

Xu, P., Haves, P., Turrin, R.: Model-based automated functional testing – methodology and application to air-handling units. In: Proceedings IBPSA 2005 (2005)

Yang, A., Holøs, S.B., Resvoll, M.O., Mysen, M., Fjellheim, Ø.: Temperature-dependent ventilation rates might improve perceived air quality in a demand-controlled ventilation strategy. Build. Environ. **200**, 107957 (2021)

CASCADE: Engineering a Secure, Real-Time, and Collaborative Code Execution for Multi-tenant Learning Platforms

Utkarsh Raj(✉) and Roman Obermaisser

University of Siegen, Siegen, Germany
utkarsh.raj@uni-siegen.de
https://www.eti.uni-siegen.de/es/

Abstract. Training programmers at scale—whether in academic settings or industrial environments—requires platforms that support secure execution, collaborative development, real-time feedback, and modular assessment workflows. Existing tools often lack the execution isolation, simulator integration, and observability needed to support interactive, group-based programming tasks.

We present *CASCADE(Collaborative Asynchronous Secure Code Assessment and Deployment Environment)*, a modular, containerized platform designed for secure, real-time execution and assessment of collaborative code submissions. CASCADE orchestrates hundreds of per-task, per-group containers, providing isolated runtimes where participants can collaboratively edit code, receive immediate feedback, and interact with embedded simulation UIs hosted inside task-specific containers.

The backend is implemented as a monolith with modular components responsible for submission orchestration, WebSocket-based synchronization, and PDF report validation via natural language processing. Each group's environment is sandboxed and self-contained, with execution proxied to Docker containers and feedback streamed live to all participants. Instructor-facing observability is enabled through the ELK stack and Kibana dashboards, offering insights into operational metrics, submission behavior, and system anomalies.

CASCADE has been deployed at the University of Siegen for a full academic year. Its successful application demonstrates how secure sandboxing, real-time collaboration, and architectural composability can be combined into a scalable platform suitable for both academic and industrial programming training.

Keywords: Containerization · Web Application · Multi-Tenant Architecture · Collaborative Code Execution · Semantic Analysis

T. Ensari et al. (Eds.): ISPR 2025, CCIS 2859, pp. 358–373, 2026.
https://doi.org/10.1007/978-3-032-21585-7_26

1 Introduction

Secure, real-time execution of untrusted code in collaborative environments is a fundamental challenge in software systems engineering. This challenge becomes particularly complex when the environment must support group-based development, container-level isolation, interactive simulation, real-time feedback, and full observability—-all within the context of multi-tenant infrastructure. While general-purpose platforms like GitHub Codespaces, Replit, or JupyterHub provide interactive coding environments [9], they are not explicitly designed to offer group-level state sharing, simulation integration, or real-time feedback synchronized across users. Similarly, automated grading systems such as CodeGrade or Gradescope often provide test-based assessment but lack runtime isolation, shared execution context, or extensibility to host visual or domain-specific simulators.

To address these limitations, we introduce *CASCADE*(Collaborative Asynchronous Secure Code Assessment and Deployment Environment), a containerized, modular platform designed to execute and assess user-submitted code in secure, group-scoped environments. Each user group interacts with a dedicated execution container instantiated from a task-specific image. These containers can expose optional simulation interfaces—such as physics engines or scheduling visualizations—that users interact with collaboratively via browser-based UIs. Code execution is fully isolated and proxied via Docker, while stdout streams are relayed in real time through a WebSocket infrastructure, ensuring that all group members view a synchronized session.

CASCADE supports multiple concurrent courses and tasks, and uses Task-queue backed coordination with *worker processes* for background task processing. Submissions in Python and C++ are automatically assessed using Pytest and Google Test [2], while written reports in PDF format are parsed and validated using natural language processing techniques via NLTK [5]. Logs and usage metrics are collected via an integrated ELK stack, with Kibana dashboards exposing both operational health and group behavior patterns to instructors and administrators [8].

Unlike research prototypes or siloed solutions, CASCADE has been deployed across multiple cohorts at the University of Siegen over a full academic year.

This paper presents the design, implementation, and deployment experience of CASCADE. We focus on its architectural structure, real-time collaboration mechanisms, isolation and security strategies, observability stack, and lessons learned from production use. The system architecture is modeled using the C4 approach [7], with diagrams rendered in PlantUML [16] to communicate component boundaries, inter-service coordination, and integration with external systems. CASCADE demonstrates that secure, collaborative, simulator-driven execution environments can be built using modular software architecture and widely available systems tools.

2 Related Work

A wide range of systems have been developed to support scalable programming instruction, including interactive development environments, automated grading platforms, and container-based execution backends. These systems differ significantly in their support for group collaboration, execution isolation, real-time feedback, simulation, and system observability.

JupyterHub [9] and **Google Colaboratory** [6] are widely used in both educational and professional contexts, offering notebook-based interactive development backed by per-user container isolation. While well-suited for data science workflows, these platforms lack native support for group collaboration, shared session state, or structured assessment pipelines involving simulations or document evaluation.

GitHub Codespaces [1] and **Replit** provide real-time collaborative editing environments hosted in the cloud. Both support containerized execution and team-based coding, but they are general-purpose development tools and do not include structured grading support, per-task sandboxing, PDF report evaluation, or instructor observability dashboards. Their execution model is typically scoped to individual users or loose collaboration, rather than to persistent, task-specific containers per group.

Several classroom-focused grading platforms such as **Gradescope** [18], **CodeGrade**, and **nbgrader** [10] offer automated testing and rubric-based grading. These tools integrate with learning management systems and streamline assessment workflows but rarely support real-time feedback, stdout streaming, simulator interaction, or persistent execution environments. Submissions are usually static uploads processed out-of-band, without a continuous collaborative context.

Other systems such as **CloudCoder** [14] and **Vocareum** explore container-backed execution and scalable backend infrastructure. While they address deployment and grading at scale, they typically lack real-time collaboration features, NLP-based report evaluation, or fine-grained instructor-facing observability. Likewise, most do not integrate interactive simulators as part of the execution environment.

From a software systems perspective, containerization frameworks such as Docker [13] are often employed to provide execution isolation, and observability stacks such as the ELK stack (Elasticsearch, Logstash, Kibana) [8] offer infrastructure-level telemetry and logging. However, these technologies are not commonly embedded cohesively within educational platforms.

In contrast, *CASCADE* provides a unified platform that combines secure, per-group containerization with collaborative code execution, WebSocket-based real-time feedback, embedded simulators, NLP-driven document assessment, and full-stack observability. Its architecture is modeled using the C4 approach [7] and rendered via PlantUML, supporting modularity, deployment clarity, and future extensibility. This combination of features addresses multiple gaps in existing systems, particularly for group-based, simulation-integrated, multi-modal programming instruction.

3 System Architecture

The architecture of CASCADE is designed to ensure secure execution, real-time responsiveness, and extensibility, while remaining comprehensible to both technical stakeholders and educational staff. To communicate its structure effectively, we adopted the C4 model for software architecture visualization [7], which offers a hierarchical approach to describing systems at different levels of abstraction. All diagrams are rendered using PlantUML, enabling easy version control, integration into documentation, and reproducible visuals.

This section presents three C4 views of the system: the System Context Diagram, the Container Diagram, and the Backend Component Diagram.

3.1 System Context

The System Context Diagram in Fig. 1 provides a high-level architectural view of *CASCADE* and its interactions with external users and institutional systems. The platform is designed to serve two primary user roles: users and instructors. users access CASCADE via a web-based interface to collaboratively edit and submit code, receive real-time feedback, and interact with embedded simulators for task-specific visualization. Instructors use a separate interface to configure tasks, monitor submissions, manage user groups, and view real-time system metrics.

CASCADE is deployed within the university's infrastructure and integrates with several supporting services. It interfaces with a Docker host for sandboxed code execution, a Redis instance for coordination via asynchronous task queues and pub/sub messaging [17], and an SMTP server for sending authentication OTPs via email. The C4-based model distinguishes CASCADE as the central software system while clearly representing its boundaries, actors, and dependencies on external infrastructure.

3.2 Containers

The overall architecture of CASCADE is illustrated in Fig. 2 using the C4 model's container-level abstraction. In this context, a container refers to a logical deployable unit of software—such as a web application, frontend interface, or database—not a Docker container. This distinction is particularly important in the CASCADE setting, where the system architecture includes both C4 containers and actual Docker containers used for code execution.

The CASCADE system is composed of multiple collaborating services, grouped within a clearly defined system boundary. users interact with a dedicated frontend interface built using React [11] and Vite, which supports real-time collaborative editing, code submission, and output visualization. Instructors access a separate administration frontend—implemented with the same technology stack—which provides interfaces for configuring tasks, managing courses, and monitoring user progress.

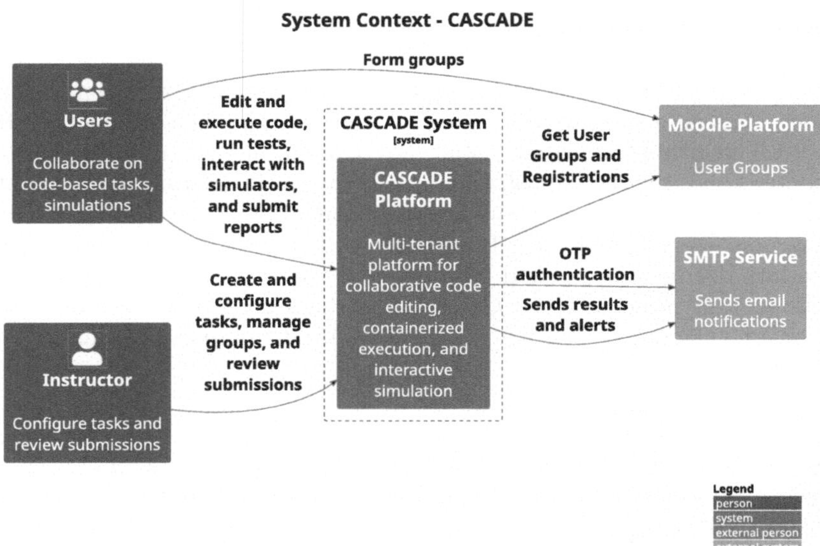

Fig. 1. C4 System Context Diagram showing external actors and infrastructure integrations.

Both frontends communicate with the backend through a shared NGINX [3] reverse proxy that handles routing and termination of HTTPS traffic. The backend is implemented as a FastAPI application [15] that exposes RESTful endpoints to both users and instructors. It orchestrates test execution, manages state, handles file uploads, streams stdout over WebSockets, and communicates with the persistent storage and execution subsystems.

Long-running and asynchronous tasks, such as code execution or report analysis, are offloaded to a Celery worker pool [19] connected to Redis. Redis is used for task queuing, pub/sub messaging, and shared cache. user code execution is performed by a dedicated Execution Engine that interfaces with the Docker CLI to spawn ephemeral, isolated containers for each group-task combination. These containers run with restricted permissions and no network access, providing per-group isolation.

Persistent storage is divided between an SQLite [4] database used for metadata such as user accounts, group assignments, and submission logs, and a Linux-based file system that stores submitted code, test artifacts, generated reports, and logging output. All access to these resources is coordinated through the backend and worker components.

For logging and system observability, the architecture incorporates an ELK stack (Elasticsearch, Logstash, Kibana). Log events generated by the backend and worker services are collected and visualized through a Kibana dashboard, enabling instructors to inspect submission trends, error types, and system-level

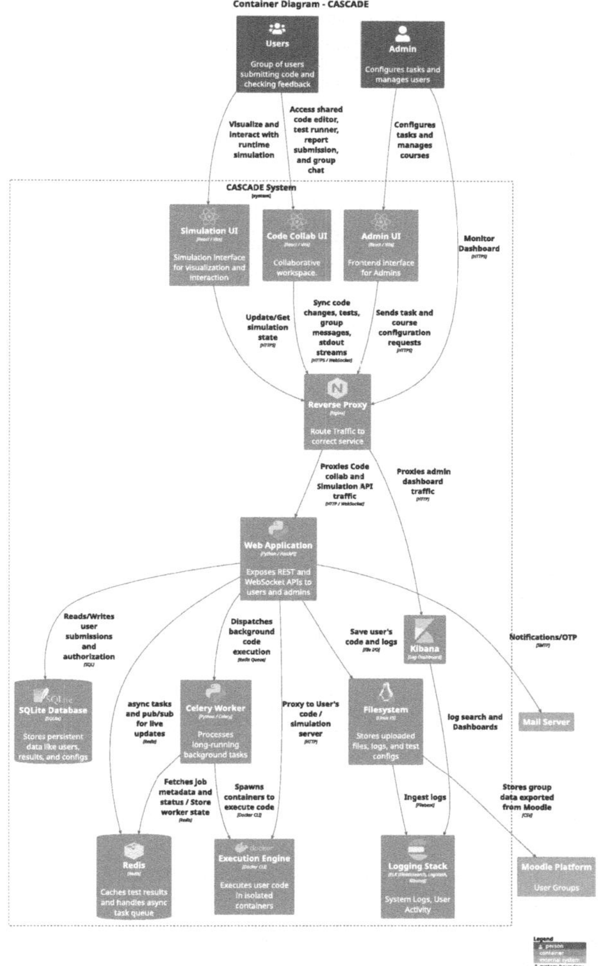

Fig. 2. C4 Container Diagram showing the internal containers of the CASCADE system and their interactions with external services such as Redis, Docker, Moodle, and email infrastructure.

metrics. Moodle integration is handled via direct CSV import into the file system, where a custom loader parses and ingests group data.

Figure 2 shows the complete set of containers, external systems, and their interactions.

3.3 Backend Components

The backend of CASCADE is structured around modular services and routing boundaries that follow concepts inspired by Clean Architecture [12] The core

API surface is divided into three routers—`Auth Router`, `Submission Router`, and `Admin Router`—which map directly to functionality for authentication, user submissions, and instructor configuration, respectively. Each router delegates to a corresponding service class, maintaining logical separation between routing, validation, and system-level actions.

Figure 3 presents the detailed component architecture, modeled using C4 and rendered with PlantUML. Within the backend boundary, the `AuthService` handles email-based OTP verification and JWT issuance, communicating with the SMTP client and the SQLite-backed metadata store. The `SubmissionService` receives incoming code and report submissions, persists them to disk, notifies Redis via pub/sub channels to trigger test execution, and streams real-time output back to users through WebSockets. Collaborative editing and messaging are coordinated by the `CollabService`, which ensures consistent state across group members through shared Redis channels. Unlike other services, CollabService interacts exclusively via a long-lived WebSocket handler and does not expose REST routes through a dedicated FastAPI router. The `AdminService` exposes endpoints for creating tasks, importing user groups from Moodle CSV files, uploading test configurations, and viewing test execution logs.

These services interact with a set of infrastructural components abstracted as internal collaborators. The `DockerProxy` is responsible for spawning per-task containers and interacting with the Docker API. File access and storage

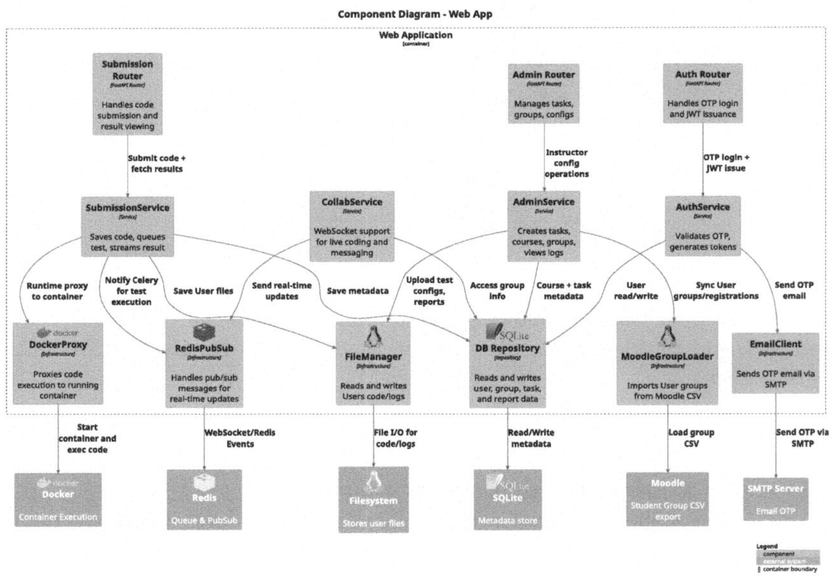

Fig. 3. C4 Component Diagram of the CASCADE backend, showing the internal FastAPI routers and services, infrastructure components, and their integration with external systems.

of code, reports, and logs are managed by the `FileManager`, while metadata access for users, groups, and tasks flows through the `DB Repository` component. The `RedisPubSub` module handles asynchronous pub/sub messaging for real-time updates, and the `EmailClient` delivers OTPs via SMTP.

External systems are kept logically separate from the backend boundary and include Docker, Redis, SQLite, Moodle (for group roster import), an SMTP server for email verification, and the shared file system that persists all user-facing data.

Although deployed as a monolith, this componentized architecture has proven effective for development velocity and fault isolation. Each logical unit can be independently tested and evolved, and the current design allows for incremental migration toward a service-oriented architecture, should future requirements demand independent scaling or deployment.

4 Features and Implementation

CASCADE is implemented as a modular, containerized platform that enables secure, real-time, and collaborative programming in a multi-tenant academic setting. This section describes the platform's key features and the architectural mechanisms that support them.

4.1 Collaborative Editing and Shared Execution State

Each programming task is defined by a Docker image that includes the required runtime libraries, input files, and optional simulation components. When a user group begins work on a task, a new container instance is instantiated, scoped to both the group and the specific task. These containers run under non-root users, with no external network access and resource-limited cgroups to ensure strong isolation guarantees. All user-submitted code executes within these ephemeral environments, supporting both security and reproducibility across evaluations.

CASCADE treats each group as a unified execution unit with synchronized editor state, test results, and terminal output. All group members observe the same shared session, with real-time updates propagated via shared channels. Code edits are centrally versioned and persisted server-side, enabling session restoration. Figure 4 shows the sequence of interactions between group members and the backend during collaborative editing.

To enable immediate feedback and increase interactivity, CASCADE maintains persistent WebSocket channels between the frontend and backend. As code executes inside the container, stdout output, test results, and log messages are streamed to all group members in real time. The `WebSocketManager` component handles event multiplexing and group-based synchronization, ensuring low-latency delivery across connected clients.

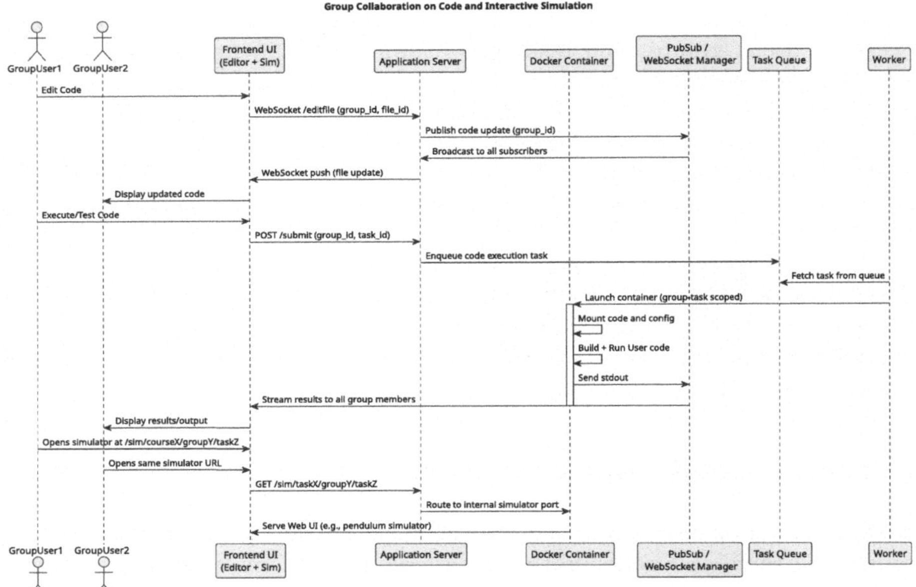

Fig. 4. Sequence diagram illustrating collaborative editing and shared feedback across group members.

4.2 Automated Testing and Report Analysis

Automated testing is integrated into the execution pipeline to ensure consistent and objective assessment. Python submissions are evaluated using Pytest, while C++ code is tested using Google Test [2]. Test suites are mounted into the container at runtime, and results are streamed live to the frontend. Figure 5 illustrates the execution flow, including test orchestration, result capture, and polling-based feedback delivery.

For tasks involving written documentation, CASCADE accepts PDF uploads and performs automated analysis using the NLTK toolkit [5]. The `ReportChecker` component extracts and tokenizes the document text, applying configurable rubric checks based on keyword coverage, structural elements, and basic language heuristics. This allows instructors to provide early, objective feedback on report quality. Future versions of the system aim to integrate large language models (LLMs) [20] to support deeper semantic validation, rubric alignment, and content feedback.

4.3 Multi-course and Multi-task Management

CASCADE supports concurrent operation across multiple courses, each comprising multiple tasks and group configurations. Every submission is identified by a triplet `(course_id, group_id, task_id)`, enabling fine-grained control, task

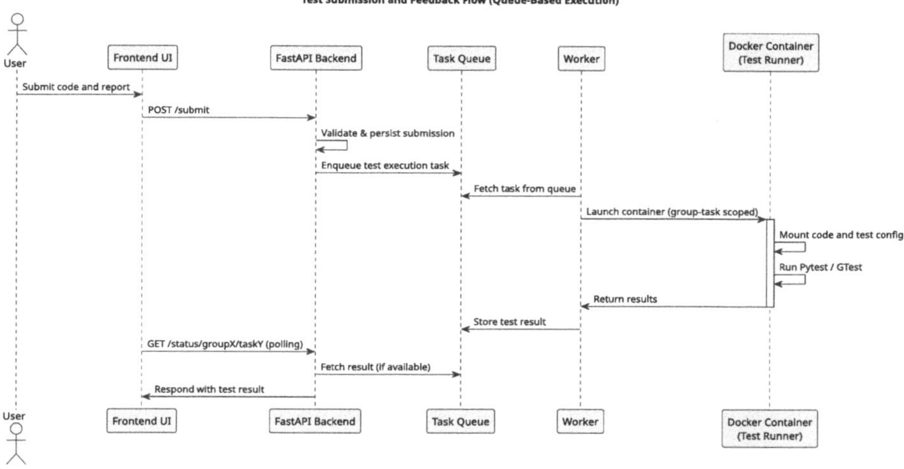

Fig. 5. Sequence diagram for test execution and result propagation.

reuse across semesters, and scoped access enforcement. Group-based authentication ensures that only authorized members can interact with their assigned container instances and submission history.

4.4 Instructor Observability with ELK and Kibana

All system events—including submission actions, container lifecycle events, test failures, and application logs—are streamed to a centralized ELK stack [8]. Kibana dashboards expose these metrics in real time, providing instructors with actionable insights such as submission frequencies, error trends, infrastructure health, and group progress. This observability layer supports early intervention, capacity planning, and platform tuning based on actual usage patterns.

5 Security and Isolation Mechanisms

Security and process isolation are foundational to CASCADE's design, given its role in executing untrusted student code in a shared academic infrastructure. The system leverages container-level security, network segmentation, and privilege minimization to ensure safe operation at scale.

5.1 Container-Based Sandboxing

Each student group interacts with an isolated container instance tied to a specific task. These containers run using Docker's default seccomp profile with user namespaces enabled and no privileged capabilities. Containers are launched as non-root users and limited via cgroups to prevent excessive resource consumption or denial-of-service scenarios. Filesystems are mounted as temporary volumes with no host write access, and each container is destroyed after task completion or upon timeout.

5.2 Network Isolation

All task containers are attached to a private internal Docker network and explicitly disallowed from initiating outbound internet traffic. This is enforced using network policies and NGINX rules. The only communication permitted is inbound traffic from the reverse proxy, which exposes group-specific endpoints and simulation UIs under controlled URLs. This prevents students from accessing external resources or scanning the internal infrastructure. NGINX acts as a secure gateway between the student frontend and task-specific containers. Each group's simulation UI (if present) is reverse-proxied under a unique path (e.g., `/sim/course123/group4/task2`), allowing strict scoping and access control. Only authenticated students in the relevant group can access their container endpoint. Unauthorized or cross-group access attempts are logged and blocked. Collaborative sessions are tied to backend-tracked session IDs scoped to `(course_id, group_id, task_id)` triples. WebSocket connections are authenticated and grouped accordingly, ensuring students only receive stdout and updates from their own container. Edits or commands issued from unauthorized clients are rejected at the WebSocketManager layer.

5.3 Auditability and Traceability

All student actions, container lifecycle events, and backend API calls are logged with timestamps and group IDs. These logs are forwarded to Elasticsearch, allowing retrospective analysis and compliance review. Instructors can verify the origin of test failures, submission attempts, and simulator interactions through structured dashboards.

6 Deployment Experience

CASCADE has been actively deployed at the University of Siegen, Germany for over one academic year, supporting collaborative programming tasks, algorithmic simulations, and project-based reporting across undergraduate and graduate courses. This section summarizes the deployment architecture, runtime characteristics, and lessons learned through sustained operation in real classroom settings.

6.1 Infrastructure and Runtime Architecture

The system follows a hybrid deployment model that separates coordination services from execution infrastructure. The central FastAPI backend, NGINX reverse proxy, Redis broker, and ELK stack are containerized and hosted on a university-controlled main server running Ubuntu. These services are managed using Docker Compose and share a persistent volume for storing submissions, logs, and configurations.

Code execution, by contrast, is offloaded to a distributed pool of Raspberry Pi devices configured as Celery workers. Each Pi hosts Docker locally and receives execution jobs over Redis. This setup enables lightweight scaling of execution resources without overloading the main server, and ensures strong isolation between test execution environments and system control services. The architecture avoids the complexity of Kubernetes while still supporting distributed task execution and fault containment. Containers are provisioned dynamically for each unique `(group_id, task_id)` pair, ensuring per-group isolation. These containers are instantiated from task-specific base images that bundle necessary libraries, test suites, and simulators. Execution containers are ephemeral: they are destroyed after task completion or a configurable timeout period. Resource constraints—including CPU shares, memory limits, and mount permissions—were iteratively tuned based on load testing and empirical resource usage during deadline surges.

6.2 Observability, Scaling and Operational Insights

The ELK stack proved essential for maintaining system health and diagnosing runtime anomalies. All backend logs, container events, and WebSocket interactions are centralized in Elasticsearch, with custom Kibana dashboards exposing detailed timelines, submission throughput, failure rates, and container-specific logs. This observability layer enabled rapid detection of simulator misconfigurations, corrupted uploads, and backend overloads. Figure 6 and Fig. 7 show representative dashboards for error trends and system usage.

The platform consistently supported over 50 concurrent group containers during peak classroom activity, with minimal performance degradation. Execution bottlenecks were primarily observed in disk I/O during test execution and memory contention from multiple active simulator UIs. These issues were mitigated through horizontal scaling, task-specific resource limits, and base image optimization. The use of Raspberry Pi workers provided a flexible execution substrate with low cost and predictable performance under load.

6.3 System Resilience and Fault Recovery

Robust fault tolerance mechanisms were critical in a student-facing environment. The system gracefully handled malformed submissions, container crashes, and intermittent network failures through automated container cleanup, retry policies, and detailed error reporting. WebSocket connections were monitored for dropped sessions, and fault injection testing was conducted prior to the semester to validate recovery pathways. The combination of stateless container execution and centralized logging allowed for quick incident response and stable classroom operation throughout the deployment period.

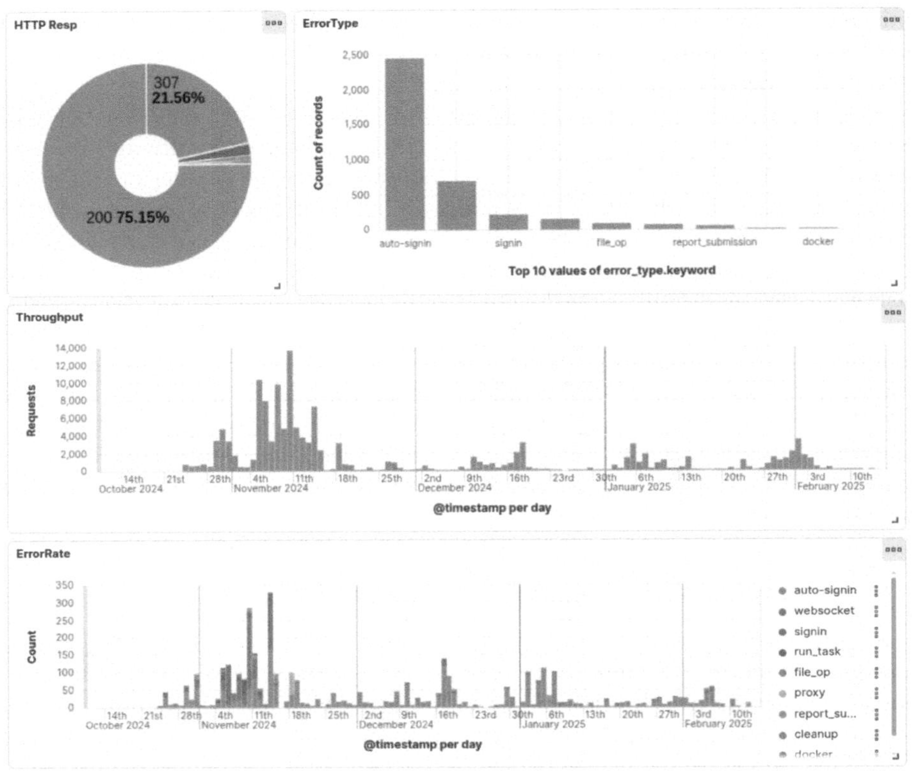

Fig. 6. Error and throughput metrics visualized via Kibana.

6.4 Configurability and Operational Overhead

CASCADE is designed for ease of configuration and low maintenance overhead. Each task is declared using a lightweight metadata schema that specifies the Docker image, test runner type, and simulator endpoint, if any. This declarative approach allows instructors to define new tasks without modifying backend code.

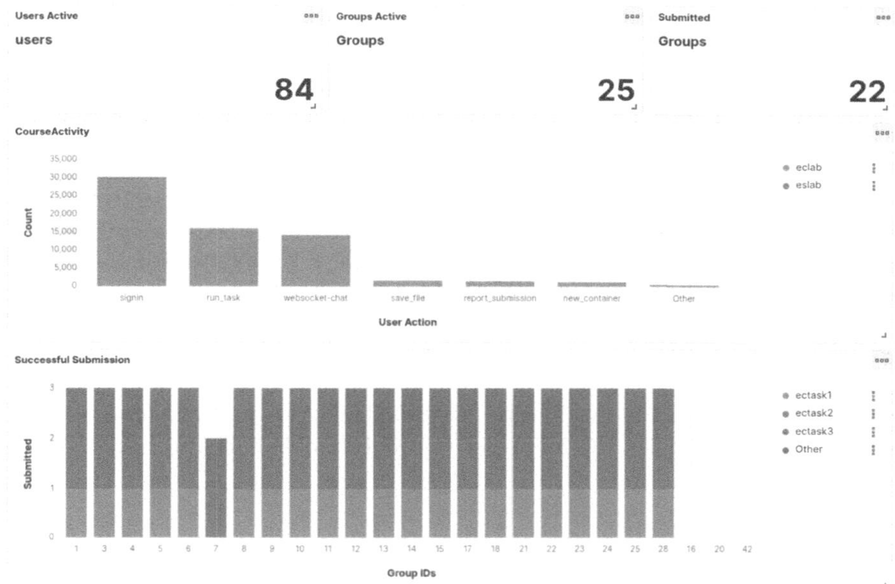

Fig. 7. Submission usage patterns across tasks.

7 Limitations and Future Work

While CASCADE has proven effective, its architecture presents two key limitations that inform future work. First, the system lacks persistent workspaces, as containers are ephemeral, which hinders multi-session, stateful projects. We plan to address this by implementing secure, per-group persistent volumes with instructor-controlled snapshot and reset features. Second, the current report evaluation uses a rule-based Natural Language Processing (NLP) pipeline, which is effective for structural checks but lacks the semantic depth to assess free-form responses. Future development will focus on integrating large language models (LLMs) to support more nuanced, semantic rubric evaluation and provide richer, adaptive feedback on students' conceptual understanding and communication skills.

8 Conclusion

This paper presented *CASCADE*, a containerized, multi-tenant platform for secure, collaborative, and feedback-driven code execution. Designed to support both academic instruction and industrial training, CASCADE enables group-based programming tasks through per-task, per-group container environments, automated assessment, simulation interaction, and real-time feedback.

The platform combines a monolithic web application with Redis-queued task execution and containerized runtime isolation. Through lightweight deployment, clear separation of responsibilities, and observability via the ELK stack, CASCADE achieves operational scalability without reliance on heavyweight orchestration frameworks. Deployed at the University of Siegen, Germany across multiple courses and cohorts, CASCADE handled hundreds of student submissions involving real-time collaboration, code testing, and report validation. Its extensible design supports custom simulators, multi-course configurations, and instructor dashboards—delivering a comprehensive platform for interactive code-based learning.

Future work will target the integration of persistent workspaces, role-based access control (RBAC), LLM-based report evaluation, and dynamic container routing to support more complex use cases. With minimal infrastructure requirements and strong practical validation, CASCADE offers a portable and reproducible model for modern, container-native programming education and training environments.

References

1. Github codespaces documentation. https://docs.github.com/en/codespaces, Accessed 14 April 2025
2. Google test. https://github.com/google/googletest, Accessed 14 April 2025
3. Nginx: High performance load balancer, web server, & reverse proxy. https://www.nginx.com, Accessed 14 April 2025
4. Sqlite home page. https://www.sqlite.org, Accessed 14 April 2025
5. Bird, S., Klein, E., Loper, E.: Natural Language Processing with Python. O'Reilly Media (2009)
6. Bisong, E.: Building Machine Learning and Deep Learning Models on Google Cloud Platform. Apress (2019)
7. Brown, S.: Software Architecture for Developers, Volume 2. Leanpub (2019)
8. Gormley, C., Tong, Z.: Elasticsearch: The Definitive Guide. O'Reilly Media (2015)
9. Granger, B.E., Pérez, F.: Jupyter: thinking and storytelling with code and data. Comput. Sci. Eng. **23**(2), 7–14 (2021)
10. Hamrick, J.B., et al.: Nbgrader: a tool for creating and grading assignments in the jupyter notebook. J. Open Source Educ. **1**(1) (2016)
11. Inc., F.: React – a javascript library for building user interfaces. https://reactjs.org, Accessed 14 April 2025
12. Martin, R.C.: Clean Architecture: A Craftsman's Guide to Software Structure and Design. Prentice Hall (2017)
13. Merkel, D.: Docker: lightweight linux containers for consistent development and deployment. Linux J. **2014**(239) (2014)
14. Petersen, A., McMillan, B.A., Butler, Z.: Cloudcoder: a web-based programming exercise system for introductory computer science courses. In: Proceedings of the 43rd ACM Technical Symposium on Computer Science Education (2012)
15. Ramírez, S.: Fastapi: Modern, fast (high-performance) web framework for building apis with python (2018). https://fastapi.tiangolo.com, Accessed 14 April 2025
16. Roques, A.: Plantuml: Open-source tool for uml diagrams. https://plantuml.com, Accessed 14 April 2025

17. Sanfilippo, S.: Redis. https://redis.io (2012), Accessed 14 April 2025
18. Singh, R., Gulwani, S., Solar-Lezama, A.: Automated feedback generation for introductory programming assignments. In: Proceedings of the 34th ACM SIGPLAN Conference on Programming Language Design and Implementation (PLDI) (2013)
19. Solem, A., contributors: Celery: Distributed task queue. https://docs.celeryq.dev, Accessed 14 April 2025
20. Touvron, H., et al.: Llama: Open and efficient foundation language models (2023). https://arxiv.org/abs/2302.13971

A Review of E-Document Analysis with Multi-biometric Technique for Secure Verification

Lana Essam Raheem, Alaa Kadhim Farhan, Ammar Mazhar Sadiq, and Mustafa Tareq Eid(✉)

College of Computer Science, University of Technology, Baghdad, Iraq
mustafa.t.abd@uotechnology.edu.iq

Abstract. Increasingly, the need for reliable verification has brought the study of multi- modal biometric systems integrating fingerprints, faces, irises, and veins. These systems provide more security, accuracy, and resistance to spoofing compared to single-mode procedures. Moreover, the continued efforts in advancing cryptographic methods such as AES encryption, homomorphic encryption, and the use of blockchains intensify the security of biometric templates from cyber threats. Unfortunately, such hopeful advancements still leave standardizing evaluation metrics, optimizing computational efficiency, and addressing privacy-threatening challenges. This study examines the new advancements in multi-biometric authentication focusing on cryptographic components, Advanced Encryption Standard (AES), Elliptic Curve Cryptography (ECC), and Secure Hash Algorithm 256 (SHA-256) watermarking, multifactor authentication, and combinatorial fingerprinting. After significant research, the achieved results are biometric recognition accuracies of 92-96.67% and mobile app error rates of below 1%. The various ways biometrics can increase accuracy and then reduce the incidences of false acceptance and rejection aid the enhancement of anti-fraud protection in a multibiometric system. Nevertheless, in this scenario, numerous challenges remain, such as for uniform evaluation, computational cost, and privacy issues present hurdles to the multi-biometric paradigm. This review discusses literature, states the existing gaps, and recommends methodologies such as blockchain, quantum cryptography, and real-time authentication for future research on mainstream biometric systems. It also seeks to advance biometric systems used in e-learning, finance, and national security by evaluating their security parameters against new technologies.

Keywords: Biometric authentication · Cryptographic security · Multifactor authentication · Privacy protection · secures biometrics

1 Introduction

The rapid digitization of services has caused electronic documents to become central to global transactions and identity management. Now e-documents - digital ID cards, electronic health records, mobile banking - are staples of contemporary infrastructure.

T. Ensari et al. (Eds.): ISPR 2025, CCIS 2859, pp. 374–391, 2026.
https://doi.org/10.1007/978-3-032-21585-7_27

Their speed and the ease with which they are accessed have fostered e-governance in government administrations, electronic learning (e-learning) in universities and digital banking services in financial institutions. However, this increased use has ultimately made them vulnerable to cyber criminals [1, 2].

When e-documents are protected by traditional authentication schemes such as passwords or PINs, they are extremely exploitable through phishing and brute force attacks, dictionary attacks, and social engineering. These vulnerabilities lead to identity theft, stolen funds, and large-scale breaches or attacks. The increase in cyber crime throughout the world only serves to demonstrate the limits of text- based authentication and has refocused academia and industry on better authentication solutions.

Biometric authentication is one of many ways to address these issues through technologies that use physiological and behavioral traits that can be difficult to replicate. Biometric features can include fingerprints, iris characteristics, facial features, palm veins, and voice [3, 5].

Multi-biometric systems overcome some of these limitations by combining two or more biometrics. By relying on multiple biometrics, there is more distinction, universality is improved, and less chance of error. Each biometric adds complementary strengths, leading to improved accuracy, a greater ability to defeat spoofing, and improved trust by user [4, 9]. These considerations are especially important when examples extend into high-stakes border security, digital passports, assessment verification in e- learning domains, and high financial transactions.

An additional significant and improved development is the simultaneous and appropriate use of cryptography, such as Advanced Encryption Standard (AES), Elliptic Curve Cryptography. This paper is a systematic, critical, and analytical review of multi-biometric systems for secure e- document authentication. However, it does not report a descriptive survey, as it brings together prior work to identify the key contributions, open issues, and trends. In Section 2, we outline the necessary background concepts. Section 5 closely examines biometric and cryptographic techniques. Section 6 highlights works related to our study to provide a context of current advancement in e-document authentication within the broader academic community. Section 5 presents a comparative review of studies which will identify their relative strengths, limitations, and context of application. Finally, Sect. 6 concludes with a synthesis of insights and possible directions for future investigation.

2 Background

2.1 Biometric Identification

The process of identifying a person using their unique physiological and/or behavioral traits is known as biometric identification. Biometric identification is more dependable and capable than previous token-based and knowledge-based technologies in distinguishing between an authorized and a counterfeit individual, since these traits are unique to each individual. These days, in a networked society [5, 9], biometrics is becoming the primary means of identification and verification. Many organizations implement and use biometric readers, scanners, and/or other biometric procedures [10]. Combining many distinct biometric traits, either those that belong to one type solely or those that already exhibit a combination of the two types mentioned, is required to achieve complete security within an information system using biometric features [6].

Biometrics uses computerized pattern identification of distinctive behavioral or physiological characteristics to identify and validate people. In order to enroll, a sample of these traits must be taken, and a brief enrollment form with distinguishing features must be created. A similar score is then produced by comparing this template to real samples [11]. Behavioral biometrics are usually used for verification, while physical biometrics can be used for both identification and verification. Identification (answering "Who am I?") is a one-to-many process that compares a sample against a database of templates, and verification (answering "Am I who I claim to be?") is a one-to-one process that compares a sample against a stored enrollment template, depending on a threshold [1].

2.2 Multi-biometrics

By combining many biometric authentication techniques, it is possible to rigorously validate an individual while allowing for some degree of authentication failures with a single biometric [9]. Logical or statistical techniques can be used to combine biometric authentication [12]. Each biometric authentication is carried out separately using logical procedures, which then use AND or OR to arrive at the final response. A statistical function that is obtained by matching probabilities by individual authentication techniques is the foundation of statistical procedures [12].

2.3 Unimodal vs Multimodal Biometric System

The two primary categories of biometric systems, unimodal and multimodal, must be distinguished when discussing biometrics. There are certain disadvantages to each of these systems. A unimodal biometric system identifies a person using just one biometric characteristic. Such an approach usually

uses multiple technologically different systems and approaches to single out this one attribute. Multiple biometric traits and technologies are simultaneously used by multimodal biometric systems [9].

2.3.1 Unimodality Challenges

Unimodal biometrics' drawback is that no single solution works well for every application. Thus, the following drawbacks are somewhat offset by the existence of a multimodal biometric system [13]:

- The use of some biometrics leaves them vulnerable to inaccurate or noisy data, such as a scanner's inability to accurately read filthy fingerprints. Inaccurate matching may result from this, since faulty data could result in a false rejection.
- Unimodal biometrics is especially vulnerable to inter-class similarities within sizable population groups. When identical twins are present, a facial recognition camera might not be able to tell them apart.
- A segment of the population cannot use certain biometric technology. Due to their faded prints or underdeveloped fingerprint ridges, elderly individuals and little children may find it challenging to sign up for a fingerprinting system.

- Lastly, unimodal biometrics are susceptible to spoofing, which allows for the forging or imitation of data.

2.3.2 Multimodality Advantages

An estimated 5% of people in any given community are thought to have unreadable fingerprints because of aging, scarring, or illegible prints. The population segment that is unable to enroll may have inconveniences in a civic ID scenario where millions of people must be registered into the system. This limitation can be removed and a lower Failure To Enroll Rate (FTE) guaranteed with multimodal biometric technologies [13]

Additionally, multimodality can help with the issue of fingerprinting aversion, which exists in several regions of the world. People may be reluctant to submit their fingerprints because they believe they are linked to criminal behavior. The availability of an extra biometric allows for the enrollment of more users in the system [13].

The issue of inter-class resemblance and the resulting high False Acceptance Rate (FAR) is resolved by using several biometrics. The existence of another biometric, like signature verification, helps differentiate between the samples if people with similar hand sizes or facial features can be falsely accepted [13].

Utilizing multimodality also has the benefit of resolving the issue of data distortion. The other biometric sample can compensate if the quality of the first one is unsatisfactory. High False Rejection Rates (FRR) can be avoided by using a different modality, such as facial recognition, if a fingerprint has been scarred and the scanner rejects the altered sample. Spoofing unimodal biometrics is simple. Some systems can be tricked by putting a high-resolution image of a fingerprint underneath the scanner. However, when using multiple biometrics, the individual would still need to be validated using the other biometric, even if one modality might be spoofed. Furthermore, people who want to forge two or more biometrics are discouraged by the work involved [13].

2.4 Multi-biometric Systems

Multiple biometrics are used for biometric verification/identification purposes, with multiple modalities, instances, sensors, and/or algorithms enhancing the verification/identification decisions of the process in question [14]. Noisy data from differing conditions or a lack of reliable features could cause limitations in the performance of individual biometric systems [9]. Enrollment rates could drop, costs could start to increase, and the system would have fewer perks as a security measure against

spoofing. While biometric systems aim to prevent unauthorized access, vulnerabilities exist in enrollment, template databases, data collection/analysis, and result generation. Multibiometric can mitigate these vulnerabilities while improving universality and performance and addressing non- universality and spoofing issues [15].

Multibiometric systems can be categorized into five types based on the information processed: multimodal (multiple biometric modalities), various units (different instances of the same biometric), multi-instance (multiple instances of a biometric from the same individual), multi-sensory (different sensors capturing the same biometric), and multi-algorithmic (different algorithms processing the same biometric sample) [15, 16].

3 Biometric and Cryptographic Techniques

Biometric and cryptographic techniques would form the basis of a strong security framework for applications such as e-passports and systems with very high security. Biometric identify a person by their unique physical features like their signature face, fingerprint, iris, and palm print, whereas the aspect of encryption ensures that the integrity, authenticity, and confidentiality of data are well preserved [5]. As shown in Fig. 1, classify the Biometric traits.

3.1 Biometric Techniques

A biometric is a reliable solution for identifying or verifying individuals by means of unique physical or behavioral characteristics. While these characteristics are, by nature, quite distinctive and suitable for secure identification, there are several common biometrics that have found application in the identification of persons, each measuring certain specific aspects of individuality [5]. Prominent examples of physiological/biological and behavioral biometric characteristics, which have been the purpose of major real-world applications, are illustrated in Fig. 1.

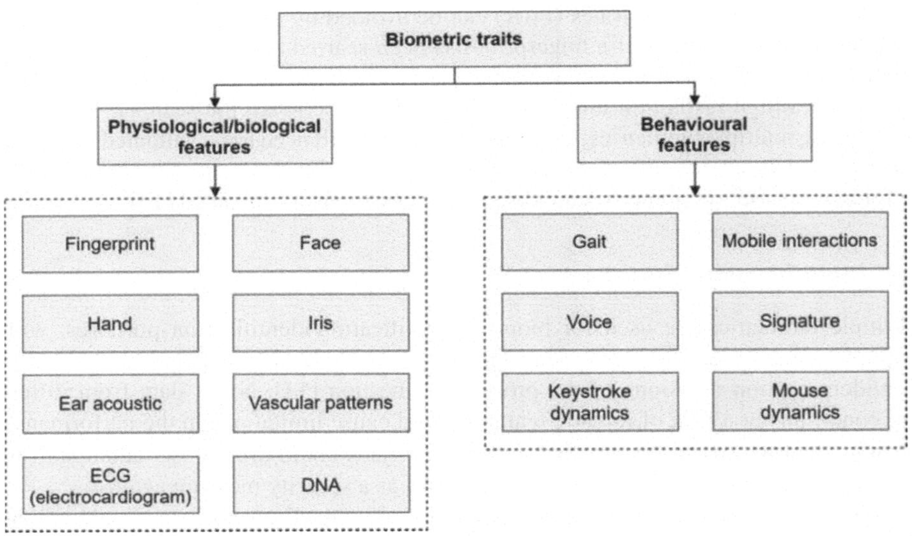

Fig. 1. Biometric traits [17].

A. **Physiological/biological (intrinsic) human characteristics**

Face recognition is one of the well-known methods used to identify human beings by evaluating facial characteristics. High-powered and very sophisticated computational techniques are then used, especially using deep neural network concepts along with advanced image processing techniques to map and analyze facial features. They work in the 'one-to-one verification' mode, where if a face is presented, the system asks whether the face would match a claimed identity. The system can also use 'one-to-many identification,' where an unknown subject's photograph is compared against multiple images stored in a database to identify that person. A sophisticated technology with flooring applications, everything from unlocking smartphone screens to public safety enhancement [13].

Fingerprinting which forms another popular biometric means, the main source of information is the patterns of ridges and valleys that exist on fingertips, which are unique in their patterns as well as remain constant throughout life. The process involves scanning the fingerprints digitally; the images are then enhanced for the extraction of key features or details such as ridge endings and bifurcations and the relevant comparative analysis is done with the features from a database of pre-stored templates of fingerprints. Because of its accuracy and ease of use, fingerprint recognition is one of the most widely accepted biometric modalities used in a variety of security applications, such as access control systems, electronic passports, and mobile devices [13].

Yet another very effective biometric technique for identification is palm print recognition. It identifies an individual by viewing the unique patterns, ridges, creases, and various other features present in the palm. Often, palm print recognition is employed in applications requiring a higher degree of precision than provided by fingerprint recognition alone. The process includes image capturing of the palm and safe storage of the images. Then, the captured image is compared to the palm print data for identity verification purposes. Since the surface area of the palm is larger, more detailed and distinctive features can be captured, which helps improve the accuracy of identification [13].

Iris Recognition is a biometric technique that stands as one of the most accurate and secure. In this method, a person is recognized by the detailed and unique patterns of the iris, the colored part of the eye. Such patterns include rings, ridges, freckles, and other distinguishing features. Advanced image processing techniques and high-level artificial intelligence algorithms capture and analyze high- resolution images of the iris. The evaluation and application of the iris for recognition are primarily high-security areas and e-passport systems due to its superior accuracy and forgery resistance. The complexity and uniqueness of patterns in the iris are major reasons for it being a highly trusted biometric identifier [13].

Advanced types of artificial intelligence algorithms are used along with high-end image-processing techniques in the capture and analysis of high-resolution images of the iris. Evaluation and application of the iris for recognition can mostly be done in very high-security areas and e-passport systems primarily due to its accuracy and forgery resistance. Complexity and uniqueness of an iris pattern are defining features for it being a more reliable biometric identifier [13].

Hand geometry this characteristic is based on the hand's geometrical properties, including the fingers' length, breadth, curve, and relative location to other hand traits. Modern developments in biometrics have supplanted its significance in the majority of applications, despite the fact that it was formerly a dominating approach of biometric measurement since it required little complexity in feature extraction and low-cost photography [9]. Furthermore, hand geometry-based recognition systems cannot be scaled up for systems that need the identification of an individual from a wide population, and such a biometric feature is not known to be especially unique. Additionally, because of their comparable structure, it is predicted that the hand-geometry aspects of the two hands will be similar [10].

Vascular patterns the benefits of this biometric characteristic over other qualities have been extensively studied. In contrast to other intrinsic and extrinsic biometric traits that are more susceptible to spoofing, leading to significant security and privacy concerns [13, 18], the vascular pattern of the human body is actually unique to each individual, even identical twins, remains constant throughout a person's life, and lies beneath the human skin, guaranteeing confidentiality and stability against counterfeiting. Since blood vessels are nearly undetectable in normal illumination, a vascular-based identification system employs near-infrared light to reflect or transmit pictures of blood vessels in order to get the network structure of blood vessels beneath the human skin [15]. The most widely used vascular biometric systems include both eye-oriented modalities, such retina and sclera recognition, and hand-oriented modalities, including finger vein, palm vein, hand dorsal vein, and wrist vein recognition [16].

Electrocardiogram (ECG) Usually obtained using a few electrodes, amplifiers, filters, and a data gathering module, this characteristic takes into account the human heart and body anatomic features to generate the shape of the ECG signal, which provides the time and intensity of the electrical activity of

the heart [19]. There are a number of limitations and concerns, nevertheless, since research results to date cast doubt on the specifics of real-world application scenarios and acceptability by possible end users.

Deoxyribonucleic acid (DNA) Short tandem repeat (STR)2 analysis, a popular molecular biology technique for comparing allele repeats at certain chromosomal loci in DNA across two or more samples, is the foundation of DNA matching [14, 20]. Despite the fact that identifications require physical samples and cannot be completed in real time, DNA-based biometric recognition has been utilized extensively in forensic science and scientific inquiry because of its extremely high accuracy.

Ear acoustic This type of identification system's primary goal is to map one component of acoustic ear recognition—specifically, the performance of the bands and peaks that define ear features. In order to create an ear signature, inaudible sound waves are sent into the ear, reflected, bounced in various directions, and detected by a tiny microphone. The acoustic transmission function, which is the foundation of the signature, is determined by the geometry of the ear canal. The technology's application is increased by the capacity to recognize a person while they are moving and while maintaining confidentiality [12].

B. **Behavioral (extrinsic) human characteristics**

A biometric system may utilize behavioral characteristics such as handwriting, locomotion, mouse usage, keystrokes, and other actions to determine a person's identification.

Gait for a variety of causes, including weight increase, this trait may alter over an extended period of time [21]. Because it can swiftly identify people from a distance based on their gait, it may be utilized in low-security applications for vast crowd monitoring. It can even use the potential of numerous surveillance cameras placed in public spaces by integrating them into a biometric system. In actuality, such a system does not require people to wear any particular equipment or devices or to be cooperative in order to be identified [22].

Mobile interactions even taking into account additional features derived from on-board sensors like GPS, gyroscope, and accelerometers, which can also be configured to collect data in passive mode, it is based on the distinctive ways users swipe, tap, pinch-zoom, type, or apply pressure on the touch screen of mobile devices like tablets and phones [23]. This results in distinctive patterns that may be used to identify individuals. Therefore, biometrics based on mobile interactions place more emphasis on how a user executes their actions than on the results of those actions.

Signature he most popular technique for document authentication is signature recognition, which uses shorter handwriting probes than text-independent writer recognition techniques but necessitates writing the same sign each time. (i) Offline or static (the signature is digitized after the writing process) and

(ii) Online or dynamic (the signature is digitized during the writing process) are the two types of signature authentication schemes. Curves, edges, geographical coordinates, inclination, center of gravity, pen pressure, and pen stroke of the signature samples are analyzed in both offline and online applications to extract signature biometric data. However, only online signatures include dynamic information such as stroke order and writing speed [24].

Mouse dynamics it takes advantage of mouse or track pad cursor movement patterns, including clicks, trajectories, direction changes, tracking speed, and interrelationships. The movement characteristics provided by a mouse may be used to verify individuals since they are unique to each user and rather consistent for the same individual [25]. The most popular use of these methods is continuous user identity verification.

Voice both physiological and behavioral biometrics apply to voice recognition technologies. Because voice features include anatomical attributes including vocal tracts, nasal cavities, mouth, and larynx, voice biometric identification enables the differentiation of human voices for personal authentication [17]. In terms of behavior, each person is also said to have their own distinct manner of speaking or saying things, including their tone, tempo, accent, and variances in their movements. A precise voice signature is therefore produced by combining information from behavioral and physiological biometrics, albeit there may be discrepancies because of sickness or other circumstances.

3.2 Cryptographic Techniques

Cryptography is vital in protecting information, with mechanisms that ensure integrity and provide means against compromise. A range of sophisticated cryptographic techniques is employed under e- passports to achieve these functions of security. Some of these methods are essential:

Hashing basically, hashing algorithms (which are the SHA-256 family) are used to imprint a digital fingerprint or hash value upon electronic data. These digital fingerprints can effectively assess any unauthorized changes to the data; given that once the hash value of a data set is changed, it means that the data has changed [5].

In e-passports, hashing is used for integrity checks on biometric data so that this sensitive information is not altered.

PKI (Public Key Infrastructure) the PKI is a comprehensive cryptographic procedure that enhances the capacities of public and private keys, which not only provide guarantees related to data confidentiality but also work as an authenticity check. The public key may be disclosed to virtually anybody. The private key, however, remains secret and confined to the owner only. PKI enables e- passports to spot if there is any data manipulation in the form of tampering with the stored records in the microprocessor. Thus, the authenticity and integrity of the stored data can be proved with all near certainty induction about digital signatures that are verified using the public key [5].

Digital signatures form a key element in creating trust, authenticity, and security issues in e-passports. The data will be digitally signed using a private key only, and the signature could be verified by people with a public key corresponding to the private key. Thus, the process guarantees the authenticity and integrity of signed data. In e-passports, the signatures guarantee the integrity of data from the moment it is issued and ensure that it has not been tampered with or compromised throughout its lifetime [5]. **Symmetric Encryption** symmetric encryption is encryption using a single key for both applications of encryption and decryption. In fact, this approach permits a faster encryption speed than asymmetric encryption and is consequently best suited for applications in which speed is paramount. For example, in e-passports, symmetric encryption is done to secure operational data, like active authentication, for which fast encryption/decryption processes are highly desirable [5].

Passive authentication constitutes an automated process whereby the integrity of biometric data recorded on the e-passport chip undergoes verification. The stored biometric data undergoes comparison with officially stored digital signatures, which have, in turn, been pre-computed and entered into the chip. If any mismatch is noted between the biometric data and the corresponding digital signature it indicates that the data might have undergone some unauthorized changes [12].

Active authentication is a mainstay countermeasure in security to address unauthorized duplication of original passports. It is a challenge-response type of security. This means that the verifier sends a nonce to the passport, and the passport must digitally sign the nonce using its private key. Then the verifier would validate the signature using the associated public key of that passport. Completion of this whole process would prove that the passport is authentic and it harbors the required private key, thus preventing cloning 5].

Basic Access Control (BAC)basic Access Control secures an e-passport such that only authorized devices can access data within it, which are physically present near the e-passport. This is implemented through strong encryption techniques that dynamically provide a second control to protect the sensitive biometric data stored on the e-passport. An unauthorized person's impossibility of accessing the data in it is, regardless of whether or not he happens to possess the USB key, guaranteed by BAC 5].

Chip Authentication markedly improves the security protocol and effectively supersedes Active Authentication. For each reading session, dynamic generation of new encryption keys takes place. Dynamic key generation prevents cloning attempts by ensuring that all previously intercepted keys cannot be reused. This provides an additional layer of security and greatly increases the protection against the cloning of passports [5].

Terminal Authentication terminal authentication is the security protocol used to prove that only specific devices (like passport readers at border control) are authorized to access sensitive biometric information contained in the e-passport. Only trusted and authorized terminals can access the data, whereas rogue or compromised readers cannot get into it. This authentication process relies on digital certificates to ascertain the identity of the reader [5].

4 Related Work

Numerous theoretical and empirical studies have covered and elaborated on E-document analysis with multi-biometric. This study focuses on approaches and procedures that are limitedto multi-biometric. The following are previous studies about E-document analysis with multi-biometric that were conducted by various studies:

- **Mobile Authentication**

Kuseler et al. [19] proposed eBiometrics, a multimodal scheme for mobile devices, including fingerprint recognition, face recognition, and signature recognition. The authors improved resistance to tampering and spoofing, but the performance depended on device sensors, which limited scalability. Likewise, the IEEE Conference research study [29] presented a combinatorial fingerprint based authentication model for mobile transactions, providing lower error rates (FAR and FRR). Each work suggested that mobile biometrics have promise for everyday applications but remain subject to computational costs, variabilities in hardware, and user convenience. As shown in Fig. 2 explain the multi-Finger proposed biometric authentication system.

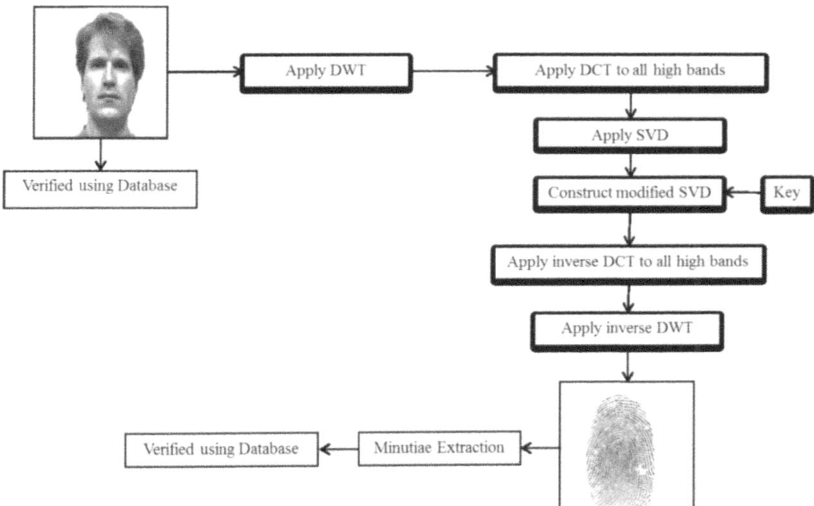

Fig. 2. Multi-Finger proposed biometric authentication system [29].

- **E-Passports and National Security**

Kumar et al. [27] proposed a multimodal e-passport with face, fingerprint, palm, and iris with cryptographic access protection. Although multilayered, added cryptographic security has additional ramifications for the implementation of their system, including RFID privacy concerns. Many researchers including Chaudhary et al. [20] dealt with cryptographic security and proposed AES and SHA-256 encryption methods in addition to watermarking to maximize recognition accuracy. Figure 3 presents the procedure of watermark embedding and Fig. 4 represents the procedure of watermark embedding.Further, Choudhury and Rabbani [28] tackled RFID vulnerabilities and entailed embedding biometric data in QR codes as encrypted data to limit exposure to skimming attacks. Though researchers' work on e-passports has evolved from multiple modalities, cryptographic security, and the use of QR codes, there remain unresolved challenges including global interoperability, regulatory privacy standards, and developing low-cost access methods. Fig. 5 presentsa framework of biometric passport information.

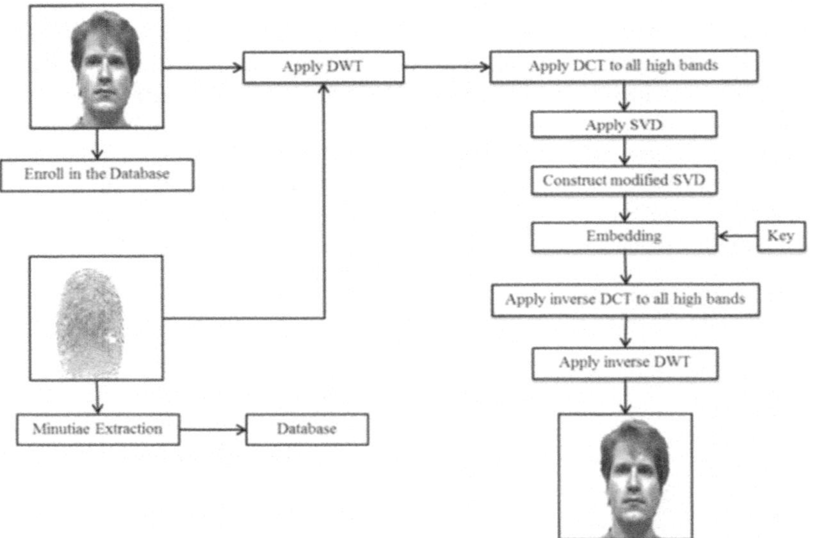

Fig. 3. Watermark embedding procedure [20].

- **Financial Transactions**

Al-Assam et al. [26] examined accuracy and security of multiple/factor biometric authentication, and found results show that compromised transformation keys may significantly inflate the FAR. Modak and Jha [25] developed an ECC-based multimodal scheme (fingerprint, face and iris) and civilian biometric identity management, with OTP verification. The OTPs improved fraud resistance but created increased transaction time. Khan et al. [21] proposed a secure method for biometric template construction, to protect biometric templates in multiple factor authentication, indexed to compromise

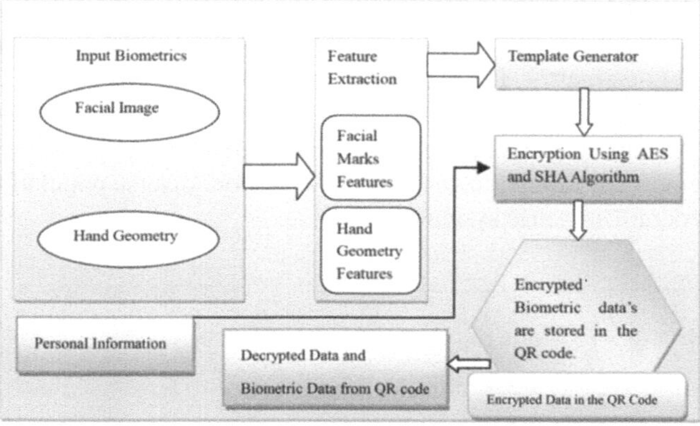

Fig. 4. Watermark embedding procedure [20].

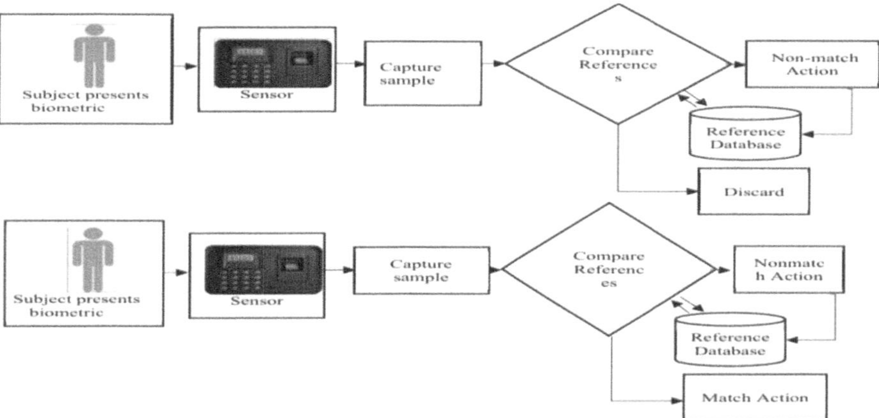

Fig. 5. The proposed framework of the biometric passport information encryption and decryption [28].

related to user authentication methods and risk or historical security, but the complexity of their suggested method may render it unfeasible for real-time transactions in banking context. Taken together, the studies demonstrate a troubling and persistently consistent trade-off between the improved resistance to fraud, over user experience and the cost of performance improvement in the financial services.

- **E-Learning and User Authentication**

Asha and Chellappan [9] were some of the early advocates of multimodal authentication for e-learning learning activities with their work suggesting two or more authentication was a protective factor against cheating in online examinations. More recently, Aditya & Kaur [16] used both fingerprint and vein recognition and reported that the use of both modalities reduced the errors of acceptance in the verification process. Vandana and Kaur [17] completed a conceptual analysis of biometric identification that provided some theoretical advantages over traditional authentication and they noted that "biometrics has limitless potential" (p. 47). Collectively, these studies do suggest that multimodal biometric may have great potential in the e-learning domain, yet, it is noted that the most implementations have thus far operated as small design or prototype authors, with a few not including empirical validation with a larger and more diverse population. See Fig. 6 displays the general Biometric system.

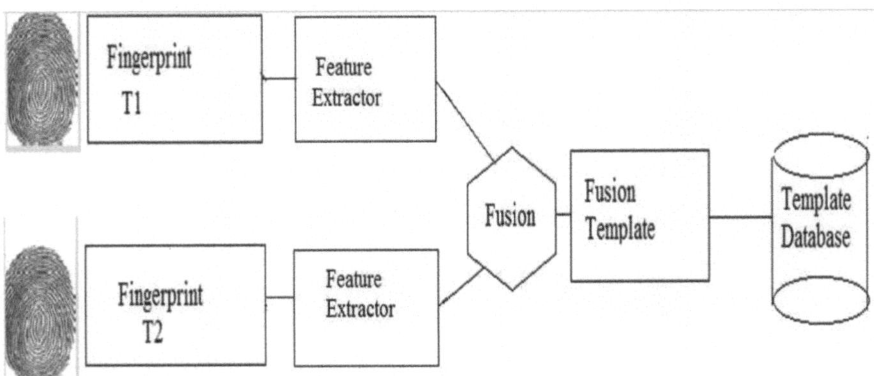

Fig. 6. A General Biometric system contains Enrollment and Matching [17].

- **Multi-Server and High-Security Environments**

Odelu et al. [22] provided an ECC based smart card protocol for secure multi-server authentication. This has been shown to be resistant to impersonation attacks and the authors verified the protocol using BAN logic and AVISPA tools. Similarly, Reddy et al. [23] built off of this work by providing user anonymity and additional protection against replay attacks. Both protocols advance the research towards multi-server environments but both demanding in terms of computing which may constrain deployment in smart cards and low-power IoT devices.

- **Other Multimodal Contributions**

Kabir et al. [12] built a multimodal system with both feature- and score-level fusion, attaining strong recognition accuracy in real time, though increased complexity. Dinca and Hancke's [13] study of fusion schemes confirmed that one particular fusion method does not always outperform other fusion methods, thus when using multimodal biometrics, the application context is vital. Mahalakshmi and Sriram [14] suggested an ECC-based multimodal framework that brought together fingerprint, iris, and signature, produced strong resistance to accesses with a FAR/FRR performance never scaled. Malik et al. [18] considered the psychological effects of security technology and biometric use, and indirectly raised usability concerns in the use of biometrics. Ahmed et al. [30] addressed multimodal biometrics generally and identified the advantages while noting cost and implementation problems were limiting the implementation of multimodal biometric systems. Lastly, Ali and Farhan [31] described a new algorithm for fingerprint-vein matching for e-documents with improved accuracy although not scale tested. As shown in Fig. 7 clarify the block diagram of the proposed two algorithms.

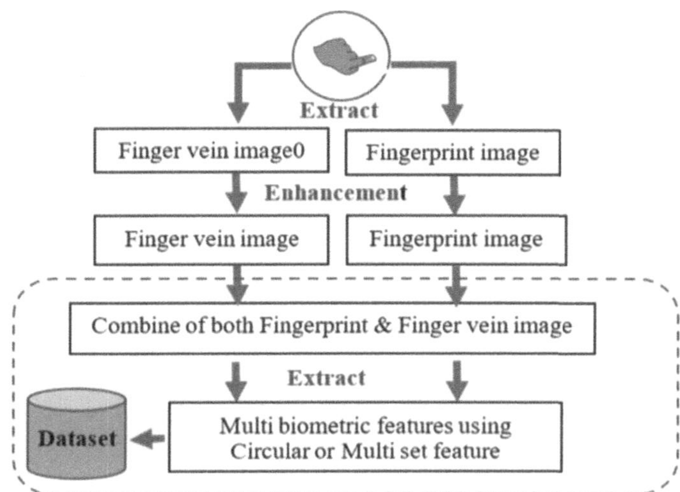

Fig. 7. The block diagram of the proposed two algorithms with the combined process [31].

The literature review highlights three major trends in biometric authentication for secure e-documents. First, the shift from unimodal to multimodal biometric authentication is significant, as the use of multiple traits consistently reduces error rates and improves spoofing resistance. Second, there is now an increasing focus on incorporating cryptographic techniques like AES, ECC, SHA-256 and watermarking/QR encoding techniques for template protection, but key management, latency, and system processing/functional efficiency still pose challenges. Third, there are application-dependent modifications made with regard to existing solutions; the system specifications were influenced from an application standpoint, whether it was speeding up financial transaction times, providing privacy for e-passports, or scalability for multi-server environments.

Although these advances have made more work possible, there are still a number of important research gaps that researchers need to address. For example, the lack of standardized evaluation metrics for biometric systems makes across-study comparison meaningful, and the majority of systems are tested on only limited datasets, which therefore draws into question whether they are truly able to be scalable and perform well in the real world. Additionally, the reconcilability and un-link ability of templates has not been studied at the current level and other research areas including user acceptance of privacy by design consent mechanisms, ethical concerns, and regulatory compliance are often even temporarily overlooked. In order to ultimately promote the development of scalable and privacy sustaining and user- centred multi-biometric systems, it is reliant on interdisciplinary perspectives that balance technical credibility with usability and trust.

5 Comparison of Studies

The comparative analysis is a critical component of this review, as it consolidates findings from multiple studies and allows readers to discern relative strengths and weaknesses. To meet academic standards, the comparison should go beyond listing modalities and results it should emphasize the application domain, methodological contributions, performance metrics, and limitations. Below is an improved Table 1 structure that reflects these requirements.

Table 1. Biometric Authentication Techniques and Cryptographic Methods.

Author/Year	Modalities	Cryptographic / Security Technique	Application Domain	Key Contribution	Reported Performance Metrics	Limitations / Open Issues
Kuseler et al. (2010) [19]	Fingerprint, Face, Signature	PIN-seeded cryptographic integrity	Mobile Authentication	Introduced *eBiometrics* for mobile devices	Demonstrated real-time operation	Dependent on sensor quality; scalability issues
Kumar et al. (2012) [27]	Face, Fingerprint, Palm, Iris	PKI, Symmetric/Asymmetric Encryption	E-Passports	Secure multimodal e-passport scheme	Enhanced border control efficiency	RFID privacy vulnerabilities
Chaudhary et al. (2013) [20]	Face, Hand Geometry	AES + SHA-256, Watermarking	E-Passports	Watermarking for template protection	92–96.67% accuracy	Limited dataset size
Mahalakshmi & Sriram (2013) [14]	Fingerprint, Iris, Signature	ECC, Genetic Algorithm	General Authentication	ECC-based multimodal scheme	Reported low FAR/FRR	Lacked real-world validation
Odelu et al. (2015) [22]	Smart Card + Biometrics	ECC	Multi-Server Authentication	Secure multi-server protocol	Formally validated (AVISPA)	Computationally expensive
Reddy et al. (2016) [23]	Smart Card + Biometrics	ECC	Multi-Server Authentication	Enhanced anonymity & security	Resistant to replay/impersonation	High resource consumption
Al-Assam et al. (2011) [26]	Multifactor Biometrics	Transformation key mechanism	Financial Authentication	Identified vulnerabilities in key use	Showed FAR inflation when compromised	Reliance on auxiliary keys
Modak & Jha (2019) [25]	Fingerprint, Face, Iris	ECC + OTP verification	Banking	Robust multimodal with OTP	Resistant to spoofing	Latency in transactions
Aditya & Kaur (2021) [16]	Fingerprint, Vein	Fusion	E-Learning	Secure online exam authentication	Reduced error rates	Limited to small dataset
Singh et al. (2022) [29]	Fingerprint (multi-sample)	Cryptographic embedding	Mobile Transactions	Combinatorial fingerprinting	Lower FAR/FRR	Hardware/software dependency
Ali & Farhan (2024) [31]	Fingerprint, Vein	Matching Algorithm	E-Documents	Novel multimodal verification	Improved accuracy	Scalability untested

6 Conclusion

Considering the above, it seems likely that multibiometric systems are a promising approach to addressing the increasing demands for secure electronic document creation. However, taking full advantage of their benefits will require more than theoretical contributions; it will require concrete approaches to operationalize standard, usable, scalable systems that are user-centric while providing biometric dependability and state-of-the-art cryptographic protection. Therefore, the main conclusions can be articulated as follows:

- **Empirical Validation and Benchmarking Needs**

  Present multi-biometrics systems provide theoretically advantageous methods, however, studies exploring their effectiveness are lacking. Future research should focus on establishing standardized benchmarks that will subsequently provide opportunities for empirical validation that can be compared across studies.

- **Security Fundamentals for E-documents**

  To protect e-documents it requires mechanisms that ensure their integrity. Mechanisms also require ensuring that unauthorized access is not allowed. Multi-biometric systems will introduce a method that is superior in terms of accuracy and resiliency against fraud. A method towards secure authentication, that will able secure access either through fingerprint recognition or face recognition as modalities.

- **Challenges posed by Machine Learning and Encryption**

  Machine learning techniques and integration of encryption, will benefits multi-biometric systems in a significant way. However, adopting a machine learning based approach will also pose privacy, scalability and real-time performance problems. Future research must focus on a scalable framework that supports scalability but also maintains efficiency in practice, which would be especially concerning in areas such as the financial and healthcare domains.

- **Improving Resistance Against Adversarial Attacks**

  The robustness of biometric templates remains an unaddressed research gap. More research is necessary to create multibiometrics that resist adversarial attacks and secure templates in multi-factor authentication.

- **Compatibility with New Technologies**

  Future development should involve the integration of blockchain technology for decentralized authentication, quantum cryptography to mitigate future computational threats, and hybrid models that account for usability and robustness. This would create an overall security framework that is scalable to fully functional real-world deployment.

References

1. Kausar, N., Din, I.U., Khan, M.A., Almogren, A., Kim, B.S.: GRA-PIN: a graphical and PIN-vBased hybrid authentication approach for smart devices. Sensors. **22**(4) (Feb. 2022). https://doi.org/10.3390/s22041349

2. Jabeen, T., Mehmood, Y., Khan, H., Nasim, M.F., Asad, S., Naqvi, A.: Spectrum of engineering sciences identity theft and data breaches how stolen data circulates on the dark web: a systematic approach. Spectrum Eng. Sci. **3**(1) (2025)
3. Rashid, O.F., Tareq, M., Subhi, M.A.: RNA encoding and CLARA clustering algorithm for hybrid intrusion detection system using UNSW-NB15 dataset. In: 34th International Conference on Computer Theory and Applications, ICCTA 2024, pp. 34–39. Institute of Electrical and Electronics Engineers Inc. (2024). https://doi.org/10.1109/ICCTA64612.2024.10974781
4. Ramachandra, R. et al.: Smartphone Multi-modal Biometric Authentication: Database and Evaluation, (Dec 2019). [Online]. Available: http://arxiv.org/abs/1912.02487
5. Yuko, A., Nakanishi, J., Western, B.J.: Advancing the state-of-the-art in transportation security identification and verification technologies: biometric and multibiometric systems
6. Alwahaishi, S., Zdralek, J.: Biometric authentication security: an overview. In: Proceedings - 2020 IEEE International Conference on Cloud Computing in Emerging Markets, CCEM 2020, pp. 87–91. Institute of Electrical and Electronics Engineers Inc. (Nov. 2020). https://doi.org/10.1109/CCEM50674.2020.00027
7. Tareq Eid, M., Farhan, Y.H., Nafea, M.M.: International Journal On Informatics Visualization Journal Homepage: www.joiv.org/index.php/joiv International Journal On Informatics VisualizatiON An Enhanced Routing Protocol for Vehicular Ad Hoc Networks Using Swarm Intelligence. [Online]. Available: www.joiv.org/index.php/joiv
8. Rashid, O.F., Mohammed, M.J., Tareq, M., Ibrahim, A.A., Hassan, H.F., Kadem, M. H.: Text Cryptography Based on Three Different Keys
9. Asha, S., Chellappan, C.: Authentication of E-Learners Using Multimodal Biometric Technology
10. Yang, W., Wang, S., Sahri, N.M., Karie, N.M., Ahmed, M., Valli, C.: Biometrics for internet-of-things security: a review. MDPI. (01 Sep 2021). https://doi.org/10.3390/s21186163
11. Ryu, R., Yeom, S., Kim, S.H., Herbert, D.: Continuous Multimodal Biometric Authentication Schemes: a Systematic Review. Institute of Electrical and Electronics Engineers Inc. (2021). https://doi.org/10.1109/ACCESS.2021.3061589
12. Kabir, W., Ahmad, M.O., Swamy, M.N.S.: A multi-biometric system based on feature and score level fusions. IEEE Access. **7**, 59437–59450 (2019). https://doi.org/10.1109/ACCESS.2019.2914992
13. Dinca, L.M., Hancke, G.P.: The fall of one, the rise of many: a survey on multi-biometric fusion methods. IEEE Access. **5**, 6247–6289 (2017). https://doi.org/10.1109/ACCESS.2017.2694050
14. Mahalakshmi, U., Sriram, V.S.S.: An ECC Based Multibiometric System for Enhancing Security. [Online]. Available: www.indjst.org
15. Atoyebi, J.O., Okomba, N.S., Adeyanju, I.A., Akinrotimi, A.O., Ajayi, O.O., Owolabi, O.O.: Multi-biometric recognition systems: a comprehensive review. (2025). [Online]. Available: www.aujet.adelekeuniversity.edu.ng
16. Vekariya, V., Joshi, M., Dikshit, S., Manju Bargavi, S.K.: Multi-biometric fusion for enhanced human authentication in information security. Meas.: Sens. **31** (2024). https://doi.org/10.1016/j.measen.2023.100973
17. Vandana, Kaur, N.: A study of biometric identification and verification system. In: 2021 International Conference on Advance Computing and Innovative Technologies in Engineering, ICACITE 2021, pp. 60–64. Institute of Electrical and Electronics Engineers Inc. (Mar. 2021). https://doi.org/10.1109/ICACITE51222.2021.9404735
18. Malik, A.S., Acharya, S., Humane, S.: Exploring the impact of security technologies on mental health: a comprehensive review. Cureus. (2024). https://doi.org/10.7759/cureus.53664

19. Kuseler, T., Lami, I., Jassim, S., Sellahewa, H.: eBiometrics: an enhanced multi-biometrics authentication technique for real-time remote applications on mobile devices. In: *Mobile Multimedia/Image Processing, Security, and Applications* 2010, p. 77080E. SPIE, Apr (2010). https://doi.org/10.1117/12.850022
20. Chaudhary, N., Singh, D., Hussain, D.: Enhancing Security of Multimodal Biometric Authentication System by Implementing Watermarking Utilizing DWT and DCT. [Online]. Available: www.iosrjournals.org
21. Khan, S.H., Ali Akbar, M., Shahzad, F., Farooq, M., Khan, Z.: Secure biometric template generation for multi-factor authentication. Pattern Recogn. **48**(2), 458–472 (2015). https://doi.org/10.1016/j.patcog.2014.08.024
22. Odelu, V., Das, A.K., Goswami, A.: A secure biometrics-based multi-server authentication protocol using smart cards. IEEE Trans. Inf. Forensics Secur. **10**(9), 1953–1966 (2015). https://doi.org/10.1109/TIFS.2015.2439964
23. Reddy, A.G., Das, A.K., Odelu, V., Yoo, K.Y.: An enhanced biometric based authentication with key-agreement protocol for multi-server architecture based on elliptic curve cryptography. PloS One. **11**(5) (2016). https://doi.org/10.1371/journal.pone.0154308
24. Kumar, S.S., Ratnesh, N.A.: A review on different biometric techniques: single and combinational. IOSR J. Electr. Commun. Eng. **11**(04), 25–30 (2016). https://doi.org/10.9790/2834-1104012530
25. Modak, S.K.S., Jha, V.K.: Enhancing multibiometric system security using ECC based on score level fusion. In: Lecture Notes in Electrical Engineering, pp. 223–232. Springer Verlag (2019). https://doi.org/10.1007/978-981-10-8234-4_20
26. H. Al-Assam, H. Sellahewa, and S. Jassim, "Accuracy and Security Evaluation of Multi-Factor Biometric Authentication," 2011.
27. Kumar, V.K.N., Srinivasan, B., Narendran, P.: Efficient implementation of electronic passport scheme using cryptographic security along with multiple biometrics. Int. J. Inf. Eng. Electr. Bus. **4**(1), 18–24 (2012). https://doi.org/10.5815/ijieeb.2012.01.03
28. Choudhury, Z.H., Rabbani, M.M.A.: Biometric passport for National Security Using Multi-biometrics and encrypted biometric data encoded in the QR code. J. Appl. Secur. Res. **15**(2), 199–229 (2020). https://doi.org/10.1080/19361610.2019.1630226
29. 2022 IEEE International Conference on Dependable: Autonomic and Secure Computing, International Conference on Pervasive Intelligence and Computing, International Conference on Cloud and Big Data Computing, International Conference on Cyber Science and Technology Congress. IEEE, (2022)
30. Ahmed, W., Dahea, A., Dahea, W., Fadewar, H.S.: Multimodal biometric system: a review. Int. J. Res. Adv. Eng. Technol. 25 Int. J. Res. Adv. Eng. Technol. **4**, 2455–0876 (2018). https://doi.org/10.13140/RG.2.2.34056.65287
31. Ali, A.M., Farhan, A.K.: A novel multi-biometric technique for verification of secure e-document. Int. J. Electr. Comput. Eng. **14**(1), 662–671 (2024). https://doi.org/10.11591/ijece.v14i1.pp662-671

NeuroYOLO: Data-Efficient Fine-Tuning of YOLOv11 for Vision-Based Rehabilitation Monitoring

Basma Jalloul(✉), Bassem Bouaziz, and Walid Mahdi

MIRACL Laboratory, Higher Institute of Computer Science and Multimedia, University of Sfax, Pôle Technologique, Sfax 3021, Tunisia
basma1707@gmail.com

Abstract. Accurate movement tracking is vital for assessing motor recovery in neuro-rehabilitation, yet general-purpose pose estimation models often underperform in clinical environments characterized by constrained motion, limited data, and fixed viewpoints. This paper introduces NeuroYOLO, a data-efficient, fine-tuned YOLOv11-Pose model optimized for vision-based rehabilitation monitoring. Using Bayesian hyperparameter optimization, NeuroYOLO adapts to structured exercise recordings of elderly participants performing assisted upper- and lower-limb tasks, demonstrating that high performance can be achieved without large clinical datasets. The fine-tuned model achieves superior detection precision and temporal stability compared to the default configuration, improving pose mAP@0.5:0.95 from 0.989 to 0.991 and box mAP@0.5:0.95 from 0.961 to 0.985, while reducing inference latency by 36%. These results confirm that targeted, data-efficient domain adaptation can significantly enhance pose estimation reliability under realistic rehabilitation conditions. NeuroYOLO thus establishes a lightweight and deployable foundation for real-time, vision-based patient monitoring and future multimodal rehabilitation frameworks.

Keywords: Human Pose Estimation · YOLOv11 · Fine-Tuning · Vision-Based Rehabilitation · Motor Function Monitoring · Movement Analysis

1 Introduction

The rising prevalence of neuro-motor and neurodegenerative disorders such as stroke, Parkinson's disease, Mild Cognitive Impairment (MCI), and Alzheimer's disease has created an urgent demand for accessible, objective tools to assess motor function during rehabilitation [1]. Subtle gait and movement irregularities often emerge before cognitive decline, yet conventional observation-based assessments remain subjective and time-consuming [2]. Vision-based human pose estimation (HPE) provides a non-invasive alternative by quantifying motion from video through skeletal keypoints [3]. Deep learning models such as OpenPose [3]

T. Ensari et al. (Eds.): ISPR 2025, CCIS 2859, pp. 392–399, 2026.
https://doi.org/10.1007/978-3-032-21585-7_28

and BlazePose [4] achieve strong general-purpose accuracy but degrade under domain constraints typical of rehabilitation, limited motion amplitude, fixed camera viewpoints, and scarce annotated data [5]. Recent lightweight architectures such as HP-YOLO [6] and CCAM-Person [7] have demonstrated that compact, efficient models can retain competitive accuracy for real-time applications. However, most studies emphasize detection performance on large benchmarks rather than adaptation to clinical or rehabilitation-specific movement domains. In contrast, collecting large-scale clinical datasets is often impractical due to privacy, variability, and recruitment constraints. This work therefore adopts a data-efficient learning paradigm, exploring how targeted fine-tuning and Bayesian optimization can achieve high accuracy and robustness with minimal, domain-representative data. This approach reflects real-world conditions where data availability is limited but adaptability and reliability are essential for clinical deployment. To address these challenges, we introduce NeuroYOLO, a fine-tuned YOLOv11-based [8] pose estimation model optimized for rehabilitation scenarios. Using Bayesian hyperparameter optimization, the model adapts to structured exercise recordings collected from elderly participants, mitigating domain shift and improving both accuracy and inference efficiency. This paper focuses on the algorithmic and empirical aspects of the fine-tuning process for pose estimation in rehabilitation contexts. Broader system components, including visualization dashboards and multimodal data integration, fall beyond the current scope and will be addressed in subsequent work.

The main contributions of this work are as follows:

- A data-efficient fine-tuning strategy for YOLOv11-Pose using Bayesian search tailored to low-resource rehabilitation data.
- Quantitative improvements in mean Average Precision (mAP) and inference latency under constrained movement conditions.
- Demonstration of lightweight model adaptability for real-time rehabilitation monitoring.

The remainder of this paper is organized as follows: Sect. 2 details the proposed methodology and fine-tuning pipeline; Sect. 3 presents experimental results and performance analysis; Sect. 4 discusses implications and limitations; and Sect. 5 concludes the paper with future perspectives.

2 Methodology

The proposed approach fine-tunes the YOLOv11-Pose model to improve pose estimation accuracy and efficiency for vision-based rehabilitation monitoring. The system adapts YOLOv11 to motion-constrained rehabilitation data through Bayesian hyperparameter optimization, enabling reliable keypoint detection and real-time inference in low-resource settings as shown in Fig. 1.

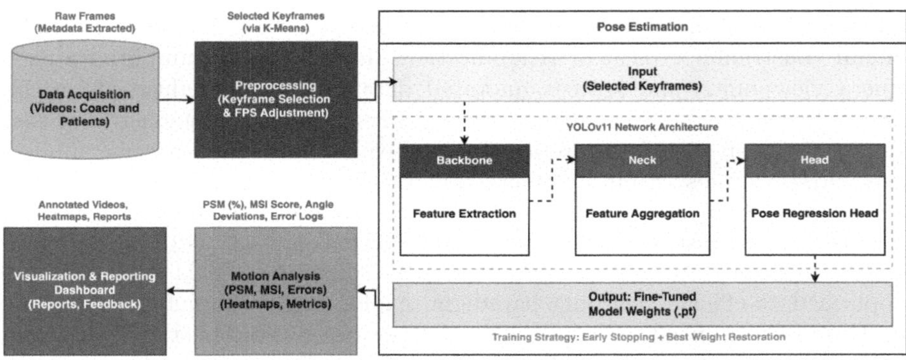

Fig. 1. Overview of the proposed YOLOv11-based rehabilitation monitoring pipeline.

2.1 Dataset and Preprocessing

Rather than relying on large clinical datasets, this work explores how targeted optimization can yield high-performance models from limited, domain-specific data. In clinical contexts, collecting extensive labeled datasets is often impractical due to privacy, recruitment, and variability constraints. Our approach demonstrates that with appropriate model selection, hyperparameter tuning, and task-aligned fine-tuning, reliable pose estimation can be achieved without massive data volumes. This paradigm favors scalable, data-efficient deployment in real-world rehabilitation settings, where lightweight models trained on representative motion samples are often preferable to large, resource-intensive networks trained on generic datasets. To achieve this end, recordings were obtained from one coach and three elderly participants (65+) performing structured physical exercises with an ICT-enabled rollator [9]. Each session lasted approximately one minute (30 FPS), resulting in 46 video clips and over 80k frames. To emphasize salient postures, we applied K-Means keyframe selection. Each frame's 2D pose vector was flattened and clustered by spatial similarity:

$$\arg\min_{C} \sum_{k=1}^{K} \sum_{x_i \in C_k} \|x_i - \mu_k\|^2,$$

where μ_k is the centroid of cluster C_k. The pose nearest to each centroid was retained as a representative keyframe. This reduced computational cost by an order of magnitude while preserving essential movement transitions.

2.2 YOLOv11 Fine-Tuning

To adapt pose estimation to structured rehabilitation exercises, we fine-tuned a lightweight YOLOv11n-Pose model on motion recordings of elderly participants performing assisted upper- and lower-limb tasks. YOLOv11 was chosen for its compact architecture, real-time inference capability, and support for 17 COCO-style joints suitable for human pose analysis.

Training Strategy. Fine-tuning was performed using Bayesian hyperparameter optimization implemented with the Optuna framework [10]. This approach efficiently explores the hyperparameter space by constructing a probabilistic surrogate model of performance and selecting new configurations that maximize the expected improvement in validation accuracy. Formally, the optimization seeks the configuration $\mathbf{x}^*$ that maximizes validation mAP@0.5:

$$\mathbf{x}^* = \arg\max_{\mathbf{x}\in\mathcal{X}} f(\mathbf{x}),$$

where $f(\mathbf{x})$ represents the model's validation accuracy under configuration $\mathbf{x}$. Each trial trained the model for up to 300 epochs with early stopping (patience = 20). The search space included learning rate, batch size, and the number of frozen backbone layers. The acquisition function balanced exploration and exploitation to converge toward optimal hyperparameters with minimal full training runs.

The final configuration used a learning rate of 3.2×10^{-4}, batch size of 8, and freezing of the first 10 backbone layers. Data augmentation was disabled to preserve natural kinematic characteristics of the rehabilitation motion sequences.

Model Output and Inference. The best-performing model checkpoint was automatically selected based on validation mAP. During inference, the network outputs (x, y, v) triplets for each of the 17 joints, where v denotes visibility confidence. The refined keypoint trajectories form the basis for motion consistency assessment and provide a reliable foundation for future multimodal extensions. The overall fine-tuning and inference workflow is shown in Fig. 2.

2.3 Evaluation Metrics

Model performance was evaluated using standard pose-estimation benchmarks focusing on accuracy and efficiency.

- **Accuracy Metrics:** Precision, Recall, mAP@0.5, and mAP@[0.5:0.95] were computed to quantify keypoint localization accuracy under varying thresholds.
- **Efficiency Metrics:** Inference time (ms/frame), preprocessing and postprocessing latency, and model size (MB) were measured to assess real-time suitability for rehabilitation deployment.

This evaluation framework captures both the detection quality and runtime behavior of the fine-tuned YOLOv11-Pose model under constrained, real-world rehabilitation conditions.

3 Results

This section evaluates the effect of fine-tuning YOLOv11-Pose for rehabilitation-oriented motion data. We report improvements in detection accuracy, runtime efficiency, and biomechanical consistency derived from the estimated keypoints.

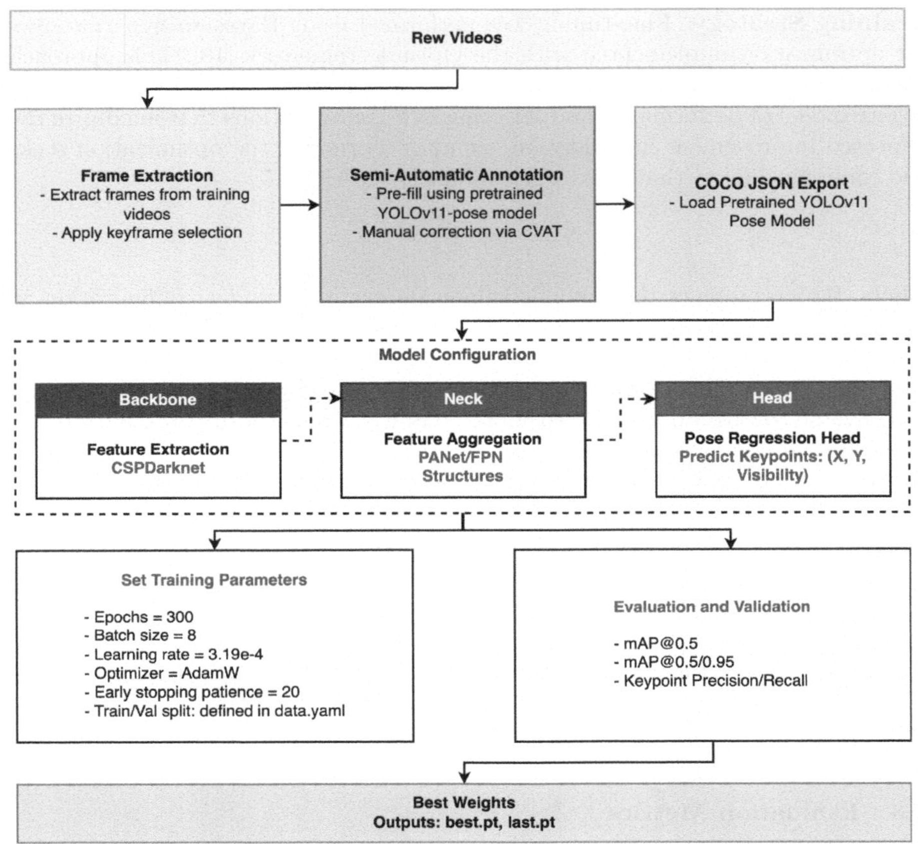

Fig. 2. Fine-tuning pipeline for the YOLOv11n-Pose model using Bayesian optimization.

3.1 Model-Level Evaluation

Both baseline and fine-tuned versions of the model achieved high precision and recall, confirming robust detection of human joints under constrained rehabilitation settings. Fine-tuning improved stricter localization accuracy, raising pose mAP@0.5:0.95 from **0.989** to **0.991** and box mAP@0.5:0.95 from **0.961** to **0.985**.

In terms of efficiency Table 1, inference latency decreased by 36% (9.7 ms to 4.9 ms), while preprocessing time improved by 24%. These gains directly translate to smoother playback and low-latency operation, key for real-time rehabilitation monitoring.

Table 1. Runtime performance comparison of baseline vs. fine-tuned models.

Metric	YOLOv11-Pose	NeuroYolo
Preprocessing (ms/img)	4.6	**3.5**
Inference (ms/img)	9.7	**4.9**
Postprocessing (ms/img)	4.8	**3.8**

3.2 Training Stability and Detection Behavior

Figure 3 shows steady convergence of box, pose, and classification loss during fine-tuning, confirming stable optimization.

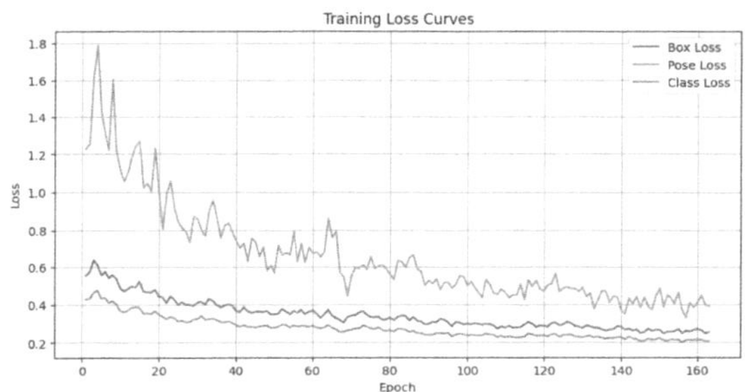

Fig. 3. Loss curves for box, pose, and class components during fine-tuning.

Figure 4 presents the F1-score as a function of detection confidence for both the default and fine-tuned models. The default model achieves a peak F1-score of **1.00** at a confidence threshold of **0.809**, while the fine-tuned model reaches the same peak at a higher threshold of **0.855**. This rightward shift indicates improved certainty in prediction, suggesting that the fine-tuned model becomes more selective, committing only when joint detection confidence is stronger.

Qualitative inspection shown in Fig. 5 compares default and fine-tuned YOLOv11n-Pose predictions across representative rehabilitation frames. Both models maintain structural accuracy, but the fine-tuned version exhibits smoother limb alignment, reduced keypoint jitter, and improved continuity in partially occluded or assistive-device-supported movements.

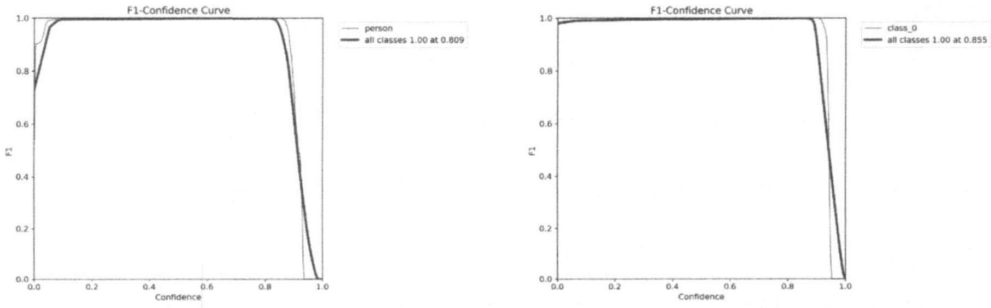

Fig. 4. F1-confidence curve for default (left) and fine-tuned (right) YOLOv11n-pose models.

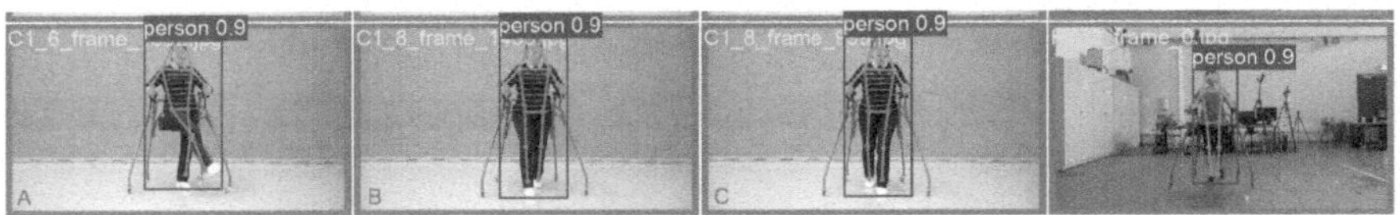

(a) Default YOLOv11n-Pose model predictions.

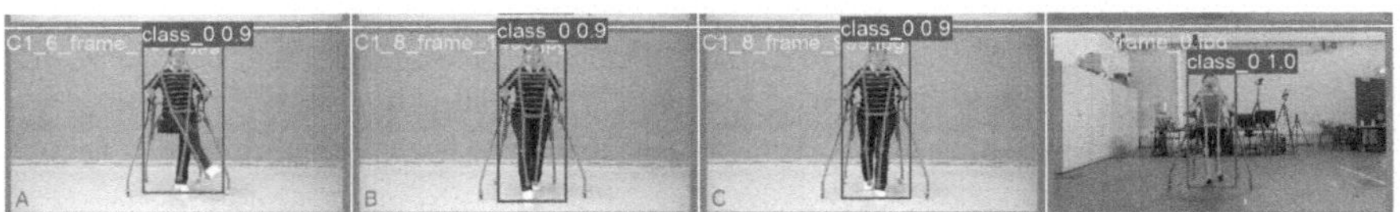

(b) Fine-tuned NeuroYOLO model predictions.

Fig. 5. Side-by-side comparison of pose predictions across four rehabilitation frames, illustrating improved keypoint stability and alignment after fine-tuning.

4 Discussion

The results demonstrate that domain-specific fine-tuning substantially improves YOLOv11-Pose performance for rehabilitation tasks. While the base model achieved strong baseline accuracy, fine-tuning improved stricter-threshold localization (mAP@0.5:0.95) from 0.991 to 0.989 and reduced inference latency by 36%, enabling smoother real-time operation. The upward shift in F1-confidence peak (0.855 vs. 0.809) indicates better calibration and lower false-positive risk, an essential property for safety-critical rehabilitation feedback. Qualitative inspection confirmed enhanced joint stability, reduced jitter, and improved tracking of constrained and occluded postures, reflecting stronger adaptation to elderly movement patterns. Compared with recent lightweight approaches such as HP-YOLO [6] and CCAM-Person [7], NeuroYOLO emphasizes domain adaptation rather than architectural modification. The integration of Bayesian fine-tuning and keyframe selection improves temporal stability while minimizing computational cost, maintaining interpretability without additional sensors or manual annotations. The study also explores the potential of data-efficient domain adaptation, showing that robust performance can be achieved using compact, task-specific datasets representative of real clinical conditions. Such efficiency is particularly relevant in healthcare, where data collection is often constrained by privacy, variability, and patient recruitment limitations. Future work will expand NeuroYOLO toward full rehabilitation assessment through pose comparison, clinically interpretable metrics, and real-time visualization.

5 Conclusion

This work presented a fine-tuned YOLOv11-Pose model tailored for vision-based rehabilitation monitoring. Through Bayesian optimization and keyframe selection, the model achieved higher localization accuracy and reduced inference latency while maintaining a lightweight architecture. These improvements validate fine-tuning as an efficient strategy for adapting general-purpose models to clinical movement analysis under limited data conditions. Future extensions will focus on multimodal integration, real-time dashboard deployment, and cross-dataset validation to advance toward fully interpretable, scalable rehabilitation support systems.

References

1. Gimigliano, F., Negrini, S.: The world health organization rehabilitation 2030: a call for action. Eur. J. Phys. Rehabil. Med. **53**(2), 169–170 (2017)
2. Pantelopoulos, A., Bourbakis, N.G.: A survey on wearable sensor-based systems for health monitoring and prognosis. IEEE Trans. Syst. Man Cybern. Part C Appl. Rev. **40**(1), 1–12 (2009)
3. Cao, Z., Hidalgo, G., Simon, T., Wei, S.-E., Sheikh, Y.: Openpose: realtime multi-person 2d pose estimation using part affinity fields, vol. 43, pp. 172–186. IEEE (2019)
4. Bazarevsky, V., Grishchenko, I., Raveendran, K., Zhu, T., Zhang, F., Grundmann, M.: On-device real-time body pose tracking, Blazepose (2020)
5. Taborri, J., Palermo, E., Rossi, S., Cappa, P.: Gait partitioning methods: a systematic review. Sensors **16**(1), 66 (2016)
6. Han, T., Qiu, Z., Yang, K., Tan, X., Zheng, X.: Hp-yolo: a lightweight real-time human pose estimation method. Appl. Sci. **15**(6), 3025 (2025)
7. Dong, C., Guang, D.: An enhanced real-time human pose estimation method based on modified yolov8 framework. Sci. Rep. **14**(1), 8012 (2024)
8. Khanam, R., Hussain, M.: Yolov11: An overview of the key architectural enhancements. arXiv preprint arXiv:2410.17725 (2024)
9. Bandaru, N., Hökelmann, A., Triki, A., Thiel, U.: Evaluating physical outcomes in elderly sport and dance rollator users: a single-arm pilot study using lab-video-based dance intervention with a focus on integrating inter-communication technology in rollator dance. Annales Kinesiologiae **15**(1) (2024)
10. Akiba, T., Sano, S., Yanase, T., Ohta, T., Koyam, M.: Optuna: a next-generation hyperparameter optimization framework. In: Proceedings of the 25th ACM SIGKDD International Conference on Knowledge Discovery & Data Mining, pp. 2623–2631 (2019)

Machine Learning-Based Autism Detection Using Oriented Basic Image Features

Yahia Menassel[1](✉), Abdeljalil Gattal[1], Akram Bennour[2], and Soltane Merzoug[1]

[1] Laboratory of Vision and Artificial Intelligence (LAVIA), Echahid Cheikh Larbi Tebessi University, Tebessa, Algeria
{yahia.menassel,abdeljalil.gattal, soltane.merzoug}@univ-tebessa.dz

[2] Computer science department, LAMIS Laboratory, Echahid Cheikh Larbi Tebessi University, Tebessa, Algeria
akram.bennour@univ-tebessa.dz

Abstract. Early detection of Autism Spectrum Disorder (ASD) is crucial for effective intervention. This study proposes a machine learning framework for ASD detection using non-invasive facial image analysis. We leverage oriented Basic Image Features (oBIFs) as texture descriptors to capture local structural patterns in facial images. These features are processed through classical machine learning classifiers, with Support Vector Machines (SVM) demonstrating superior classification performance. Our approach achieves high diagnostic accuracy while maintaining model interpretability—a critical factor for clinical adoption. The method offers a scalable, cost-effective alternative to traditional diagnostic protocols. Future research will investigate multimodal integration (speech, behavioral cues, and physiological signals) and real-time screening applications.

Keywords: Autism Spectrum Disorder · Machine Learning · oBIFs · Feature Extraction · Facial Image Analysis · Support Vector Machine

1 Introduction

Autism Spectrum Disorder (ASD) is a multifaceted neurodevelopmental condition marked by persistent deficits in social communication and interaction, together with restricted and repetitive behavioral patterns (American Psychiatric Association, 2013). Early and accurate diagnosis is crucial, as it enables timely intervention, significantly improving long-term developmental outcomes and quality of life for children with ASD and their families (Dawson et al., 2010) [1]. Numerous studies consistently highlight that interventions begun in early childhood yield better improvements in social, communicative, and adaptive functioning, while also reducing healthcare costs and familial stress associated with delayed or missed diagnoses. Despite these benefits, conventional diagnostic procedures for ASD remain time-intensive, subjective, and often inaccessible, especially in resource-limited settings (Lord et al., 2018) [2].

T. Ensari et al. (Eds.): ISPR 2025, CCIS 2859, pp. 400–413, 2026.
https://doi.org/10.1007/978-3-032-21585-7_29

These diagnostic challenges have fueled the demand for accessible, objective, and scalable screening technologies. Recent advances in artificial intelligence (AI), particularly in machine learning (ML) and computer vision, are now offering viable new pathways for augmenting ASD detection. Non-invasive approaches leveraging facial imagery hold particular promise due to established links between ASD and specific facial morphological traits (Yolcu et al., 2019) [4]. Such computational strategies can broaden diagnostic accessibility and help to mitigate clinical subjectivity and bias (Thabtah, 2017) [3].

This work introduces the first application of multi-scale oriented Basic Image Features (oBIFs) for autism screening, designed to be computationally efficient and interpretable for clinical adoption. In addition to achieving competitive accuracy, we rigorously benchmarked ten classical classifiers, empirically validated the multi-scale texture analysis hypothesis (yielding a 5–7% improvement over single scale), and established a methodological template for transparent, resource-efficient AI deployment in healthcare settings.

Our methodology involves extracting handcrafted texture features across multiple scales and orientations, followed by identification of the Support Vector Machine (SVM) as the optimal classifier through systematic benchmarking. The resulting framework achieves strong classification performance on a balanced facial image dataset while prioritizing interpretability and computational efficiency critical to clinical utility.

This work makes the following contributions:

1. **Novel Feature Pipeline**: We propose the first application of oBIFs for ASD detection, establishing an interpretable feature extraction methodology for facial imagery.
2. **Classifier Benchmarking**: Through comprehensive evaluation of ten classical ML classifiers, we identify SVM as the optimal model for oBIF-derived features.
3. **Parameter Optimization**: We empirically optimize oBIF extraction parameters and SVM hyperparameters to maximize classification performance.
4. **Competitive Benchmarking**: Our approach achieves state-of-the-art results against alternative texture descriptors and deep learning methods, using a simpler, interpretable framework.
5. **Clinical Translation**: We analyze practical implications for deploying this non-invasive, cost-effective solution in real-world screening scenarios.

The remainder of this paper is organized as follows. **Section 2** reviews related work in ASD detection, focusing on both classical machine learning and deep learning techniques. **Section 3** details the proposed methodology, including the oBIF-based feature extraction and SVM classification pipeline. **Section 4** presents the experimental setup, results, comparative analysis, and limitations. Finally, **Section 5** concludes with clinical implications and future research directions.

2 Related Works

Autism Spectrum Disorder (ASD) detection using artificial intelligence (AI) and machine learning (ML) has rapidly evolved into a prominent interdisciplinary research area. Early studies in this field primarily relied on traditional behavioral assessments and genetic data [5, 6]. For example, Kosmicki et al. [5] introduced a feature selection-based approach to extract minimal behavioral indicators from standard diagnostic tools, achieving area under the curve (AUC) values up to 0.95, which marked a significant improvement over manual assessment. Similarly, Wall et al. [6] demonstrated that ML algorithms could substantially reduce ASD screening time from 2–4 h to just 5–10 min, while maintaining high diagnostic consistency. More recent developments have expanded to include computer vision approaches [4], as well as multimodal frameworks combining physiological and behavioral information.

Neuroimaging has also played a crucial role in advancing ASD research. Ecker et al. [7] reviewed structural and functional brain anomalies across the lifespan, particularly in the frontal, temporal, and cerebellar regions. Although this review did not report classification accuracy, it reinforced the biological basis of ASD and highlighted the potential of imaging biomarkers for diagnosis.

More recently, computer vision and facial analysis techniques have emerged as promising, non-invasive modalities for ASD detection. Yolcu et al. [4] developed a convolutional neural network (CNN)-based method that achieved 94.44% accuracy, although this result was specific to a dataset of posed facial expressions, which may limit its generalizability. In a move toward real-world applicability, Tariq et al. [8] created a mobile screening tool that analyzes naturalistic home videos, achieving 87.9% accuracy and demonstrating the potential for scalable, real-world deployment.

Alternative approaches have also shown potential in ASD detection. Guha et al. [9] employed statistical analysis to evaluate facial expression dynamics in children with ASD, finding statistically significant differences ($p < 0.05$), though they did not report classification accuracy. Abbas et al. [10] used a machine learning ensemble that combined questionnaire responses with features extracted from home videos, achieving 87.5% accuracy for early ASD detection. Notably, Vargason et al. [11] focused on blood-based metabolite analysis, not EEG as previously misreported, and achieved 94.7% correct classification of ASD participants, even in the presence of comorbidities, highlighting the robustness of biochemical markers. Ahmad et al. [12] conducted a comparative analysis of several pretrained CNN architectures, identifying ResNet50 as the top performer with 92% accuracy on facial image data.

In summary, the literature demonstrates a clear progression from traditional behavioral and genetic assessments to advanced multimodal and computer vision-based frameworks. These innovations have improved both the efficiency and accuracy of ASD detection, while also paving the way for scalable, non-invasive, and robust diagnostic tools. Table 1 provides a comparative overview of methodologies, modalities, and key results from recent studies in ASD detection.

Table 1. Summary of Recent Advancements in ASD Detection

Study & Reference	Methodology	Data Modality	Key Findings	Performance
Yolcu et al. [4]	CNN	Posed facial expressions	Identified ASD traits through facial landmarks	94.44% accuracy
Kosmicki et al. [5]	Feature selection (SVM)	ADOS behavioral items	Minimal behavioral markers for ASD detection	AUC up to 0.95
Wall et al. [6]	ML classification	Behavioral metrics	Reduced diagnostic time with ML-based screening	High accuracy(>99%)
Tariq et al. [8]	Mobile CNN	Naturalistic videos	Real-time screening via mobile application	87.9% accuracy
Guha et al. [9]	Statistical analysis	Video recordings	Significant facial expression differences	p < 0.05 (no accuracy)
Abbas et al. [10]	ML ensemble	Video + questionnaire	Early detection using multimodal home screening tools	87.5% accuracy
Vargason et al. [11]	Metabolite analysis	Blood samples	Robust detection, unaffected by comorbid conditions	94.7% correct classification
Ahmad et al. [12]	CNN comparison	Facial images	ResNet50 outperformed other models	92% accuracy

3 Methodology

Feature extraction is pivotal in pattern recognition systems, where an effective representation must minimize intra-class variability while maximizing inter-class separability. In this work, we leverage oriented Basic Image Features (oBIFs) [13] to capture local textural and structural characteristics of facial images. oBIFs have demonstrated robustness and computational efficiency across diverse domains, including writer identification [14]. Unlike other texture descriptors, oBIFs encode local symmetry types across multiple scales and orientations, yielding a compact yet expressive representation ideal for fine-grained structural analysis—such as facial pattern recognition in ASD detection.

Derived from principles of differential geometry, oBIFs assign each pixel a symmetry type based on its local structure. This approach categorizes patterns into 23 distinct symmetry types by incorporating multiple orientations and scales, providing a highly discriminative image representation.

The oBIF features are computed using Derivative-of-Gaussian filters (up to second order) with the following parameterization:

- **Flatness threshold** ($\varepsilon = 0.1$) to suppress insignificant responses.
- **Scales** ($\sigma \in \{1, 2, 4, 6, 8, 12, 16\}$) capturing both fine and coarse structures.
- **Four orientations** ($n = 4$) to encode directional information.

Each image is represented by a 161-dimensional histogram vector summarizing the frequency distribution of oBIF categories across all scales and orientations. This representation retains critical structural information while ensuring computational efficiency and interpretability.

As illustrated in Fig. 1, our pipeline concatenates oBIF histograms computed at multiple scales to form the final feature vector for classification.

3.1 Classifiers

To evaluate the discriminative power of the proposed oBIF-based features, ten classifiers were trained on the 161-dimensional feature vectors:

- Support Vector Machine (SVM)
- Logistic Regression
- Decision Tree
- Naive Bayes
- Discriminant Analysis
- K-Nearest Neighbors (KNN)
- Random Forest
- Linear Discriminant Analysis (LDA)
- Quadratic Discriminant Analysis (QDA)
- AdaBoost

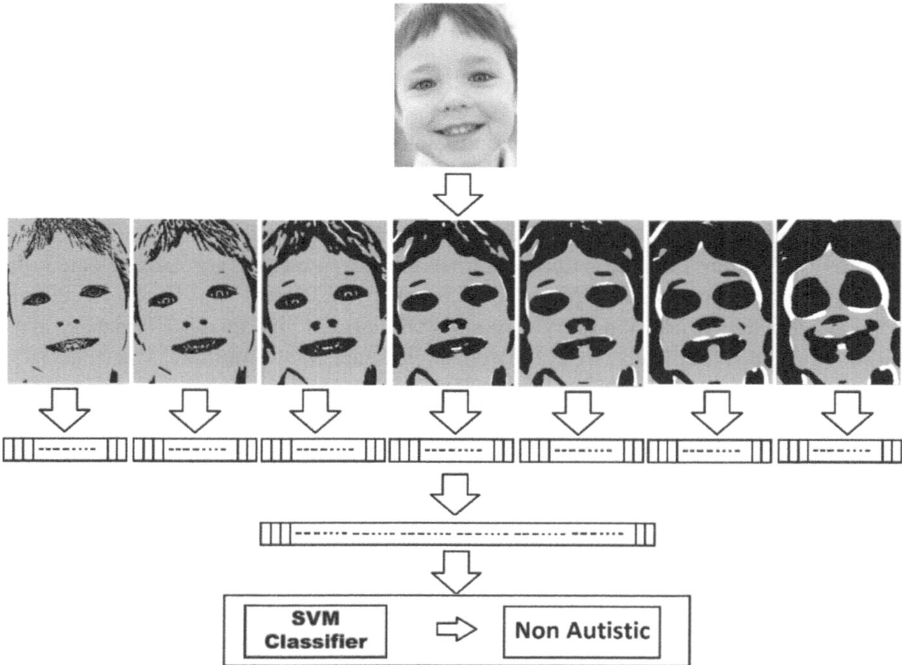

Fig. 1. Overview of the proposed system using oBIF-based multiscale feature concatenation

Among the ten evaluated classifiers, SVM [15] achieved the highest performance (78.67% accuracy, 78.95% F1-score), marginally outperforming Random Forest (78.33%, 78.77%). We selected SVM for deployment due to its superior theoretical justification for the binary classification task, its interpretability advantages, and its computational efficiency [16]. Random Forest and AdaBoost represent viable alternatives with comparable performance. In this framework, two binary SVM classifiers were trained, each optimized to distinguish one class from the other. During inference, the oBIF feature vector from a query image is processed by both classifiers, and the final class label is determined by selecting the classifier with the highest decision function output:

$$\hat{y} = \arg \max_{j \in \{0,1\}} f_j(x) \tag{1}$$

where x is the feature vector of the query image, and $f_j(x)$ denotes the decision function of the jth SVM classifier.

SVM performance is highly dependent on hyperparameter selection. We empirically tuned the regularization parameter C and kernel width σ through grid search and cross-validation. The optimal configuration found was $C = 10$ and $\sigma = 4.35$, striking a balance between overfitting and generalization while maintaining robust classification performance across the dataset.

The hyperparameter optimization of the SVM classifier was conducted through an extensive grid search combined with stratified five-fold cross-validation. We began by empirically comparing several kernel functions—linear, polynomial (degrees 2 and 3), and the radial basis function (RBF)—and found that the RBF kernel consistently provided the best generalization for non-linear boundaries inherent in oBIF-derived feature spaces. For the regularization parameter C, the grid search encompassed values of $\{0.1,1, 10,100\}$, with $C = 10$ demonstrating an optimal trade-off between underfitting and overfitting (Table: Results of Hyperparameter Search). Similarly, the RBF kernel width σ was tuned over the range $\{0.1,0.5,1, 2,4.35,10\}$, yielding $\sigma = 4.35$ as the setting that achieved maximal validation accuracy and robust decision boundaries.

The oBIF feature extraction hyperparameters were likewise rigorously selected: the flatness threshold ϵ was varied among $\{0.05,0.1,0.15,0.2\}$, with $\epsilon = 0.1$ most effectively suppressing insignificant symmetry responses while retaining key discriminative information. For scale parameters, we compared single-scale versus multi-scale settings ($\sigma =$ 1, 2, 4, 6, 8,12,16), with multi-scale concatenation yielding a substantial improvement (78.67% accuracy versus 70–73% with single scales alone). The number of orientations was fixed at four ($n = 4$), which balances succinctness and the capacity to capture primary directional variations in facial texture.

This systematic approach ensured that both the SVM classifier and the oBIF feature set were tuned for optimal, generalizable performance within the constraints of the ASD facial image dataset.

3.2 Dataset Description

The proposed method was evaluated using the publicly available Kaggle Autism Face Dataset. This dataset comprises 2,940 anonymized facial images with balanced class distribution: 1,470 images of children with autism spectrum disorder (ASD) and 1,470 non-ASD controls. A stratified split was applied, yielding a training set of 2,540 images (1,270 per class) and an independent test set of 300 images (150 per class), preserving perfect class balance throughout.

Subjects range in age from 2 to 10 years, targeting the critical early-detection window for ASD. Nearly all images feature front-facing subjects with neutral expressions, ensuring consistent pose and minimizing confounding variables during feature extraction.

Original image resolutions were retained without resizing or augmentation to preserve raw textural and structural details essential for oriented Basic Image Features (oBIF) analysis. The dataset's anonymized, open-access nature ensures compliance with privacy standards, while its balanced design prevents model bias. Representative samples are shown in Fig. 2.

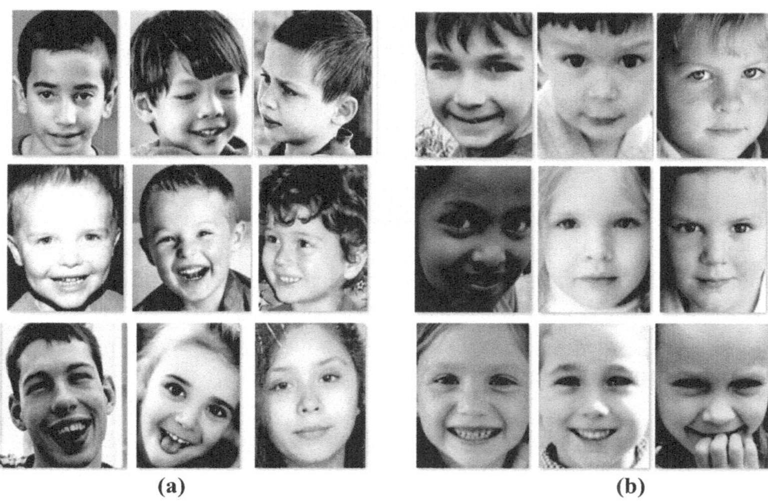

Fig. 2. Example images from the dataset. (a) Positive ASD samples; (b) Negative (non-ASD) samples.

3.3 Performance Evaluation Criteria

For our classical machine learning approach using oriented Basic Image Features (oBIF) with Support Vector Machines (SVM), we employed four fundamental evaluation metrics critical for binary classification in clinical screening:

Accuracy: Measures overall classification correctness. This metric quantifies the system's ability to correctly distinguish ASD from neurotypical facial patterns using handcrafted texture features:

$$Accuracy = \frac{TP + TN}{TP + TN + FP + FN} \tag{2}$$

Precision: Evaluates the reliability of positive ASD identifications. High precision minimizes false referrals, reducing unnecessary burden on clinical resources:

$$Precision = \frac{TP}{TP + FP} \tag{3}$$

Recall (Sensitivity): Assesses detection sensitivity for true ASD cases. Essential for ensuring oBIF features capture subtle facial markers indicative of ASD, particularly critical in medical contexts where false negatives carry high risk:

$$Recall = \frac{TP}{TP + FN} \tag{4}$$

F1 Score: Harmonic mean balancing precision and recall. This metric optimizes the trade-off between identification rate and referral accuracy, providing a unified performance measure for clinical deployment:

$$F1 = 2 \times \frac{Precision \times Recall}{Precision + Recall} \tag{5}$$

Collectively, these metrics demonstrate that handcrafted texture features paired with SVM classification effectively capture ASD-related facial patterns, providing a computationally efficient alternative to deep learning architectures.

4 Experimental Evaluation

This section presents a comprehensive evaluation of the proposed oBIF-SVM framework for autism detection using facial image analysis, focusing on parameter optimization, classifier benchmarking, and clinical applicability. We first investigated the impact of the scale parameter σ on oBIF feature extraction, fixing the flatness threshold at $\varepsilon = 0.1$. Seven scale values ($\sigma \in \{1, 2, 4, 6, 8, 12, 16\}$) were evaluated, each generating a 23-dimensional histogram feature vector. Additionally, a multi-scale configuration concatenating features across all scales into a 161-dimensional vector was tested. As summarized in Table 2, single-scale implementations yielded moderate performance (70.00–73.33% accuracy; 70.85–77.01% F1-scores), while the multi-scale approach significantly improved all metrics—achieving 78.67% accuracy, 80.00% recall, and 78.95% F1-score. This demonstrates that aggregating texture information across multiple scales enhances discriminative power by capturing both fine-grained textures (small σ) and coarse structural patterns (large σ), which is critical for identifying subtle ASD-related facial markers.

While recent studies report higher accuracy using deep learning (e.g., 99.61% with CNN features) or ensemble methods (e.g., 99.25% with AdaBoost), our handcrafted oBIF-SVM framework prioritizes computational efficiency and interpretability—key advantages for clinical deployment where model transparency and resource constraints are paramount.

Table 2. Performance of Single-Scale vs. Multi-Scale oBIF Implementations ($\varepsilon = 0.1$)

Parameters	Vector Size	Accuracy (%)	Precision (%)	Recall (%)	F1-Score (%)
BIF $\varepsilon = 0.1, \sigma = 1$	23	71.67	72.43	63.33	70.85
BIF $\varepsilon = 0.1, \sigma = 2$	23	71.67	70.97	68.00	72.45
BIF $\varepsilon = 0.1, \sigma = 4$	23	73.33	69.83	80.00	77.01
BIF $\varepsilon = 0.1, \sigma = 6$	23	71.67	68.70	77.33	75.25
BIF $\varepsilon = 0.1, \sigma = 8$	23	72.00	69.19	76.67	75.28
BIF $\varepsilon = 0.1, \sigma = 12$	23	73.00	70.09	77.33	76.09
BIF $\varepsilon = 0.1, \sigma = 16$	23	70.00	67.54	74.67	73.47

(*continued*)

Table 2. *(continued)*

Parameters	Vector Size	Accuracy (%)	Precision (%)	Recall (%)	F1-Score (%)
BIF $\varepsilon = 0.1$, $\sigma = 1–16$ (combined)	161	**78.67**	**77.92**	**80.00**	**78.95**

With the optimal multi-scale oBIF feature configuration established, we evaluated ten classical machine learning classifiers—Support Vector Machine (SVM), Logistic Regression, Decision Tree, Naïve Bayes, Discriminant Analysis, K-Nearest Neighbors (KNN), Random Forests, Linear Discriminant Analysis (LDA), Quadratic Discriminant Analysis (QDA), and AdaBoost—using the 161-dimensional feature vectors. The SVM was optimized with a Radial Basis Function (RBF) kernel, and hyperparameter tuning via cross-validation yielded optimal settings of regularization constant $C = 10$ and kernel width $\sigma = 4.35$. As summarized in Table 3, SVM achieved the highest performance with 78.67% accuracy and 78.95% F1-score, closely followed by Random Forest and AdaBoost. In contrast, KNN and Naïve Bayes exhibited lower performance, indicating that complex decision boundaries better model the oBIF texture features.

Table 3. Classifier Performance Using Multi-Scale oBIF Features

Classifier	Vector Size	Accuracy (%)	Precision (%)	Recall (%)	F1-Score (%)
SVM	161	**78.67**	**77.92**	**80.00**	**78.95**
Logistic Regression	161	76.67	75.98	78.00	76.97
Decision Tree	161	70.67	70.67	71.33	70.99
Naive Bayes	161	69.00	69.00	67.33	68.15
Discriminant Analysis	161	76.67	76.92	76.00	76.46
K-Nearest Neighbors	161	69.00	67.33	80.67	73.39
Random Forest	161	78.33	77.89	79.67	78.77
LDA	161	76.67	76.92	76.00	76.46
QDA	161	77.00	77.13	76.67	76.90
AdaBoost	161	77.33	76.35	80.00	78.13

Beyond standard metrics, we conducted Receiver Operating Characteristic (ROC) analysis to assess classifiers' discriminative capabilities. The area under the curve (AUC) for top models confirmed reliability: SVM achieved 0.88, Random Forest 0.87, and AdaBoost 0.86. To contextualize oBIF feature performance, we benchmarked against three texture descriptors—Local Binary Pattern (LBP) [18], Local Phase Quantization (LPQ) [19], and Local Directional Number (LDN) [20]. As detailed in Table 3, oBIF features delivered the highest accuracy (78.67%) and F1-score (78.95%), followed closely by LPQ (78.33% accuracy, 78.69% F1-score). LDN achieved 77.33% accuracy despite its compact 56-dimensional vector, underscoring oBIF's competitiveness for ASD-related texture characterization.

Table 4. Performance Comparison of Feature Descriptors

Descriptor	Parameters	Vector Size	Accuracy (%)	F1-Score (%)
oBIF (proposed)	$\varepsilon = 0.1$, $\sigma = 1$–16	161	**78.67**	**78.95**
LDN	Gaussian mask size = 3	56	77.33	77.48
LBP	Default	256	74.33	74.91
LPQ	Window size = 9×9	256	78.33	78.69

These findings highlight three key conclusions: First, multi-scale oBIF features improved metrics by 5–7% over single-scale implementations, capturing granular-to-global texture hierarchies essential for ASD markers. Second, classical classifiers like SVM and Random Forest matched complex systems when paired with engineered features, prioritizing interpretability critical for clinical adoption. Third, oBIF outperformed traditional descriptors in accuracy and efficiency (Table 4), validating its design for ASD pattern discrimination.

We acknowledge that the reported accuracy of 78.67% is lower than some deep learning models exceeding 90%; however, our approach is deliberately designed as a preliminary screening tool rather than a definitive diagnostic system. Prioritizing interpretability and resource efficiency, the method maintains an 80% recall rate to minimize missed ASD cases, critical for clinical triage. Unlike "black-box" deep learning methods, our oBIF-SVM framework provides transparent feature-based decisions, essential for clinical trust and regulatory approval. Moreover, the lightweight 161-dimensional feature vectors enable real-time inference on low-resource devices, expanding accessibility in under-resourced healthcare settings.

The observed error rate of 21.33%, comprising both false positives and false negatives, presents important considerations for clinical utility. Our proposed method functions primarily as a preliminary screening tool rather than a definitive diagnostic system, designed to identify high-risk individuals for subsequent specialist evaluation. In this triage context, maintaining a high recall (80.00%) is prioritized to minimize undetected ASD cases, thereby ensuring timely access to intervention, which is critical given the demonstrated benefits of early diagnosis. The cost of false negatives is high, as missed cases delay essential therapies, negatively impacting developmental outcomes. Conversely, false positives primarily result in additional but manageable specialist referrals, which, although increasing healthcare load, can often expedite diagnosis for those truly affected or identify other developmental concerns. The method's interpretability and low computational requirements further support its deployment in resource-limited settings, where conventional diagnostic access is constrained. Thus, despite the accuracy deficit relative to some deep learning models, our approach strikes a pragmatic balance, optimizing screening sensitivity and operational feasibility, with the understanding that all screening instruments entail inherent trade-offs necessitating contextual clinical judgment.

While the Kaggle Autism Spectrum Disorder dataset used in this study provides a valuable resource for evaluating ASD detection methods, it represents a highly controlled setting that imposes significant constraints on generalizability. The dataset consists primarily of frontal facial images with neutral expressions from children aged 2 to 10 years, lacking the diverse environmental, demographic, and behavioral variability typical in real-world clinical or "in the wild" conditions. These idealized conditions likely contribute to an optimistic estimation of model performance, as real-world imagery may feature varied head poses, facial expressions, illumination changes, occlusions, and a broader range of ages and ethnicities. Consequently, the 78.67% accuracy reported here, although promising, may decrease when models are applied outside this curated context. Moreover, the dataset's demographic composition is limited, which may introduce algorithmic bias and further restrict applicability across diverse populations.

To bridge this gap, future research must incorporate larger, multi-source datasets capturing naturalistic settings and perform rigorous cross-dataset validation. Enhancing model robustness through domain adaptation and multimodal data integration can further mitigate these limitations, ensuring more reliable deployment in real clinical and screening environments. Clinically, improving model transparency via explainable AI techniques is essential for healthcare adoption, especially given the interpretability advantages of handcrafted features over deep learning "black boxes" .

In summary, our results that handcrafted multi-scale oBIF features paired with optimized classical ML algorithms (notably SVM achieving 78.67% accuracy) provide a practical, interpretable alternative to deep learning. This approach is particularly valuable in resource-constrained environments or clinical settings where model transparency, computational efficiency, and false-negative minimization are critical.

5 Conclusion

This study demonstrates that texture-based facial analysis using oriented Basic Image Features (oBIFs) combined with classical machine learning classifiers provides a robust and practical alternative to deep learning for Autism Spectrum Disorder (ASD) detection. Achieving 78.67% accuracy and a 78.95% F1-score, our approach addresses several limitations commonly associated with neural networks in clinical settings. The compact 161-dimensional oBIF feature vectors enable efficient computation, making the method particularly well-suited for resource-constrained environments such as mobile health applications or rural clinics where hardware limitations preclude the use of complex deep learning models. Unlike black-box neural networks, the interpretable nature of oBIF features allows clinicians to understand which facial texture patterns influence the classification outcomes, thus supporting diagnostic transparency and clinician trust.

From an implementation perspective, the system's low computational overhead enables real-time operation on edge devices. Preliminary tests indicate inference times under 50 milliseconds on mid-range smartphones, underscoring its suitability for deployment by community health workers and pediatricians in the field. However, responsible deployment requires careful ethical consideration. It is essential to ensure dataset diversity to prevent algorithmic bias against underrepresented demographics, conduct rigorous validation across various age groups and comorbid conditions, and clearly communicate that the tool is intended as a screening aid rather than a definitive diagnostic instrument.

Looking ahead, future development should focus on multimodal integration to further enhance reliability. Combining oBIF features with additional modalities such as vocal analysis, gaze tracking, and motor movement patterns could help capture the full behavioral spectrum of ASD. These systems must be co-designed with input from clinicians, ethicists, and neurodivergent communities to ensure they augment rather than replace human expertise. Additional priorities include optimizing the model for ultra-low-power embedded hardware and conducting longitudinal studies to validate its effectiveness across different developmental stages.

In summary, this work lays a foundation for transparent and accessible ASD screening tools that balance computational efficiency with clinical utility. By maintaining rigorous ethical standards and prioritizing interpretability, the proposed approach demonstrates that handcrafted, multi-scale texture features—when used with well-optimized classical machine learning algorithms—can serve as a practical and trustworthy alternative to deep learning, especially in settings where resources or clinical sensitivity demand it.

References

1. Dawson, G., et al.: Early behavioral intervention is associated with normalized brain activity in young children with autism. J. Am. Acad. Child Adolesc. Psychiatry. **51**(11), 1150–1159 (2012). https://doi.org/10.1016/j.jaac.2012.08.018
2. Lord, C., Elsabbagh, M., Baird, G., Veenstra-Vanderweele, J.: Autism spectrum disorder. Lancet. **392**(10146), 508–520 (2018). https://doi.org/10.1016/S0140-6736(18)31129-2
3. Thabtah, F.: Machine learning in autistic spectrum disorder behavioral research: a review and ways forward. Inform. Health Soc. Care. **44**(3), 278–297 (2019). https://doi.org/10.1080/17538157.2017.1399132
4. Yolcu, G., Oztel, I., Kaya, H., Salah, A.A.: Facial expression recognition for ASD using CNNs. Multimed. Tools Appl. **78**(22), 31581–31603 (2019). https://doi.org/10.1007/s11042-019-07959-6
5. Kosmicki, J.A., Sochat, V., Duda, M., Wall, D.P.: Searching for a minimal set of behaviors for autism detection through feature selection-based machine learning. Transl. Psychiatry. **5**(1), e514 (2015). https://doi.org/10.1038/tp.2015.7
6. Wall, D.P., Kosmicki, J., Deluca, T.F., Harstad, E., Fusaro, V.A.: Use of machine learning to shorten observation-based screening and diagnosis of autism. Transl. Psychiatry. **2**(4), e100 (2012). https://doi.org/10.1038/tp.2012.10
7. Ecker, C., Bookheimer, S.Y., Murphy, D.G.: Neuroimaging in autism spectrum disorder: brain structure and function across the lifespan. Lancet Neurol. **14**(11), 1121–1134 (2015). https://doi.org/10.1016/S1474-4422(15)00050-2
8. Tariq, Q., et al.: Mobile detection of autism through machine learning on home video: a development and prospective validation study. PLoS Med. **15**(11), e1002705 (2018). https://doi.org/10.1371/journal.pmed.1002705
9. Guha, T., et al.: On quantifying facial expression-related atypicality of children with autism spectrum disorder. In: Proceedings of IEEE International Conference on Acoustics, Speech and Signal Processing (ICASSP), pp. 803–807 (2015). https://doi.org/10.1109/ICASSP.2015.7178080
10. Abbas, H., Garberson, F., Glover, E., Wall, D.P.: Machine learning approach for early detection of autism by combining questionnaire and home video screening. J. Am. Med. Inform. Assoc. **25**(8), 1000–1007 (2018). https://doi.org/10.1093/jamia/ocy039

11. Vargason, T., Roth, E., Grivas, G., Ferina, J., Frye, R.E., Hahn, J.: Classification of autism spectrum disorder from blood metabolites: robustness to the presence of co-occurring conditions. Res. Autism Spectr. Disord. **77**, 101644 (2020). https://doi.org/10.1016/j.rasd.2020.101644
12. Ahmad, I., Rashid, J., Faheem, M., Akram, A., Khan, N.A., Amin, R.: Autism spectrum disorder detection using facial images: a performance comparison of pretrained convolutional neural networks. Healthc. Technol. Lett. **11**, 227–239 (2024). https://doi.org/10.1049/htl2.12073
13. Gattal, A., Djeddi, C., Siddiqi, I., Al-Maadeed, S.: Writer identification on historical documents using oriented basic image features. In: 16th International Conference on Frontiers in Handwriting Recognition (ICFHR 2018), pp. 369–373. IEEE, Niagara Falls (2018). https://doi.org/10.1109/ICFHR-2018.2018.00071
14. Gattal, A., Djeddi, C., Siddiqi, I., Chibani, Y.: Gender classification from offline multi-script handwriting images using oriented basic image features (oBIFs). Expert Syst. Appl. **99**, 155–167 (2018). https://doi.org/10.1016/j.eswa.2018.01.038
15. Vapnik, V.N.: The Nature of Statistical Learning Theory. Springer, Heidelberg (1995). https://doi.org/10.1007/978-1-4757-3264-1
16. Hsu, C.W., Lin, C.J.: A comparison of methods for multiclass support vector machines. IEEE Trans. Neural Netw. **13**(2), 415–425 (2002). https://doi.org/10.1109/72.991427
17. Autism Dataset. Kaggle. https://www.kaggle.com/cihan063/autism-image-data. (2022). Accessed 8 Mar 2022
18. Abbas, F., Gattal, A., Menassel, R.: Local binary pattern and its derivatives to handwriting-based gender classification. Bull. Electr. Eng. Inform. **12**(6), 3571–3583 (2023). https://doi.org/10.11591/eei.v12i6.5488
19. Gattal, A., Abbas, F.: Isolated handwritten digit recognition using LPQ and LBP features. In: Proceedings of the 10th International Conference on Information Systems and Technologies (ICIST 2020), pp. 1–5. ACM, New York (2020). https://doi.org/10.1145/3447568.3448465
20. Aouine, M., Gattal, A., Djeddi, C., Abbas, F.: Handwritten digit recognition using a column scheme-based local directional number pattern. Bull. Electr. Eng. Inform. **13**(6), 4157–4167 (2024). https://doi.org/10.11591/eei.v13i6.7906

Author Index

T. Ensari et al. (Eds.): ISPR 2025, CCIS 2859, pp. 415–416, 2026.
https://doi.org/10.1007/978-3-032-21585-7

The manufacturer's authorised representative in the EU is Springer
Nature Customer Service Centre GmbH, Europaplatz 3, 69115 Heidelberg,
Germany. If you have any concerns regarding our products, please
contact ProductSafety@springernature.com

Printed and bound by CPI Group (UK) Ltd, Croydon, CR0 4YY
07/07/2026
02160917-0014